Animal Welfare from a Cross-Cultural Perspective

Guest Editors

Kristine Hill
Michelle Szydlowski
Jes Hooper
Sarah Oxley Heaney

MDPI

Basel • Beijing • Wuhan • Barcelona • Belgrade • Novi Sad • Cluj • Manchester

Guest Editors
Kristine Hill
Department of Philosophy and Social Sciences
University of Hradec Králové
Hradec Králové
Czech Republic

Michelle Szydlowski
Department of Biology
Miami University
Oxford
USA

Jes Hooper
Department of Social and Political Sciences
University of Exeter
Devon
UK

Sarah Oxley Heaney
Department of Social and Political Sciences
University of Exeter
Devon
UK

Editorial Office
MDPI AG
Grosspeteranlage 5
4052 Basel, Switzerland

This is a reprint of the Special Issue, published open access by the journal *Animals* (ISSN 2076-2615), freely accessible at: https://www.mdpi.com/journal/animals/special_issues/43L9ML67V1.

For citation purposes, cite each article independently as indicated on the article page online and as indicated below:

Lastname, A.A.; Lastname, B.B. Article Title. *Journal Name* **Year**, *Volume Number*, Page Range.

ISBN 978-3-7258-4159-2 (Hbk)
ISBN 978-3-7258-4160-8 (PDF)
https://doi.org/10.3390/books978-3-7258-4160-8

Contents

animals

MDPI

Editorial

Rethinking Animal Welfare in a Globalised World: Cultural Perspectives, Challenges, and Future Directions

Sarah Oxley Heaney [1,*], Michelle Szydlowski [2], Kristine Hill [3] and Jes Hooper [1]

1 Department of Social and Political Sciences, Philosophy, and Anthropology (SPSPA), Faculty of Humanities, Arts and Social Sciences, University of Exeter, Devon EX4 4QJ, UK
2 Department of Biology, College of Arts and Science, Miami University, Oxford, OH 45056, USA
3 Department of Philosophy and Social Sciences, Philosophical Faculty, University of Hradec Králové, 500 02 Hradec Králové, Czech Republic
* Correspondence: sarah@anthrozoo.me

1. Introduction

In an increasingly interconnected world, human and other-than-human-animal welfare has emerged as a significant global concern. The welfare of these entangled lives, whether through anthropocentric, biocentric, or other-than-human-centred posthuman lenses, face challenges arising from cultural, political, and social boundaries. Urbanisation, globalisation, pandemics, and evolving ethical standards have amplified the focus on how other-than-human animals are treated across various settings. Realising that both human and other-than-human-animal societies co-evolve and interrelate, often beyond human imaginations, human societies must navigate the complexities of their relationships with other-than-human animals in the era now classified as the Anthropocene. The need for a nuanced understanding of the interconnectedness of human and other-than-human-animal welfare is becoming increasingly urgent. By "animal welfare", we refer to the welfare of *all* animals, including humans, unless specified otherwise. This acknowledges the trans-species nature of our intertwined lives in the biosphere.

2. Animal Welfare

Animal welfare is often described as the physical, mental, and emotional wellbeing of other-than-human animals. Humans arguably have a responsibility to ensure the welfare needs of domesticated and/or otherwise-captive animals are met, and some argue that those of wild-living other-than-human animals should also be considered (e.g., [1,2]). However, beliefs about what the needs of a species, or an individual, entail and what constitutes good practice can vary greatly. An understanding of different cultural perspectives is especially pertinent for researchers and animal advocates concerned with the welfare of animals situated in touristic, foreign (relative to the researcher), or culturally diverse contexts.

For this Special Issue, we welcomed manuscripts that addressed animal welfare from a cross-cultural perspective, including those considering more-than-human cultures. We were particularly interested in submissions offering theoretical contributions towards more ethical and symbiotic relations between humans and other-than-human animals. Finally, we encouraged submissions addressing the decolonization of animal welfare in academia and/or practice.

The articles in this Special Issue shed light on diverse issues surrounding other-than-human-animal welfare, from the management of community cats in urban China to the ethical dilemmas of elephant tourism in Nepal, and the controversial civet coffee production industry in Asia. Each study provides a unique perspective on how cultural contexts, governance, and human–other-than-human relationships shape our understanding of animal welfare. Despite the diversity of these issues, there is a unifying call for more humane treatment of species and individuals, necessitating cross-cultural understanding, tailored policies, and innovative solutions.

Animal welfare encompasses a broad range of concerns, often addressing the well-being of various species in human-controlled environments e.g., [3,4]. In this Special Issue, for example, Kamal and colleagues' paper *Recognising Zooeyia to Promote Companion Animal Welfare in Urban Bangladesh* and Gu and colleagues' article titled *A Survey of Public Opinion on Community Cats' General Health and Relationship Quality with Residents in Urban China* highlight the pressing need to improve the welfare of other-than-human animals living closely alongside humans [5,6]. These studies emphasise that animal welfare is not solely an ethical issue but is intrinsically linked to the health, safety, and community well-being of individuals of all species. Furthermore, the examination of animal welfare in different cultural contexts, as seen in *Mandan, Hidatsa, and Arikara Nation Perspectives on Rez Dogs on the Fort Berthold Reservation in North Dakota, U.S.A.* by Cardona and colleagues and *Ubiquitous Love or Not? Animal Welfare and Animal-Informed Consent in Giant Panda Tourism* by Fennell and Guo, underscores the diverse factors influencing human–other-than-human-animal relationships and the growing recognition of many species as sentient beings deserving of care and respect [7,8].

The aforementioned articles of Cordonal et al. and Gu et al. are supported by Rogers et al. in *Perceptions of Cross-Cultural Challenges and Successful Approaches in Facilitating the Improvement of Equine Welfare* [7–9]. Community engagement and collaboration are essential components for improving animal welfare. Cordona et al. emphasise the need for these interventions to be culturally relevant if they hope to address the welfare of reservation dogs [9]. Accentuating the importance of local partnerships and understanding cultural contexts is vital to achieving successful outcomes. Rogers et al. advocate for working alongside local communities to foster trust and collaboration as they discuss the challenges of managing animal welfare across different countries and cultural frameworks [9]. Similarly, Cui and colleagues' work, *Farm Animal Welfare Is a Field of Interest in China: A Bibliometric Analysis Based on CiteSpace,* confirms a growing interest in farm animal welfare in China, and the need for increased collaboration amongst authors and institutions within China and less anthropocentric-focused policy efforts for more effective outcomes [10].

Cultural beliefs and practices play a pivotal role in shaping attitudes toward animal welfare. For instance, the study by Kamal et al. explores the stigma surrounding companion animal ownership in Bangladesh, illustrating how cultural perceptions can impede welfare efforts. Moreover, the nuanced understanding of animal welfare can vary significantly among populations, as demonstrated in *Perceptions of Farm Animal Sentience and Suffering: Evidence from the BRIC Countries and the United States* by Mata et al. [11] This study reveals that attitudes toward animal sentience differ not only between countries but also within them, accentuating the subtle need for intra-country, culturally sensitive approaches to animal welfare. Some articles peel off layers of complexities within human–other-than-human-animal relationships, showing how they are influenced by historical, cultural, and social factors within communities themselves. For example, Kamal et al. demonstrate how caretakers of companion animals in Bangladesh report significant social and psychological benefits, despite the societal stigma associated with their ownership [5]. Cordonal and colleagues' "rez dog" study highlights how Indigenous communities maintain deep cultural ties with free-roaming dogs, reflecting their unique perspectives on animal welfare, which are often misunderstood and clash with or are neglected by those who govern, and attempts to impose animal welfare solutions [7].

Effective governance and policy frameworks when renegotiating a one-size-fits-all approach are shown as being crucial for advancing animal welfare. *Beyond the Unitary State: Multi-Level Governance, Politics, and Cross-Cultural Perspectives on Animal Welfare* by Chaney et al. discusses how political contexts and civil society can influence the development of animal welfare policies in the UK, highlighting the importance of multi-level governance in addressing animal welfare concerns [12]. The challenges associated with implementing policies are evident in Bangladesh, which underscores the gaps in enforcing the Animal Welfare Act (2019) and the implications for companion animal welfare. Meanwhile, Cui et al. stress the Chinese institutional power dynamics within the agricultural machine—which prioritises economic aspects and certain "livestock" species—that prevent community voices from being heard [10]. These aspects compel contributors to consider the ethical and moral dimensions of animal welfare. Ethical dilemmas surrounding the use of other-than-human animals in various contexts are a recurring theme. For instance, Fennel and colleagues' *Ubiquitous Love or Not? Animal Welfare and Animal-Informed Consent in Giant Panda Tourism* and Szydlowski's article *Wicked Problems, Novel Solutions: Nepalese Elephant Tourism and Conservation* on Nepalese elephants both raise critical questions about animal-informed consent, challenging readers to consider whether giant pandas and elephants genuinely "agree" to their roles as tourist attractions [8,13].

Several authors discuss how economic factors are intricately linked to animal welfare. Cui and colleagues' study *Farm Animal Welfare Is a Field of Interest in China: A Bibliometric Analysis Based on CiteSpace* demonstrates how addressing welfare concerns can lead to economic benefits such as improved growth and product quality [10]. The dual focus on welfare and economic viability is also evident in *Wicked Problems, Novel Solutions: Nepalese Elephant Tourism and Conservation* and *Thinking with Civets: The Role of Zoos in the Decolonisation of Animal Tourism,* both of which discuss the complex relationship between economic reliance on animal tourism and the ethical treatment of animals [13,14]. *Perceptions of Cross-Cultural Challenges and Successful Approaches in Facilitating the Improvement of Equine Welfare* and *Mandan, Hidatsa, and Arikara Nation Perspectives on Rez Dogs on the Fort Berthold Reservation in North Dakota, U.S.A.* also identify challenges such as limited financial resources [7,9].

The commodification of animals in tourism is further explored in *Thinking with Civets: The Role of Zoos in the Decolonisation of Animal Tourism,* which examines how colonial legacies continue to impact animal welfare in tourism contexts [14]. Both Cardona et al.'s article on "rez dog" and Hooper's paper on civets offer a critical examination of colonial legacies and explore how decolonisation could affect the practices of care and commodification analysed in their research [7,14]. *Thinking with Civets* reveals ongoing impacts on the commodification of animals and illustrates how the history of colonialism continues to shape contemporary animal tourism practices [14].

Culture itself is being increasingly recognised as existing in various forms across species (e.g., [15–17]). Acknowledging the nuances unveiled by considering cultural differences and decolonisation for *all* animals is paramount in unpacking human–other-than-human-animal relationships. These considerations are foregrounded in Szydlowski's article, where she reflects upon the effects on *all individuals* wrapped up in animal welfare concerns. Here, Szydlowski illustrates the need to rethink current ideologies surrounding the use of other-than-human animals in the promotion of both conservation and tourism practices. Elephants serve as a nucleus for ongoing political, land use, and environmental justice debates throughout SE Asia. In range countries, such as Nepal, elephants and their caregivers face additional challenges thanks to their liminal status as tourist transport, objects of worship, subjects of conservation practice, and co-workers in forestry efforts aimed at endangered-species conservation. Arguing that finding "common ground" requires the use of novel problem-solving techniques, Szydlowski suggests that only when community members, international interests *and* endangered individuals themselves are valued equally will positive change occur. This emphasis brings attention to the bespoke welfare needs of persons, human or other-than-human, rather than creating universal, coarse remedies intended to be panaceatic but which often result in discordant outcomes.

3. Conclusions

As we confront the multifaceted challenges of interrelated animal welfare in a globalised world, it becomes clear that understanding animal welfare through culturally sensitive, ethically grounded, and historically informed lenses is paramount. While the included articles discuss the broader implications of animal welfare in the Global South, we recognise the growing need for more ethical practices that respect both cultural contexts and animal rights in both hemispheres, while being cognisant of the stratified, historically layered, and fluid boundaries that form our world.

Conflicts of Interest: The authors declare no conflict of interest.

References

1. Faria, C.; Horta, O. Welfare Biology. In *The Routledge Handbook of Animal Ethics*; Fischer, B., Ed.; Routledge: New York, NY, USA, 2019; pp. 455–466.
2. Faria, C. *Animal Ethics in the Wild: Wild Animal Suffering and Intervention in Nature*; Cambridge University Press: New York, NY, USA, 2022.
3. Baker, S.E.; Cain, R.; van Kesteren, F.; Zommers, Z.A.; D'Cruze, N.; Macdonald, D.W. Rough Trade: Animal Welfare in the Global Wildlife Trade. *BioScience* **2013**, *63*, 928–938. [CrossRef]
4. Kagan, R.; Carter, S.; Allard, S. A universal animal welfare framework for zoos. *J. Appl. Anim. Welf. Sci.* **2015**, *18* (Suppl. S1), S1–S10. [CrossRef] [PubMed]
5. Kamal, A.-H.M.; Dell, C.A.; Kang, T. Recognizing Zooeyia to Promote Companion Animal Welfare in Urban Bangladesh. *Animals* **2023**, *13*, 1523. [CrossRef]
6. Gu, X.; Zhang, Z.; Peng, G.; Ni, A.; Wang, B.; Xiong, X.; Liu, Y.; Wang, L. A Survey of Public Opinion on Community Cats' General Health and Relationship Quality with Residents in Urban China. *Animals* **2024**, *14*, 525. [CrossRef] [PubMed]
7. Cardona, A.; Hawes, S.M.; Cull, J.; Connolly, K.; O'Reilly, K.M.; Moss, L.R.; Bexell, S.M.; Yellow Bird, M.; Morris, K.N. Mandan, Hidatsa, and Arikara Nation Perspectives on Rez Dogs on the Fort Berthold Reservation in North Dakota, U.S.A. *Animals* **2023**, *13*, 1422. [CrossRef] [PubMed]
8. Fennell, D.A.; Guo, Y. Ubiquitous Love or Not? Animal Welfare and Animal-Informed Consent in Giant Panda Tourism. *Animals* **2023**, *13*, 718. [CrossRef] [PubMed]
9. Rogers, S.; Lee, N.Y.P.; White, J.; Bell, C. Perceptions of Cross-Cultural Challenges and Successful Approaches in Facilitating the Improvement of Equine Welfare. *Animals* **2023**, *13*, 1724. [CrossRef] [PubMed]
10. Cui, L.; Tang, W.; Deng, X.; Jiang, B. Farm Animal Welfare Is a Field of Interest in China: A Bibliometric Analysis Based on CiteSpace. *Animals* **2023**, *13*, 3143. [CrossRef]
11. Mata, F.; Jaeger, B.; Domingues, I. Perceptions of Farm Animal Sentience and Suffering: Evidence from the BRIC Countries and the United States. *Animals* **2022**, *12*, 3416. [CrossRef] [PubMed]
12. Chaney, P.; Jones, I.R.; Narayan, N. Beyond the Unitary State: Multi-Level Governance, Politics, and Cross-Cultural Perspectives on Animal Welfare. *Animals* **2024**, *14*, 79. [CrossRef] [PubMed]
13. Szydlowski, M. Wicked Problems, Novel Solutions: Nepalese Elephant Tourism and Conservation. *Animals* **2024**, *14*, 171. [CrossRef] [PubMed]
14. Hooper, J. Thinking with Civets: The Role of Zoos in the Decolonisation of Animal Tourism. *Animals* **2023**, *13*, 1739. [CrossRef] [PubMed]
15. Aplin, L.M. Culture and cultural evolution in birds: A review of the evidence. *Anim. Behav.* **2019**, *147*, 179–187. [CrossRef]
16. Sapolsky, R.M. Social Cultures among Nonhuman Primates. *Curr. Anthropol.* **2006**, *47*, 641–656. [CrossRef]
17. Whitehead, H.; Rendell, L. *The Cultural Lives of Whales and Dolphins*; University of Chicago Press: Chicago, IL, USA, 2014.

animals

MDPI

Essay

Thinking with Civets: The Role of Zoos in the Decolonisation of Animal Tourism

Jes Hooper

The Department of Social and Political Sciences, Philosophy, and Anthropology, The University of Exeter, Exeter EX1 2LP, UK; jh1220@exeter.ac.uk

Simple Summary: The conditions afforded to zoo animals across the globe remain diverse and disparate. This paper brings two forms of animal commodification into dialogue: (1) the management of civets in global zoos and (2) the rising trend in civet coffee production and tourism in Asia. By qualitatively analysing the entanglements between colonialism, animal welfare, and conservation in civet tourism, this paper calls for enhanced reflexivity and decolonisation of animal-based tourism. Suggestions are made on how zoos may answer this call.

Abstract: Civets belong to the family *Viverridae*, an ancient line of 'cat-like' animals. Despite their large geographic distribution across southeast Asia and parts of Africa, little scientific attention has been attributed to *Viverrids* or *Viverrid*–human relations. This paper applies the lens of civets to explore the tensive intersection between animal welfare, conservation, and colonialism within the tourism landscape. Through thinking with civets, this paper brings two forms of animal commodification into dialogue: (1) the management of civets in zoos around the globe and (2) the rising trend in civet coffee production and tourism in Asia. By qualitatively analysing the entanglements between colonialism, animal welfare, and conservation and how each impacts the lives and treatment of civets in tourism, this paper calls for enhanced reflexivity and thus the decolonisation of animal-based tourism. Suggestions are made on how zoos may progress towards the decolonisation of animal tourism, and the argument is made that zoos are well positioned and morally obligated to answer this call. By doing so, greater attention can be given to the animals whose lives are most affected by the global tourism landscape.

Keywords: *Viverridae*; zoos; animal welfare; conservation; decolonisation; animal tourism

1. Introduction

I will always remember the first time I encountered a member of the *Viverridae* family. As an eager undergraduate student studying Animal Science, I would frequently volunteer at my nearest accredited zoo. A small, family-run business, the zoo had a modest collection of exotic and domestic species. With a young primary target demographic, there was much emphasis on children's education. Fun facts were painted on signs throughout the animal collection, and children were encouraged to take part in 'Zoolympics'—activity stations placed intermittently throughout the zoo trail where children and their guardians could see if they could run as fast as a cheetah or jump as far as a kangaroo. Of the animal displays, there was an overrepresentation of round-faced and fluffy species, from domestic rabbits to red pandas—those with charismatic features that zoo visitors consider most appealing [1].

As I arrived for my shift on this particular day in 2011, the mood in the zoo felt different. There was an excited murmuration of volunteer gossip filling the air as rumours abounded about the latest zoo arrival—today was the day that the new binturong (*Arctictis binturong*) exhibit opened to the public. Having never heard of this species before, I eagerly awaited my opportunity to go and see them. As soon as it was lunch break, I diverted away from the direction of the food stalls and headed to the new binturong enclosure. Whatever image of an animal I could have conjured would certainly have been less unusual than that which met me upon my arrival. I considered the animals who looked back at me nothing short of captivating. Weighing up to 20 kg and measuring ~1.8 m from head to tail, with large paws, long shaggy grey and black fur, and elliptical pupils, binturong also possesses a powerful prehensile tail for arboreal locomotion [2]. Known affectionately as 'bearcats' given their size and anatomically cat- and bear-like features, binturong are an ideal species to engage zoo visitors [3].

Their large size and preference for height make binturong visually noticeable. Despite their crepuscular nature, binturong move around their enclosure and rest in full view of zoo visitors, and their activity patterns are unaffected by the presence or noise of spectators [4]. Their lack of direct relatedness to bears or cats makes for fun facts for enclosure signage, and their musk secretions give off a distinct pop-corn smell [5] for a multisensorial immersive zoo experience. Smell is an important sense for forming memories and arousing feelings [6], so odour may offer zoo education programmes an additional factor to contribute to long-lasting educational messages. Binturong are also formally recognised as in need of conservation, listed as 'Vulnerable to Extinction' (VU) on the IUCN Red List of Threatened Species [7], another characteristic that is favoured by zoo visitors [8]. On reflection, it was unsurprising that binturong were imbued with exoticism and mystery to me, for despite having lived alongside humans throughout history, their illusive nature had left binturong and their civet (*Viverrid*) relatives as an understudied, albeit ancient, set of species [9]. The inclusion of *Viverrids* in zoos could also provide an opportunity to enhance scientific knowledge. In all, binturong make for a sensible species selection for zoos, as curators increasingly attempt to fulfil a variety of (often conflicting) institutional objectives.

Most accredited zoos cite their primary objectives as conservation, education, recreation, and research [10], though academics have often debated the ability of zoos to fully meet these objectives [11–13]. Some scholars have condemned the keeping of endangered animals captive as nothing more than a "macabre" exercise of human curiosity [14], whereas others have noted the benefits of zoos as spaces for fostering pro-conservation behaviour [15] and contributing to the preservation of declining species [16]. Within anthropology, the zoo is notable for its geographical and theoretical positioning alongside museums [17,18], both of which have emerged as places where specimens are displayed in pursuit of human knowledge [19–21], see the zoo not as a space for animals but as human-constructed cultural institutions that disseminate social and cultural meaning through the display of live animals ([20,21] are translated to English in [22]). This article is grounded in a similar anthropological perspective. I am concerned both with the history and the future possibilities of zoos and their role within contemporary global animal tourism. I am especially interested in the tenacious intersection between animal welfare, conservation, and colonialism within the tourism landscape.

The way zoos are represented in different cultural contexts and how accredited zoos in economically advanced nations position themselves on the global stage is an area of research that has gone relatively understudied, as is the link between colonial narratives and animal welfare within tourism. As I will go on to explain in the forthcoming sections, the role of zoos throughout Europe in the 19th to 21st centuries has shaped and responded to public perceptions of and concerns for animal welfare in tourism. However, as I will demonstrate, there are many ways that accredited zoos fail to adequately address animal welfare in the global tourism and leisure space, in which they are themselves firmly situated. Nor do they always reflect upon their colonial histories when collection planning or offering recompense to economically emerging nations within endangered species home ranges. While zoos have evolved substantially from their curiosity-driven emergence to their current position as leaders in conservation, education, animal welfare, and research, it was colonialism that made the modern zoo's emergence possible [23]. It is with the lingering shadow of the zoos' colonial origins that this article is concerned, as I ask for further reflection on how zoos may continue to move beyond their colonial past. Just as [24] advocated for the decolonisation of tourism literature, this essay advocates for the decolonisation of animal tourism and argues zoos are morally and situationally well positioned to pursue this mission.

This essay contributes to the growing anthropological literature concerned with the 'species turn', the refocusing of scholarly inquiry towards the role of animals within human-dominated societies [25]. It is part of an ongoing multi-year investigation into human–civet relationships that arguably became embedded in my consciousness during my first encounter with a binturong. For the past four years, I have been conducting a multi-sited and multi-method approach to the anthrozoological investigation, a critical analysis of contemporary human–civet relationships. This article is a qualitative literature review that has formed alongside the collection of empirical data, which is presented elsewhere in the literature (see [9,26,27]). Through thinking with civets in this essay, I first address the colonial underpinnings of zoos, their advances towards improved standards of animal welfare, and the demographics of *Viverrid* species housed in the zoos. I then introduce the colonial origins of civet coffee and the emergence of civet coffee tourism before discussing the commonly associated animal welfare and conservation issues and the ways zoos are used by tourists to condone and condemn poor animal welfare practises. Finally, I conclude the article with a series of recommendations that accredited zoos could strive towards in a bid to decolonise global animal tourism. While some of the recommendations I make are theoretical and thus somewhat limited in their ability to be directly implemented, I hope that this essay provokes conversation and inspires accredited zoos from economically advanced nations to adopt greater reflexivity in their current and future practises.

2. Zoos: Past and Present

The colonial histories of zoos are well documented in anthropological and historical texts [23,28,29]. While many of the first living animal specimens entered Europe via the slave trade routes of the 1770s [30], by the 1900s, humans were also subject to zoo display as a spectacle of the primitive [31]. Certainly, entertainment and a desire to satisfy human curiosity about the exotic were prime directives of early public zoos [28]. Enclosures were built with panoptic displays designed for unobstructed onlooker observation [32], and visitors could ride on the backs of elephants or watch chimpanzees perform tea parties and drink alcohol [33]. It was in the 1950s that zoo animal welfare began to take centre stage as zoos (particularly those in Europe and North America) were pressured by the general public to improve captive conditions for the animal's experience [34]. Such a shift came in the wake of increased accessibility to television and the popular broadcasting of nature documentaries, which showcased for the first time the natural habitats from which zoo animals had been separated. Sir David Attenborough's television series 'Zoo Quest', for example, transported UK viewers to the wild to see zoo animals in their "proper setting" [35]. The imagery of wild animals immersed in natural landscapes and expressing active species-typical behaviour such as play, allogrooming, hunting, and foraging inadvertently lead viewers to question the appropriateness of zoo environments. Thus, it is notable how animal welfare and ethics are entangled. Where animal welfare is used to describe an animal's psychological and physical health [36], where their ability to cope with their environment can be measured by parameters such as intrinsic need [37], it was in questioning the ethics of whether animals were deserving of having these needs met that animal welfare became central within the zoo industry. Therefore, from the mid-20th century, naturalistic and immersive enclosures were more systematically adopted in zoos in economically advanced nations, along with a renewed commitment to prove to the public that the zoo as an institution remained a valuable societal asset for both humans and animals [10].

Today, the global modern zoo industry attracts more than 700 million annual visitors [38], and visitor satisfaction continues to be of paramount importance. Zoo visitors not only claim to rate animal welfare as a primary concern when visiting zoo collections [39], but zoo visitors are said to learn more [40], view zoos more positively [41], and commit to pro-conservation behaviour more readily [42] when they have witnessed zoo animals performing natural and desirable behaviour. Similarly, the viewing of animals expressing inactivity or stereotypic behaviour can negatively impact the experience of zoo visitors [43]. When animal welfare negatively shapes the zoo visitor's experience, this in turn risks the zoos' financial security and their ability to meet their education and conservation objectives.

The formation of global and regional accreditation programmes, such as the World Association of Zoos and Aquariums (WAZA), European Association of Zoos and Aquariums (EAZA), Association of Zoos and Aquariums (AZA), British and Irish Association of Zoos and Aquariums (BIAZA), and South East Asian Zoo Association (SEAZA), to name a few, have each emerged in response to public demand to hold zoos accountable to higher standards than are required by law in terms of institutional commitments to animal welfare, conservation, and visitor education. The accrediting bodies advocate for a holistic approach to animal welfare provisions and assessments. The WAZA defines animal welfare, for example, as "a state that is specific for every individual animal; it is how the animal experiences its world and life through its association with pleasant experiences specific for that species such as vitality, affection, safety, and excitement, or unpleasant experiences such as pain, hunger, fear, boredom, loneliness, and frustration" (WAZA, 2023). Accredited zoos are, therefore, guided by comprehensive animal welfare policies, which describe the relationships between animal health, nutrition, behaviour, mental state, and environment. Zoos are also required to assess and monitor the welfare of their animals based on these multiple indicators (see [44,45] for examples). However, animal welfare comes at a high economic price, as does ex situ conservation. Thus, the conditions afforded to zoo animals across the globe remain diverse and disparate. Where high welfare standards, including veterinary care, naturalistic enclosure design, and species-specific nutrition, are often synonymous with accredited and legislated collections, such standards are more commonplace in economically advanced nations [46]. Zoos in the United States alone spend upwards of $1 billion per year on operational costs such as animal food, veterinary care, and enclosure maintenance [47]. Artificial insemination, gamete or living animal transfer, and pre- and post-natal care take knowledge, time, and monetary resources, as does the loan of breeding animals of high conservation and political value (see [48] for giant pandas).

Through the lens of captive breeding programmes, we can also see the entanglement between animal welfare, ethics, economics, and conservation. Breeding is not only paramount for conservation, but baby animals also make zoos more attractive to paying visitors [8]. However, zoo animals are bred in captivity for a variety of reasons, from the maintenance of genetic diversity in assurance populations to providing optimal stocking densities to promote natural social behaviours [49]. While some breeding programmes facilitate conservation via species reintroduction [50,51] others are conducted to supply animals for surrogacy and mentorship purposes [52,53]. Thus, the drivers behind captive breeding can be unique and not always obvious to the visiting public. Breeding success can also be impacted by a range of factors. Mate choice [54], individual personality traits [55], environmental parameters [56], social groupings and reproductive ageing [57], and birth origin [58] are just some of the factors that can impact mate success. Additionally, many of these factors can be present for newly acquired species where species-specific knowledge is lacking and so trial-and-error practises are required [50]. Although the level of the conservation value of captive breeding can differ between programmes, animal welfare is integral to captive breeding regardless of the programme's purpose. Ref. [59] goes as far as to describe the lack of mate success as a form of "reproductive strike", a form of resistance to inadequate captive management. Indeed, if a breeding animals' welfare is compromised, then the likelihood of its breeding can also be cast into doubt [60]. However, irrespective of its importance, it remains puzzling that animal welfare is largely absent from accredited zoo mission statements [10]—perhaps its absence is in recognition of the way zoo objectives can and often do conflict in practise [61].

As an example, zoo visitors are said to be most positively impacted by their feelings of connection with zoo animals [15]. This has, in turn, resulted in the continuation of animal performances and intimate human-animal encounters such as animal feeding or 'keeper for the day' experiences. Ref. [62] found that of 1241 surveyed zoo facilities, 75% of accredited zoos offered animal-visitor experiences. All in all, animal-visitor activities were most popular in advanced economic regions, as 90% of North American zoos and 75% of European zoos offered some form of animal-visitor interaction beyond passive observation. Rates were lowest in South America, where 50% of accredited zoos offered animal-visitor experiences. Of all the recorded animal-visitor experiences on offer, 43% of accredited zoos across the world advertised hands-on petting experiences, 28% offered non-hand-feeding interactions, and 23% offered direct hand-feeding opportunities [62]. Such activities have led to concerns that close-contact human-animal interactions present mixed messages to zoo visitors and could therefore promote unethical, low-welfare interactions beyond the zoo visit [63]. Amongst these concerns is that the performative display of exotic species serves as an extension of the zoological industries' colonial histories, whereby animals serve as entertainers under the guise of educational opportunities [14]. Furthermore, although accredited zoos can house confiscated, abandoned, or rescued animals, including local native wildlife [64], the species selection in zoos is still predominately tied to entertainment-driven colonial interests. Zoos in economically advanced countries continue to prioritise the housing of charismatic endangered species, the two characteristics that zoo visitors value most [1].

Through zoo visitation and portrayals in popular media, charismatic endangered species have, however, been found to inspire pro-conservation behaviour [65]. The role of ambassador animals has been widely discussed in the literature, with concepts such as 'flagship' or 'umbrella' species undergoing several revisions. Such terms often describe animal species that can serve multiple conservation-education outcomes [66–68]. Indeed, [68] suggest that "a species' potential usefulness as a flagship for conservation could be given added value by its ability to act as a marketing surrogate for a large number of high-priority species." Where most ambassador species are endangered, [69] proposes the use of non-endangered 'proxy species' to represent endangered sister taxa with whom they share a physical resemblance. In a similar line of thought, later in his paper, an argument will be made that neither the ambassador nor its benefactor need be restricted to species formally recognised as endangered. Not only are many non-endangered taxa already in steady decline, but ambassador species could promote good animal welfare practises as well as conservation messages. First, I turn our attention back to the *Viverridae* family as a case study of population and taxonomic trends in species representation in the global zoo industry.

3. Civets in Zoos

As demonstrated in Table 1, the most popular *Viverrids* housed in zoos across the world are binturong, and approximately 60% of the world's captive binturong population is housed in zoos within advanced economies rather than in zoos within the species' natural range. (The data presented in Table 1 were grouped into Emerging and Advanced Economies as defined per country by the International Monetary Fund (see https://www.imf.org/en/Publications/WEO/weo-database/2023/April/select-country-group, accessed on 25 April 2023)) Other *Viverrid* species of conservation concern (listed as either vulnerable, 'VU', or endangered, 'EN', on the IUCN) include the Banded palm civet (*Hemingalus derbyanus*), Golden palm civet (*Paradoxurus zeylonesis*), Large-spotted civet (*Viverra magaspila*), and Owston's civet (*Chrotogale owstoni*). Only Owston's civet and Banded civet are housed in zoos in economically advanced economies and far fewer numbers than binturong. One contributing factor could be heightened rates of reproductive success for binturong, yet this raises questions as to the strategic value of captive breeding for conservation purposes when conducted so far from a species' natural habitat. While conservation concern is a likely factor under consideration when collection planning, the binturong's large charismatic appearance and crepuscular nature are more likely to contribute to their heightened representation within the zoo industry when compared to related species.

However, the question remains as to how effective binturong are as zoo ambassadors. How well do binturong in zoos represent the welfare and conservation interests of their wild counterparts and *Viverrid* relatives?

Table 1. Number of Viverrids housed in zoos recorded on the Zoological Information Management Systems (ZIMS) database as of 2023 [70]. (Table 1 comprised 365 holding collections across the globe that house *Viverrids* and use the ZIMS database to record their species data. Of these holding collections, 84% (306) were accredited by WAZA and/or regional or national associations, and only 16% (59) were unaccredited. Thus, data are lacking for unaccredited facilities, and so Table 1 represents a conservative estimate of *Viverrids* housed as part of global zoo tourism. While not all zoos across the globe use ZIMS for storing their records, the ZIMS database is the most comprehensive zoo database in the world (see https://species360.org/, accessed on 18 April 2023). Therefore, at the time of writing, the data presented here present the most accurate picture of *Viverrid* holdings currently available. The monetary total has been converted from Euros to US dollars for consistency).

Species	IUCN Status	Emerging Economies	Advanced Economies	Total
African civet, *Civettictis civetta*	LC	5	15	20
Banded linsang, *Prionodon linsang*	LC	5	0	5
Banded palm civet, *Hemingalus derbyanus*	VU	3	10	13
Binturong, *Arctictis binturong*	VU	199	288	487
Common palm civet, *Paradoxurus hermaphroditus*	LC	124	24	148
Forest Genet, *Genetta vistoriae*	LC	0	6	6
Genet (*species undocumented*)	LC	12	4	16
Golden palm civet, *Paradoxurus zeylonesis*	VU	9	0	9
Hausa genet, *Genetta thierryi*	LC	0	2	2
Large Indian civet, *Viverra zibetha*	LC	19	0	19
Large-spotted civet, *Viverra magaspila*	EN	5	0	5
Large-spotted genet, *Genetta pardina*	LC	20	19	39
Masked palm civet, *Paradoxurus larvata*	LC	36	51	87
Owston's palm civet, *Chrotogale owstoni*	EN	16	9	25
Small Indian civet, *Viverricula indica*	LC	29	1	30
Small-spotted genet, *Genetta genetta*	LC	38	41	79
Small-toothed palm civet, *Arctogalidia trivirgata*	LC	16	15	31
Spotted linsang, *Prionodon pardicolor*	LC	1	0	1
TOTAL		537	485	1022

The effectiveness of zoo ambassadors has been long debated in terms of meaningful education [71,72]. Ref. [3], however, argue that binturong are valuable species ambassadors, stating that their inclusion in accredited zoos across the world has helped to generate US $67,520 towards in situ conservation efforts. (The monetary total has been converted here from Euros to US dollars for consistency.) Indeed, ambassador species are often effective in inspiring the public to contribute financially towards conservation initiatives [68], and accredited zoos are an important societal resource for collating conservation funding. The WAZA-accredited zoos, for example, are said to collectively contribute an estimated US $350 million towards wildlife conservation each year [73,74]. In 2021, the AZA-accredited zoos reported spending over $25 million across 30 endangered species as part of the AZA 'Saving Animals from Extinction' (SAFE) programme. Its members also reported spending a combined $216.9 million on field conservation programmes in the same year, to the benefit of 954 species [75]. Thus, accredited zoos report a significant contribution towards global ex situ and in situ conservation efforts. According to [76], in the most recent peer-reviewed research investigating the regional contributions of zoos towards conservation, it was zoos and aquariums in North America and Europe that "spent the most by far on wildlife conservation (97% of expenses reported)". Thus, on this basis, it could be argued that housing endangered and charismatic species in zoos in advanced economies is a strategic benefit to global conservation. However, in reaching the sum of zoos' conservation contributions, zoo accreditation bodies rely on zoos to self-report their spending, as did Gussett and Dick, who defined wildlife conservation broadly as "in situ conservation of wild species and habitats, including related ex situ work". By this definition, zoos that reinvested in their captive facilities could claim that doing so was a form of wildlife conservation expense. Furthermore, their research highlighted that zoos in Asia received the greatest numbers of zoo visitors than either North America or Europe. It would be useful, therefore, to know how much zoos in economically advanced regions work with zoos in economically emerging ones and how much consideration visitor demographics are given when collection planning. Conservation education may be better served by targeting the demographics that have greater power to affect positive change for animals by presenting species that visitors have a greater likelihood of engaging with outside of the zoo. One of the biggest threats facing all civet species, for example, is the civet coffee production and tourism industry. Given that consumers from economically advanced nations contribute to global demand [77], the conservation and animal welfare implications of civet coffee ought to be a focal message for *Viverrid* holding collections, and yet very few species most impacted by civet coffee (such as the common palm civet, *Paradoxurus hermaphroditus*) are represented in zoological collections in regions contributing towards civet coffee consumer demand. Furthermore, preliminary research shows that few European zoos that house binturong include civet coffee as an educational message in their collections (Hooper, *unpublished thesis*), despite ethical consumerism being of interest to zoo visitors [78].

As demonstrated in Table 1, common palm civets are notable for their poor representation within zoos in advanced economies. Although perhaps one of the less aesthetically charismatic *Viverrids*, the common palm civet would be well suited to zoo representation. Common palm civets reproduce well in captivity and so would not require the captive population to be supplemented with wild animals to maintain genetic diversity and desired zoo stocking densities [27], and they can adapt well to captive settings when high welfare standards are upheld [79]. As the most used species in civet coffee production and tourism, their absence is a missed opportunity for zoo education. Thus, it may be of little surprise that tourists who visit *Viverrid* range countries are often unaware of the civet species they are most likely to come into contact with, nor are they aware of the welfare or conservation issues associated with civet coffee production [26]. As I will now demonstrate, zoos are already enmeshed in the political and tricky intersection between animal welfare, conservation, and ethics in global animal tourism, and so responding to these issues could be seen as a moral imperative.

4. Civet Coffee

Civet coffee is a luxury coffee produced through the digestive tract of the common palm civet and is said to be less bitter and smoother in taste than non-digested varieties [80]. Civet coffee has secured a lucrative commodity value through pervasive marketing claims of the product's rarity and unique production method, and its origin story has become highly fetishized throughout high-income nations despite the product itself being steeped in colonial heritage [9]. Civet coffee, also referred to in Indonesian as 'kopi luwak', is said to have been discovered in Indonesia during the time of Dutch colonial rule at the end of the 16th century [81]. Indonesian farmers were instructed by Dutch colonisers to grow and harvest coffee, though they were prohibited from consuming it themselves. As the story goes, coffee farmers observed that wild civets would enter the coffee plantations at night, where they would select only the ripest coffee cherries for consumption. The only trace of these elusive animals left behind were their faecal droppings, which were studded with partially digested coffee cherries. Forbidden to sample from their own coffee harvest and curious about the caffeinated beverage, coffee farmers turned to the civet faeces upon which the beans were removed before cleaning, drying, and roasting them [82]. Upon recognising this local phenomenon, the Dutch rulers are said to have noted the coffee's seemingly enhanced taste characteristics [83].

Rising to international fame in the early 2000s after it was featured in a series of Hollywood films and prime-time television programmes in Europe and North America [26], civet coffee's novel production method, romanticised origin story, and proclaimed rarity saw consumer demand rise exponentially [84]. Whether contemporary civet coffee is a product of authenticity, however, is debatable. Early studies confirmed that civet coffee possessed unique structural and chemical characteristics, said to be the work of the civets' digestive enzymes [80]. Indeed, civet digestion has evolved to facilitate a varied meso-carnivore diet [82]. More recent research, however, has challenged this analysis and identified flaws in the method of the original research [9,85]. Civet coffee collection methods are also widely falsified [86], and although consumers are led to believe civet coffee is produced through the collection of civet faeces by Indonesian farmers, the reality is the mass farming of caged civets to meet the demand of the global market [87]. Civet coffee's global market is currently increasing and is expected to reach US $11.75 billion by 2030 [77].

4.1. Civet Coffee and Civet Welfare

As a common species without conservation protections, common palm civets are captured from the wild to supply the civet coffee industry. While the civet coffee phenomenon originates in Indonesia, its global market growth has seen the emergence of civet coffee farms across the species' distribution, including Indonesia, Singapore, Vietnam, and India [77]. The capture of wild civets can take the form of caged capture or snaring, and so civets can often be inflicted with injuries through the process of wild harvesting [86]. Although civet coffee farming is permitted by law in several range countries, research has shown that civet coffee farms are often over the legal stocking density [86] or include endangered civet species [88,89].

On civet coffee farms, civets are housed in rows of stacked cages, usually with one animal per cage, with each cage typically no larger than a small domestic dog crate consisting of slatted or mesh floors designed for ease of faecal collection [27]. Civet welfare is highly compromised in these conditions, as civets are forced to sleep and stand on a wire [90], conditions well associated with injury and abrasions [91]. Through such confinement, civets are also forced to remain close to each other, despite members of the *Viverrid* family expressing a preference for solitary behaviours [92]. Thus, the smell, sound, and sight of other animals can cause significant psychological distress, as civets are unable to hide or increase their distance from caged neighbours. Cage-related injuries such as abrasions to paws and faces and stereotypic behaviour, including pacing and self-mutilation, are all common occurrences [26,86,90].

To obtain the most productivity from the farm practise, farmers report feeding their civets a diet either mostly or exclusively comprised of coffee cherries, and captive civets often die from lack of nutrition and caffeine toxicity [93]. Rather than seeking veterinary care for sick individuals, farmers have been known to release civets into the wild once they become ill, only for the same civet to be recaptured upon recovery [88]. Despite legislation in Indonesia specifying that farmed civets must be released after five years of coffee production, the same legislation encourages the release of civets into the civet pet trade [94]. After five years in these facilities, few civets would be likely to survive in the wild due to psychosis and injury. However, the transfer of civets from coffee farms into the pet industry only continues their exploitation. Upon retirement from civet coffee farms, civets are usually too aggressive and their physical condition too unappealing to be rehomed as companion animals. Thus, individuals may find themselves used as breeding stock instead. Preliminary research into the civet pet trade illustrates comparable conditions between pet and civet coffee-farming facilities [27].

4.2. Civet Coffee and Civet Conservation

While it is illegal in many civet coffee-producing countries to harvest *Viverrid* species protected by international conservation agreements (such as binturong, which are listed as 'vulnerable' and Owston's civet, which are listed as 'endangered' by the IUCN), these protections are poorly enforced in practise [86] and the indiscriminate capture methods (particularly snaring, which is endemic throughout Vietnam, Willcox et al. [95], puts all civet species (and other wildlife) at risk of physical and psychological harm. While conservation statuses are necessary to safeguard the most vulnerable species from unsustainable exploitation, conservation-based protections do not attribute value to the intrinsic interests of the individual and so overlook the relationship between individual animal welfare and the population health of a species. The social and ecological impacts of the overharvesting of civets are not well understood, nor is it clear how the mass capture of common palm civets from the wild for civet coffee production is impacting species population health and sustainability [87].

The assignment of species conservation statuses can also have colonial implications. Species classification is often non-reflective of how native species are perceived and engaged with by locals. The change in the conservation status of the binturong from 'Least Concern' to 'Vulnerable to Extinction' in 2008, for example, resulted in many Indonesian civet coffee farmers facing fines for housing binturong captive alongside common palm civets. As described in the ethnographic investigation of civet coffee production by Cahill [89,96], it is customary practise in Indonesia for civets of all species to be identified as one umbrella 'musang' or 'luwak' ('civet') species. Thus, where international agreements between countries recognise the difference between civet species and conservation needs, for many local people in range countries, these differences are seen as variations in civet morphology. The overnight change in legislation resulted in many civet coffee farmers receiving fines for holding binturong captive, and as word spread of the surprise inspections, many more farmers began releasing binturong from their collections [89]. While the release of wild animals may at first appear positive, there is no evidence to say that these animals were released in or near the locations where they were originally captured, and any injuries or behavioural abnormalities obtained in captivity could hinder their chances of wild survival. Thus, the welfare of these animals could have remained compromised even through their release, a phenomenon that resulted from the mismanagement of the species newly acquired conservation status. Thus, we see again how colonialism, conservation, and animal welfare are often entangled.

4.3. Colonisation of Animal Welfare in Civet Coffee Tourism

Alongside civet coffee's global fame has come a rise in civet coffee tourism. Most prominent on the popular tourist destination island of Bali, civet coffee tours facilitate tourists desire to learn about the coffee's unusual production process [26]. Tourists are guided through curated displays of coffee cherries, caged civets, civet faeces, and elderly Balinese women roasting coffee beans on an open fire. The short tour ends with a flute of endemic teas and coffees, typically served with panoramic views of the Balinese jungle and traditional rice paddies. In my previous review of 3364 Trip Advisor reviews (reported fully in [26]), I found that where some tourists claimed their civet coffee purchasing decisions were driven by curiosity or the desire to take part in a perceived 'once in a lifetime experience', others expressed a desire to support what they believed to be an authentic agricultural establishment and traditional practises. However, civet coffee tours are curated enterprises designed to exploit the colonial narrative of civet coffee's origins, thus fulfilling the tourists' expectations of traditional Balinese agri-food practises [97]. In such scenarios, animal welfare becomes secondary to tourist curiosity, with caged civets experiencing similar conditions to those in the intensive civet coffee production facilities.

Carder et al. [90] investigated 16 Indonesian civet coffee tourism sites and revealed significant animal welfare concerns present across survey locations. Display civets in tourist sites were subjected to poor housing and husbandry provisions, none of which were adequate in meeting basic animal needs as outlined by the Five Freedoms. No provisions for enrichment were observed; all civets had either limited or no access to clean water, and the animals were forced to stand and sleep on either wire mesh or slatted floors, conditions similar to those that have been reported for civet production farms [86,88]. Carder et al. [90]

also noted that tour guides openly confirmed the civet's inclusion was for display purposes only, whereas civet coffee production took place in locations non-disclosed to tourists. The secretive nature of civet coffee farming was likened by [89] to [98] 'out-of-the-way places', spaces that are free from the judgement of onlookers. Of course, civets in tourist and production facilities are intimately connected.

Without the demand for civet coffee, there would not be a heightened demand for civet capture, mass farming, or tourism. However, the relationship between consumer behaviour and issues of animal welfare and conservation is not always well understood by visitors to tourist facilities that house living animals. In fact, visitors to animal-based tourism (including zoos and other wildlife interaction experiences) overwhelmingly fail to recognise the negative implications of the experience for animals. The analysis of Moorehouse et al. (2015) of tourist perceptions of wildlife-tourist attractions found only 7.8% of tourists recognised signs of poor animal welfare or negative conservation impacts. My research into civet coffee tourism found similarly low levels. In total, 11.65% of English-speaking national and international tourists raised animal welfare concerns, with only 1.16% noting the possible conservation implications, all of whom had visited facilities where wild-caught civets were either caged or sedated to enable cage-free photographs [26].

Zoos in economically advanced nations can be used by tourists as a mechanism to either condone or condemn the animal welfare implications of civet coffee tours. For the latter, zoos are used by tourists as vehicles of judgement towards the Balinese tour guides for providing captive conditions less suitable for promoting animal welfare than would be experienced by animals in zoos within economically advanced regions [26]. The recognition that captive animal welfare can differ between animal tourism facilities shows that modern zoos have been somewhat successful at 'bringing in' humans to what visitors largely perceive as animal-dominated landscapes [99], even though most zoos in economically advanced regions of the world occur in urban spaces [32]. The rebranding of zoo 'specimens' as 'ambassadors' combined with immersive naturalistic 'enclosures' or 'habitats' rather than 'cages' [100] further acts as a semantic dilution of the historic colonial rhetoric of zoos, and so tourists do not always question the ethics of keeping animals captive. Although advances in zoo infrastructure have improved animal welfare in many zoos across the world, the co-opting of animal welfare as a mechanism for cultural judgement simply highlights the ongoing issue of colonialism within the tourism industry. In such cases, tourists overlook the economic requirements of high welfare practises and, in their judgements, are insensitive to the colonial rhetoric that comes with assumptions of animal cruelty as inherent to a particular culture. Most notably, such commentary is indicative of a lack of reflexivity from tourists who do not always recognise their role as the primary target audience. This issue thus illustrates the need for greater efforts to be made by contemporary zoos in economically advanced nations to be transparent with their visitors about their industries history, the interplay between tourism and colonialism, and how tourists themselves can contribute to positive animal welfare within the tourism landscape.

My research has also shown, however, that accredited zoos can be used by tourists as a way to overlook the welfare implications of the civets' captive conditions [26]. In such cases, tourists will claim to see no ethical difference between Balinese civet coffee tours and accredited zoos, as both hold animals' captive for display purposes. Thus, for these tourists, animal welfare is not a priority consideration for captive management. The perceived societal acceptance of animals used for public display (granted, this is but one objective of the modern zoo environment) overlooks both the complexities involved in providing positive captive animal welfare and the wider conservation impact of unregulated wild capture that is synonymous with the civet coffee industry. In highlighting the utility of the animal's captivity, tourists attempt to negate cognitive dissonance caused by the possible harms incurred by the animals' capture and display. Where tourists commented on there being little difference between animals housed on display in coffee tours and those on display in zoos (see [26]), questions must be raised as to the impact of how modern zoos portray their animals to visitors and the wider public. Indeed, close human-animal encounters are effective marketing tools. Of the tourists who reported being able to pet, hold, feed, or have photographs taken with civets, many went on to recommend the experience to others [26].

In their attempts to meet their educational aims, accredited zoos can muddy the boundaries between education, recreation, and conservation [61]. Inappropriate portrayals of zoo animals can ultimately lead to zoo visitors (both those who visit in person and those who engage with online entertainment-driven marketing campaigns [101]), developing a poor understanding of captive animal welfare. Thus, zoos risk contributing to the endorsement of unethical, unregulated animal-based tourism. Given the frequency at which animal-visitor interactions are facilitated by accredited zoos the world over [62], it is therefore somewhat unsurprising that few tourists consider the negative implications of feeding animals when abroad, whether in free-ranging, semi-captive, or captive settings [102]. Wildlife selfies are equally problematic for wildlife in tourism [103], and tourists may not be able to recognise the animal welfare or conservation implications of paying a fee to pose with wild animals when similar close contact experiences are available from accredited zoos in their home countries. Accrediting bodies have written guidelines for their members when offering animal-visitor interactions, which include measures to ensure that animal welfare and conservation messaging are upheld through the encounter [104]. However, not all messaging is effectively obtained by the public, either when delivered to the visitor partaking in the interaction [105,106] or via marketing media that can lack educational context [107,108]. Given the predicted intensification of demand for human-animal encounters in tourism [36], accredited zoos ought to be cognizant of their formative role as educators for the promotion of positive animal welfare and conservation within global tourism.

5. Towards Decolonisation of Animal Welfare in Tourism

While it is not possible for zoos to completely decolonise, given that the fundamental attribute of zoos is to care for animals in a captive setting, zoos could offer visitors steppingstones towards more ethical human-wildlife encounters within the remit of tourism and, more generally, as consumers within capitalist systems. Indeed, as noted by [63], in relation to the issue of zoos continuing to offer a merger between entertainment and education, "old views are not simply consigned to the dustbin of history when new ones emerge. Instead, they tend to linger, overlapping with newly emergent ideas". With 700 million people said to visit accredited zoos globally each year [38], contemporary zoos are in a powerful position to advocate for greater reflexivity to be adopted by tourists and the tourism industry concerning the treatment of animals.

First, however, the accredited zoo industry must recognise and reflect on current practices that overlap with the industries' colonial origins; origins that accredited zoos have made substantial strides to move beyond. For one, the ways accredited zoos engage visitors with animals (particularly the feeding and petting of wild animals) ought to be re-evaluated to ensure tourists are clear on the distinctions between high- and low-welfare animal attractions when faced with similar opportunities in unregulated facilities. Zoos could also offer more tangible ways for visitors to reflect on their behaviour and their impact on species protection by building on previously successful ethical consumerism campaigns. Zoo visitors are more likely to adopt pro-conservation behaviour during their visit than after [42], and some zoos have reported highly successful engagement rates when bringing ethical consumption into the zoo space. Smartphone recycling stations at gorilla exhibits to educate visitors about coltan mining [109] and sustainable palm oil educational initiatives at orangutan exhibits [78] have shown innovation within the zoo industry towards effective education concerning the topics of sustainability and ethical consumerism. However, these initiatives are not synonymously adopted throughout accredited zoo facilities, and many species (including civets) could contribute to these initiatives. Thus, a more nuanced species selection process could be adopted. By doing so, finite resources could be better optimised in terms of achieving more of the zoo industries contemporary objectives while also addressing the overrepresentation of endangered or otherwise charismatic species [69,110].

The first suggestion I make, therefore, towards the decolonisation of animal welfare and conservation within animal tourism is for the selection of zoo species to be made based upon a holistic assessment of the species' ability to meet education, conservation, and animal welfare objectives. The repatriation of endangered species to their countries of origin should therefore occur where either: the species in question is unlikely to be returned to the wild from their present location outside of their range country; the captive population is too small or genetically unsustainable to be managed in captivity without the further introduction of wild-captured individuals; insufficient funds are being redirected from the holding zoo to in situ conservation initiatives; or the dominant conservation threat can be more effectively addressed by engaging local communities in the species native range countries. Thus, while not all endangered species would be phased out of economically advanced zoos, the zoo industry would be increasingly accountable for their contribution to the species' protection while optimising the limited resources available to them.

In part to address possible concerns related to the disparate animal welfare conditions between zoos in economically advanced and emerging economic regions, where some of these species would be returned, programmes should be established between the current holding collections in colonial nations and accredited zoos in range of countries. These programmes should serve to support non-accredited zoos throughout the species range to deliver higher standards of animal welfare and education so that they can achieve accreditation. Thus, a global cooperative of zoos should contribute to sustainable tourism development [111]. In doing so, zoos would better serve the interests of animal welfare and conservation as a global animal tourism network, thus reaching more animals and more human visitors across the world. A unified global approach to the issue of animal welfare in tourism is needed, as there is currently no global governing body for the various forms of wildlife tourism attractions on offer [112]. While it is not to say that all modern zoos are failing to support in situ initiatives or falling short of their animal welfare obligations, it must be recognised that not all accredited zoos are fulfilling their potential in meeting these objectives [113]. Nor do zoos adequately respond to the issue of poor animal welfare within global tourism. A more holistic framework for collection planning in cooperation with zoos in species home ranges would thus assist zoos globally in meeting such objectives while decolonising animal tourism from within.

To ensure the welfare of animals during repatriation efforts, I recommend a phased programme be implemented for individuals whose welfare may be compromised by travelling or whose survivability might be reduced by their transfer. For example, this approach would apply if the animal were elderly, unwell, or where there was an insufficiency of captive facilities available in their range country that could maintain comparatively high levels of welfare. In place of the species removed from zoos within non-range nations, their non-endangered proxy relatives should be housed instead (either once the original individual has died, been retired from the public-facing collection, or after repatriation). Thus, zoos would begin transitioning to a proxy species approach, a form of surrogate defined as "a non-threatened species that are morphologically similar [to an endangered species] but not otherwise represented in zoos" [69]. A proxy approach would allow more zoo animals to fulfil the role of conservation ambassadors while keeping the rarest species within their range countries. I would, however, extend the definition of proxy species to include the representation of species either already formally recognised as endangered, those in steady decline, or those otherwise threatened by poor welfare practises. Thus, proxy species could provide both conservation and animal welfare education messages.

A greater representation of common palm civets in zoos within economically advanced regions would also offer further opportunities to educate zoo visitors about their role in global animal welfare issues. The lack of common palm civets in zoos in economically advanced nations means the only place to see common palm civets and learn about civet coffee in-person is through visiting locations in Asia where civets are extracted from the wild and housed in conditions with proven welfare implications [88,90]. The display of common palm civets in zoos in economically advanced regions could decrease the allure of the species exoticism associated with civet coffee. Education messages concerning the role of consumerism in global animal welfare and conservation issues would provide visitors with tangible ways to contribute towards sustainability through ethical consumer practises. Indeed, zoo visitors have reported a lack of tangible actions as a barrier to adopting pro-conservation behaviour beyond donating money [114]. Furthermore, the broadening of educational messages in zoos to include global animal welfare concerns would optimise the educational potential of ambassador animals. Of course, in the case of civets, this would continue to require zoos to carefully consider how the species is portrayed so as not to increase the novelty appeal of civet coffee or to promote close-contact interactions, which could be harmful to zoo residents and other animals in tourism.

My second suggestion for the decolonisation of animal tourism, therefore, is for zoos in economically advanced nations to reflect on their portrayals of wild animals to their zoo visitors. The growing interest in the relationships held between humans and animals in tourism [114–118], and the shift towards what Caton [119] calls the "moral turn" in tourism scholarship towards research that advocates for sustainable tourism, is indicative of a theoretical shift toward a species turn already underway in Anthropological practise [25]. As long-standing features of human societies and icons of animal tourism across the globe, modern zoos must respond to the call to enhance the welfare of animals in the tourism landscape more broadly than within the confines of the zoo walls. As of yet, the conservation and animal welfare messages presented to tourists have been inconsistently delivered as zoos increasingly rely on close contact with animals experience, such as petting, feeding, and posing with wild animals, to attract zoo visitors [62]. Should zoos be considered of paramount importance for the conservation of species and the pro-conservation education of the public, then zoos across the world require additional support to meet these objectives. Indeed, in the age of the Anthropocene, the sixth mass extinction largely due to humans [120], the role of zoos in global conservation ought not to be understated [113]. If zoo institutions were less dependent on revenue generated by zoo visitors, zoos would have less need to monetise their animal residents. Thus, the commodity value of zoo animals could fully progress from entertainer to conservation and animal welfare education ambassador. To this end, national governments and local authorities must support zoos financially, recognising the increasing need to respond to the current ecological crisis.

Finally, I argue that contemporary zoos in high-income nations ought to be transparent with their visitors about the advances made for animal welfare and conservation since the industries' emergence in the colonial period. While the repatriation of endangered species to their native countries and the cessation of animal encounters and performances would aid with this objective, for full educational benefits, zoo visitors should be informed about the progressively evolving nature of animal welfare and conservation science. Of particular focus should be the role of the public in demanding higher welfare standards for zoo animals, to encourage personal reflexivity towards ethical consumer and tourist behaviour.

6. Conclusions

In this paper, I have brought two forms of animal commodification into dialogue, namely the management of civets in zoos and the rising trend in civet coffee production and tourism. Through an analysis of data held on the global database 'Zoological Information Management Systems' (ZIMS), I have shown that charismatic, rare, and endangered *Viverrids* are most commonly represented in zoos within economically advanced regions of the world, far from the species natural range countries. ZIMS records also showed that common species, such as the common palm civet, were rarely present in zoos in economically advanced non-range countries, despite the numerous educational benefits the species could provide to zoo visitors. Although the IUCN lists the Common palm civet as a species of Least Concern [121], the civet coffee industry threatens the species both in terms of its conservation and individual welfare. The Common palm civets are enrolled as coffee producers in intensive low-welfare farms and as display animals in civet coffee tourism sites across Asia, industries fuelled by an ever-growing global consumer demand. Thus, common palm civets could educate zoo visitors about the conservation and welfare implications of the civet coffee industry and provide zoo visitors with tangible solutions to these and similar species issues through sustainable consumerism and tourism.

Further analysis of tourist perceptions of civet coffee tourism in Bali revealed the interplay between accredited zoos and unregulated, low-welfare, animal-based attractions. Tourists would either use zoos to justify or condemn civet coffee tourism, and many saw little difference between unregulated facilities and accredited zoos, as both offer a range of close-contact experiences with wild animals. A common theme across tourist perceptions of civet coffee tourism was a colonial narrative that is harmful to both humans and animals. These findings, therefore, highlight the need for further efforts to be made to decolonise global animal tourism. By critiquing the ways that some accredited zoo practises can be reflective of the industry's colonial past, I have argued that accredited zoos are well positioned within the global tourism landscape to lead efforts towards the decolonisation of animal tourism. Indeed, the modern zoo has already made significant strides in moving past its colonial origins.

In all, I make several theoretical suggestions on how zoos could move towards the decolonisation of animal tourism from within. My suggestions include the careful repatriation of rare species to their home countries and the use of proxy species to educate tourists about ethical consumerism and tourism. The animal welfare considerations of repatriation are discussed, along with the need for zoos in economically advanced nations to work collectively with zoos in the range of countries of repatriated animals to ensure high levels of welfare, conservation, and education are upheld on a global scale. In the immediate future, I ask the zoo industry to critically evaluate the current applications of zoo animals as educational entertainers. The cessation of certain animal-visitor interactions in accredited zoos (such as petting and feeding wild animals) would set a clearer example for tourists so they could more easily recognise the distinctions between high and low-welfare animal tourism. Finally, I propose zoos offer greater transparency to their visitors about the progress already made since the zoo industry's colonial emergence.

Funding: This research received no external funding.

Acknowledgments: Special thanks go to staff from Owston's Civet EEP, Species360, and Prof Hurn for supporting my interest in civets and their captive management over the past four years of research. It is important to note that all views and proposals made in this paper are my own and do not represent the views or opinions of those in the aforementioned organizations. I also offer thanks to the reviewers of this article, whose expertise and eye for detail have strengthened my analysis.

Conflicts of Interest: The author declares no conflict of interest. As one of the Guest Editors for this Special Issue ('Animal Welfare from a Cross-Cultural Perspective'), this manuscript was not handled by any of the Guest Editors.

References

1. Carr, N. Star Attractions and Damp Squibs at the Zoo: A Study of Visitor Attention and Animal Attractiveness. *Tour. Recreat. Res.* **2016**, *41*, 326–338. [CrossRef]
2. Veron, G.; Debruille, A.; Kayser, P.; Fernandez, D.A.P.; Bourgeois, A. Genetic Diversity and Structure of the Binturong *Arctictis binturong* (Carnivora: Viverridae)-Status of the Elusive Palawan Binturong and Implications for Conservation. *Zool. J. Linn. Soc.* **2020**, *188*, 302–318. [CrossRef]
3. Bourgeois, A.; Kayser, P.; Debruille, A.; Veron, G. Binturong *Arctictis binturong* Conservation: The Relationship between the Zoo Community and ABConservation for an Integrated Conservation Programme in Palawan, Philippines. *Int. Zoo Yearb.* **2020**, *54*, 120–130. [CrossRef]
4. Finch, K.; Leary, M.; Holmes, L.; Williams, L.J. Zoo Closure Does Not Affect Behavior and Activity Patterns of Palawan Binturong (*Arctictis binturong whitei*). *J. Zool. Bot. Gard.* **2022**, *3*, 30. [CrossRef]

5. Greene, L.K.; Wallen, T.W.; Moresco, A.; Goodwin, T.E.; Drea, C.M. Reproductive Endocrine Patterns and Volatile Urinary Compounds of *Arctictis binturong*: Discovering Why Bearcats Smell like Popcorn. *Sci. Nat.* **2016**, *103*, 1–11. [CrossRef]
6. Glachet, O.; Moustafa, A.A.; Gallouj, K.; El Haj, M. Smell Your Memories: Positive Effect of Odor Exposure on Recent and Remote Autobiographical Memories in Alzheimer's Disease. *J. Clin. Exp. Neuropsychol.* **2019**, *41*, 555–564. [CrossRef] [PubMed]
7. Willcox, D.H.A.; Chutipong, W.; Gray, T.N.E.; Cheyne, S.; Semiadi, G.; Rahman, H.; Coudrat, C.N.Z.; Jennings, A.; Ghimirey, Y.; Ross, J.; et al. *Arctictis binturong*. *IUCN Red List Threat. Species* **2016**, *1*, 337–380. [CrossRef]
8. Carr, N. Ideal Animals and Animal Traits for Zoos: General Public Perspectives. *Tour. Manag.* **2016**, *57*, 37–44. [CrossRef]
9. Hooper, J.; Linna, M.; Kassinen, S.; Salvage, J. Technologies, Bodies and Faecal Matters: Embodied Empathy with Coffee Producing Civets. *Transpositiones J. Transdiscipl. Intermedial Cult. Stud.* **2022**, *1*, 73–94.
10. Roe, K.; McConney, A.; Mansfield, C.F. The Role of Zoos in Modern Society—A Comparison of Zoos' Reported Priorities and What Visitors Believe They Should Be. *Anthrozoos* **2014**, *27*, 529–541. [CrossRef]
11. Falk, J.H.; Reinhard, E.M. *Why Zoos & Aquariums Matter: Assessing the Impact of a Visit to a Zoo or Aquarium*; Association of Zoos and Aquariums: Silver Spring, MD, USA, 2007; pp. 1–24.
12. Marino, L.; Lilienfeld, S.O.; Malamud, R.; Nobis, N.; Broglio, R. Do Zoos and Aquariums Promote Attitude Change in Visitors? A Critical Evaluation of the American Zoo and Aquarium Study. *Soc. Anim.* **2010**, *18*, 126–138. [CrossRef]
13. Falk, J.H.; Heimlich, J.E.; Vernon, C.L.; Bronnenkant, K. Critique of a Critique: Do Zoos and Aquariums Promote Attitude Change in Visitors? *Soc. Anim.* **2010**, *18*, 415–419. [CrossRef]
14. Malamud, R. The Problem with Zoos. In *The Oxford Handbook of Animal Studies*; Kalof, L., Ed.; Oxford University Press: Oxford, UK, 2017; pp. 396–410. [CrossRef]
15. Grajal, A.; Luebke, J.F.; Kelly, L.A.D.G.; Matiasek, J.; Clayton, S.; Karazsia, B.T.; Saunders, C.D.; Goldman, S.R.; Mann, M.E.; Stanoss, R. The Complex Relationship between Personal Sense of Connection to Animals and Self-Reported Proenvironmental Behaviors by Zoo Visitors. *Conserv. Biol.* **2017**, *31*, 322–330. [CrossRef] [PubMed]
16. Clarke, A.G. The Frozen Ark Project: The Role of Zoos and Aquariums in Preserving the Genetic Material of Threatened Animals. *Int. Zoo Yearb.* **2009**, *43*, 222–230. [CrossRef]
17. Gurian, E. Threshold Fear. In *Reshaping Museum Space*; Routledge: Abingdon, UK, 2005; pp. 217–228.
18. Dick, G. Natural History Museums, Zoos, and Aquariums. In *The Future of Natural History Museums*; Routledge: Abingdon, UK, 2017; pp. 155–167.
19. Anderson, K. Culture and Nature at the Adelaide Zoo: At the Frontiers of 'Human' Geography. *Environment* **2018**, *20*, 255–274. [CrossRef]
20. Zelenka, J.; Paskova, M. *Výkladový Slovník Cestovního Ruchu*; Ministry for Regional Development: Prague, Czech Republic, 2002.
21. Vaníček, J. Comparison of Visitors' Profile in Selected Zoological Gardens in the Czech Republic. *COT Bus.* **2012**, *1*, 70–72.
22. Nekolný, L.; Fialová, D. Zoo Tourism: What Actually Is a Zoo? *Czech J. Tour.* **2018**, *7*, 153–166. [CrossRef]
23. Szczygielska, M. Elephant Empire: Zoos and Colonial Encounters in Eastern Europe. *Cult. Stud.* **2020**, *34*, 789–810. [CrossRef]
24. Pratt, M.L. *Imperial Eyes: Travel Writing and Transculturation*; Routledge: Abingdon, UK, 1992.
25. Kirksey, S.E.; Helmreich, S. The Emergence of Multispecies Ethnography. *Cult. Anthropol.* **2010**, *25*, 545–576. [CrossRef]
26. Hooper, J. Cat-Poo-Chino and Captive Wildlife: Tourist Perceptions of Balinese Kopi Luwak Agrotourism. *Soc. Anim.* **2022**, *1*, 1–20. [CrossRef]
27. Hooper, J. Contamination: The Case of Civets, Companionship, COVID, and SARS. *J. Appl. Anim. Welf. Sci.* **2022**, *25*, 167–179. [CrossRef] [PubMed]
28. Song, Z. Between the Past and Future. Antwerp Zoo and the 19th Century Belgium. *Tech. Soc. Sci. J.* **2021**, *17*, 235–243. [CrossRef]
29. Sampaio, M.B.; Schiel, N.; Souto, A. From Exploitation to Conservation: A Historical Analysis of Zoos and Their Functions in Human Societies. *Ethnobiol. Conserv.* **2020**, *9*, 1–32. [CrossRef]
30. Coleman, D. Animal Studies Journal Menageries and Museums: John Simons' The Tiger That Swallowed the Boy (2012) and the Lives and Afterlives of Historical Animals. *Anim. Stud. J.* **2013**, *2*, 114–132.
31. Abbattista, G. *Beyond the 'Human Zoos' This Contribution Discusses Some Aspects of What Is Conventionally Termed as 'Living Human' or 'Ethnic Exhibitions': An Expression Denoting the Mainly Nineteenth-Early Twentieth Century Western Practice of Putting on Display*; Polistampa: Firenze, Italy, 2015; pp. 1–16.
32. Braverman, I. *Zooland: The Institution of Captivity*; Stanford University Press: Redwood City, CA, USA, 2012; p. 87.
33. Rothfels, N. *Savages and Beasts: The Birth of the Modern Zoo*; JHU Press: Baltimore, MD, USA, 2002.
34. Hosey, G.; Melfi, V.; Pankhurst, S. *Zoo Animals: Behaviour, Management and Welfare*, 2nd ed.; Oxford University Press: Oxford, UK, 2009.
35. Dibley, B.; Hawkins, G. Making Animals Public: Early Wildlife Television and the Emergence of Environmental Nationalism on the ABC. *Continuum* **2019**, *33*, 744–758. [CrossRef]
36. Carr, N.; Broom, D.M. *Tourism and Animal Welfare*; CABI: Oxford, UK, 2018.
37. Mason, G.J.; Latham, N. Can't Stop, Won't Stop: Is Stereotypy a Reliable Animal Welfare Indicator? *Anim. Welf.* **2004**, *13*, 57–69. [CrossRef]
38. Moorhouse, T.P.; Dahlsjö, C.A.; Baker, S.E.; D'Cruze, N.C.; Macdonald, D.W. The Customer Isn't Always Right—Conservation and Animal Welfare Implications of the Increasing Demand for Wildlife Tourism. *PLoS ONE* **2015**, *10*, e0138939. [CrossRef]

39. Warsaw, D.; Sayers, J. The Influence of Animal Welfare Accreditation Programmes on Zoo Visitor Perceptions of the Welfare of Zoo Animals. *Res. Artic. J. Zoo Aquarium Res.* **2020**, *8*, 2020.
40. Altman, J.D. Animal Activity and Visitor Learning at the Zoo. *Anthrozoos* **1998**, *11*, 12–21. [CrossRef]
41. Godinez, A.M.; Fernandez, E.J.; Morrissey, K. Visitor Behaviors and Perceptions of Jaguar Activities. *Anthrozoos* **2013**, *26*, 613–619. [CrossRef]
42. Godinez, A.M.; Fernandez, E.J. What Is the Zoo Experience? How Zoos Impact a Visitor's Behaviors, Perceptions, and Conservation Efforts. *Front. Psychol.* **2019**, *10*, 1746. [CrossRef] [PubMed]
43. Miller, L.J. Visitor Reaction to Pacing Behavior: Influence on the Perception of Animal Care and Interest in Supporting Zoological Institutions. *Zoo Biol.* **2012**, *31*, 242–248. [CrossRef] [PubMed]
44. EAZA. EAZA Standards for the Accommodation and Care of Animals in Zoos and Aquaria. European Associastion of Zoos and Aquaria. 2019; p. 19. Available online: http://www.eaza.net/assets/Uploads/Standards-and-policies/Standards-for-the-Accommodation-and-Care-of-Animals-2014.pdf (accessed on 10 April 2023).
45. WAZA. WAZA Code of Ethics and Animal Welfare. In Proceedings of the 58th Annual Meeting, of the World Association of Zoos and Aquariums, San José, CR, USA, 19 November 2003; p. 5.
46. Agoramoorthy, G.; Hsu, M.J. Use of Nonhuman Primates in Entertainment in Southeast Asia. *J. Appl. Anim. Welf. Sci.* **2005**, *8*, 141–149. [CrossRef]
47. Primack, R.B. *Essentials of Conservation Biology*, 3rd ed.; Sinauer Associates: Sunderland, MA, USA, 2002.
48. Collard, R.C. Panda Politics. *Can. Geogr.* **2013**, *57*, 226–232. [CrossRef]
49. Meehan, C.L.; Mench, J.A.; Carlstead, K.; Hogan, J.N. Determining Connections between the Daily Lives of Zoo Elephants and Their Welfare: An Epidemiological Approach. *PLoS ONE* **2016**, *11*, e0158124. [CrossRef]
50. Chrulew, M. Saving the Golden Lion Tamarin. In *Extinction Studies: Stories of Time, Death and Generations*; Rose, D.B., van Dooren, T., Chrulew, M., Eds.; Columbia University Press: New York, NY, USA, 2017; pp. 217–222.
51. Zhang, Y.; Wang, F.; Cui, Z.; Li, M.; Li, X.; Ye, X.; Yu, X. Can We Reestablish a Self-Sustaining Population? A Case Study on Reintroduced Crested Ibis with Population Viability Analysis. *Avian Res.* **2021**, *12*, 1–10. [CrossRef]
52. Pukazhenthi, B.S.; Wildt, D.E. Which Reproductive Technologies Are Most Relevant to Studying, Managing and Conserving Wildlife? *Reprod. Fertil. Dev.* **2004**, *16*, 33–46. [CrossRef]
53. Biggins, D.E.; Hanebury, L.R.; Miller, B.J.; Powell, R.A. Black-Footed Ferrets and Siberian Polecats as Ecological Surrogates and Ecological Equivalents. *J. Mammal.* **2011**, *92*, 710–720. [CrossRef]
54. Martin-Wintle, M.S.; Shepherdson, D.; Zhang, G.; Zhang, H.; Li, D.; Zhou, X.; Li, R.; Swaisgood, R.R. Free Mate Choice Enhances Conservation Breeding in the Endangered Giant Panda. *Nat. Commun.* **2015**, *6*, 4–10. [CrossRef]
55. Masilkova, M.; Boukal, D.; Ash, H.; Buchanan-Smith, H.M.; Konečná, M. Linking Personality Traits and Reproductive Success in Common Marmoset (*Callithrix jacchus*). *Sci. Rep.* **2022**, *12*, 1–12. [CrossRef]
56. Blais, B.R.; Wells, S.A.; Poynter, B.M.; Koprowski, J.L.; Garner, M.M.; Allard, R.A. Adaptive Management in a Conservation Breeding Program: Mimicking Habitat Complexities Facilitates Reproductive Success in Narrow-Headed Gartersnakes (*Thamnophis rufipunctatus*). *Zoo Biol.* **2022**, *41*, 346–353. [CrossRef] [PubMed]
57. Ludwig, C.; Dehnhard, M.; Pribbenow, S.; Silinski-Mehr, S.; Hofer, H.; Wachter2, B. Asymmetric Reproductive Aging in Cheetah (*Acinonyx jubatus*) Females in European Zoos. *J. Zoo Aquarium Res.* **2019**, *7*, 87–93.
58. Farquharson, K.A.; Hogg, C.J.; Grueber, C.E. A Meta-Analysis of Birth-Origin Effects on Reproduction in Diverse Captive Environments. *Nat. Commun.* **2018**, *9*, 1055. [CrossRef] [PubMed]
59. Guenther, K.M. *The Lives and Deaths of Shelter Animals: The Lives and Deaths of Shelter Animals*; Stanford University Press: Redwood City, CA, USA, 2020.
60. Greggor, A.L.; Vicino, G.A.; Swaisgood, R.R.; Fidgett, A.; Brenner, D.; Kinney, M.E.; Farabaugh, S.; Masuda, B.; Lamberski, N. Animal Welfare in Conservation Breeding: Applications and Challenges. *Front. Vet. Sci.* **2018**, *5*, 1–6. [CrossRef] [PubMed]
61. Fernandez, E.J.; Tamborski, M.A.; Pickens, S.R.; Timberlake, W. Animal-Visitor Interactions in the Modern Zoo: Conflicts and Interventions. *Appl. Anim. Behav. Sci.* **2009**, *120*, 1–8. [CrossRef]
62. D'Cruze, N.; Khan, S.; Carder, G.; Megson, D.; Coulthard, E.; Norrey, J.; Groves, G. A Global Review of Animal–Visitor Interactions in Modern Zoos and Aquariums and Their Implications for Wild Animal Welfare. *Animals* **2019**, *9*, 332. [CrossRef]
63. Carr, N. 5 Zoos and Animal Encounters to Touch or Not to Touch, That Is the Question. In *Wild Animals and Liesure: Rights and Welfare*; Routledge: Abingdon, UK, 2011; pp. 80–95.
64. EAZA. Policy and Legislation. Available online: https://www.eaza.net/about-us/areas-of-activity/policyandlegislation/ (accessed on 10 April 2023).
65. Jepson, P.; Barua, M. A Theory of Flagship Species Action. *Conserv. Soc.* **2015**, *13*, 95–104. [CrossRef]
66. Favreau, J.M.; Drew, C.A.; Hess, G.R.; Rubino, M.J.; Koch, F.H.; Eschelbach, K.A. Recommendations for Assessing the Effectiveness of Surrogate Species Approaches. *Biodivers. Conserv.* **2006**, *15*, 3949–3969. [CrossRef]
67. Branton, M.; Richardson, J.S. Assessing the Value of the Umbrella-Species Conceptfor Conservation Planning with Meta-Analysis. *Conserv. Biol.* **2011**, *25*, 9–20. [CrossRef]
68. Macdonald, E.A.; Hinks, A.; Weiss, D.J.; Dickman, A.; Burnham, D.; Sandom, C.J.; Malhi, Y.; Macdonald, D.W. Identifying Ambassador Species for Conservation Marketing. *Glob. Ecol. Conserv.* **2017**, *12*, 204–214. [CrossRef]

69. Kerr, K.C.R. Zoo Animals as "Proxy Species" for Threatened Sister Taxa: Defining a Novel Form of Species Surrogacy. *Zoo Biol.* **2021**, *40*, 65–75. [CrossRef]
70. ZIMS Species Holdings. Species360 Zoological Information Management System. Available online: https://species360.org/ (accessed on 18 April 2023).
71. Whitehouse-Tedd, K.; Spooner, S.; Scott, L.; Lozano-Martinez, J. Animal Ambassador Encounter Programmes in Zoos: Current Status and Future Research Needs. In *Zoo Animals: Husbandry, Welfare and Public Interactions. Animal Science, Issues and Research*; Berger, M., Corbett, S., Eds.; Nova Science Publishers: New York, NY, USA, 2018; pp. 89–140.
72. Spooner, S.L.; Farnworth, M.J.; Ward, S.J.; Whitehouse-Tedd, K.M. Conservation Education: Are Zoo Animals Effective Ambassadors and Is There Any Cost to Their Welfare? *J. Zool. Bot. Gard.* **2021**, *2*, 4. [CrossRef]
73. Gusset, M.; Dick, G. The Global Reach of Zoos and Aquariums in Visitor Numbers and Conservation Expenditures. *Zoo Biol.* **2011**, *30*, 566–569. [CrossRef]
74. Barongi, R.; Fisken, F.A.; Parker, M.; Gusset, M. (Eds.) *Committing to Conservation: The World Zoo and Aquarium Conservation Strategy*; WAZA Executive Office: Barcelona, Spain, 2015; p. 69.
75. AZA. 2020 Annual Report on Conservation and Science. 2021. Available online: www.aza.org/annual-report-on-conservation-and-science (accessed on 17 April 2023).
76. Gusset, M.; Dick, G. 'Building a Future for Wildlife'? Evaluating the Contribution of the World Zoo and Aquarium Community to in Situ Conservation. *Int. Zoo Yearb.* **2010**, *44*, 183–191. [CrossRef]
77. Research, S. Kopi Luwak Coffee Market Size Is Projected to Reach USD 10 Billion by 2030, Growing at a CAGR of 4.9%: Straits Research. Global Newswire. Available online: https://www.globenewswire.com/en/news-release/2022/10/04/2528177/0/en/Kopi-Luwak-Coffee-Market-Size-is-projected-to-reach-USD-10-Billion-by-2030-growing-at-a-CAGR-of-4-9-Straits-Research.html (accessed on 23 November 2022).
78. Pearson, E.L.; Lowry, R.; Dorrian, J.; Litchfield, C.A. Evaluating the Conservation Impact of an Innovative Zoo-Based Educational Campaign: "Don't Palm Us Off" for Orang-Utan Conservation. *Zoo Biol.* **2014**, *33*, 184–196. [CrossRef] [PubMed]
79. Hooper, J. *Civets in Society: What Can the Family Viverridae Teach Us about Disappearance in the Anthropocene*; The University of Exeter: Exeter, UK, 2023.
80. Marcone, M.F. Composition and Properties of Indonesian Palm Civet Coffee (Kopi Luwak) and Ethiopian Civet Coffee. *Food Res. Int.* **2004**, *37*, 901–912. [CrossRef]
81. Bickley, J. The History of Kopi Luwak. Gayo Kopi. Available online: https://gayokopi.com/history-of-kopi-luwak/?unapproved=54473&moderation-hash=0c64ec7b218a3bd26ae847c8c30ee324#comment-54473 (accessed on 1 February 2022).
82. Nakashima, Y.; Inoue, E.; Inoue-Murayama, M.; Sukor, J.R.A. Functional Uniqueness of a Small Carnivore as Seed Dispersal Agents: A Case Study of the Common Palm Civets in the Tabin Wildlife Reserve, Sabah, Malaysia. *Oecologia* **2010**, *164*, 721–730. [CrossRef] [PubMed]
83. Kopi Luwak Direct. Shop Kopi Luwak Direct [Online]. 2020. Available online: https://kopiluwakdirect.com/kopi-luwak-taste/ (accessed on 1 February 2022).
84. Muzaifa, M.; Hasni, D.; Rahmi, F. What Is Kopi Luwak? A Literature Review on Production. In *Earth and Environmental Science*; IOP Publishing: Bristol, UK, 2019; Volume 365, p. 12041.
85. Ifmalinda, I.; Setiasih, I.S.; Muhaemin, M.; Nurjanah, S. Chemical Characteristics Comparison of Palm Civet Coffee (Kopi Luwak) and Arabica Coffee Beans. *J. Appl. Agric. Sci. Technol.* **2019**, *3*, 280–288. [CrossRef]
86. Thi, M.T.; Gray, R.; Pham, T.; Thuy, H.T.; Le Kim, L.; Nhat, L.C.; Nguyen, T. Commercial Civet Farming Practices and Conservation Impacts on Wild Civet Populations in Central Vietnam. 2022; pp. 1–27. Available online: https://assets.researchsquare.com/files/rs-1806075/v1/0d558491-d773-48ae-9ae5-602b2d258753.pdf?c=1658227465 (accessed on 1 February 2022).
87. Shepherd, C.R. Observations of Small Carnivores in Jakarta Wildlife Markets, Indonesia, with Notes on Trade in Javan Ferret Badger Melogale Orientalis and on the Increasing Demand for Common Palm Civet Paradoxurus Hermaphroditus for Civet Coffee Prod. *Small Carniv. Conserv.* **2012**, *47*, 38–41.
88. Lynn, G.; Rogers, C. Civet Cat Coffee's Animal Cruelty Secrets. 2013. Available online: https://www.bbc.co.uk/news/uk-england-london-24034029 (accessed on 1 February 2022).
89. Cahill, C.W. *Feral Natures and Excremental Commodities: Purity, Scale, and the More-than-Human in Indonesia*; UC Irvine: Irvine, CA, USA, 2017.
90. Carder, G.; Proctor, H.; Schmidt-Burbach, J.; D'cruze, N. The Animal Welfare Implications of Civet Coffee Tourism in Bali. *Anim. Welf.* **2016**, *25*, 199–205. [CrossRef]
91. Lewis, E.; Boyle, L.A.; O'Doherty, J.V.; Brophy, P.; Lynch, P.B. The Effect of Floor Type in Farrowing Crates on Piglet Welfare. *Irish J. Agric. Food Res.* **2005**, *44*, 69–81.
92. Hunter, L. *Carnivores of the World*; Princeton University Press: Princeton, NJ, USA, 2019; Volume 117.
93. World Animal Protection. Civet Coffee: Campaigning for Cage-Free. 2020. Available online: https://www.worldanimalprotection.org.uk/campaigns/animals-wild/civet-coffee-kopi-luwak (accessed on 1 February 2022).
94. Peraturan Menteri Pertanian. Permentan Tentang Cara Produksi Kopi Luwak Melalui Pemeliharaan Luwak Yang Memenuhi Prinsip Kesejahteraan Hewan Nomor 37 Tahun 2015. Available online: https://peraturan.bpk.go.id/Home/Details/160544/permentan-no-37permentankb12062015-tahun-2015 (accessed on 1 February 2022).

95. Willcox, D.; Taylor, O.; Jimerson, J.; Underwood, G.; Nguyen, V. Owston's Civet *Chrotogale owstoni*: A Priority for Conservation Breeding. *Int. Zoo Yearb.* **2020**, *54*, 113–119. [CrossRef]
96. Cahill, C. [Un]becoming a Resource: Translating the Nature of Civets in Indonesia. *Ethnos* **2020**, *85*, 100–117. [CrossRef]
97. Hooper, J. Fill Your Cup: A Multpspecies Exploration of Balinese Kopi Luwak Agrotourism. In *Tourism, Heritage and Commodification of Non-Human Animals: A Posthumanist Reflection*; Lopez, A., Venegas, Q., Kline, C., Eds.; CABI: Wallingford, UK, 2023; In Press.
98. Tsing, A.L. *In the Realm of the Diamond Queen: Marginality in an Out-of-the-Way Place*; Princeton University Press: Princeton, NJ, USA, 1993.
99. Grazian, D. Where the Wild Things Aren't: Exhibiting Nature in American Zoos. *Sociol. Q.* **2012**, *53*, 546–565. [CrossRef]
100. Pennisi, L.; Lackey, N.Q.; Holland, S.M. Can an Immersion Exhibit Inspire Connection to Nature and Environmentally Responsible Behavior? *J. Interpret. Res.* **2017**, *22*, 2. [CrossRef]
101. Carr, N.; Cohen, S. The Public Face of Zoos: Images of Entertainment, Education and Conservation. *Anthrozoos* **2011**, *24*, 175–189. [CrossRef]
102. Orams, M.B. Feeding Wildlife as a Tourism Attraction: A Review of Issues and Impacts. *Tour. Manag.* **2002**, *23*, 281–293. [CrossRef]
103. Carder, G.; Plese, T.; Machado, F.C.; Paterson, S.; Matthews, N.; McAnea, L.; D'cruze, N. The Impact of 'Selfie' Tourism on the Behaviour and Welfare of Brown-Throated Three-Toed Sloths. *Animals* **2018**, *8*, 216. [CrossRef]
104. WAZA. *WAZA Guidelines for Animal-Visitor Interactions*; WAZA: Barcelona, Spain, 2020; pp. 1–6.
105. Spooner, S.L.; Jensen, E.A.; Tracey, L.; Robert, A. Evaluating the Effectiveness of Live Animal Shows at Delivering Information to Zoo Audiences Evaluating the Effectiveness of Live Animal Shows at Delivering Information to Zoo Audiences. *Int. J. Sci. Educ.* **2021**, *1*, 1–11.
106. Whitehouse-tedd, K.M.; Lozano-martinez, J.; Reeves, J.; Martin, J.H.; Prozesky, H.; Lozano-martinez, J.; Reeves, J. Assessing the Visitor and Animal Outcomes of a Zoo Encounter and Guided Tour Program with Ambassador Cheetahs. *Anthrozoos* **2022**, *35*, 307–322. [CrossRef]
107. Llewellyn, T.; Rose, P.E. Education Is Entertainment? Zoo Science Communication on YouTube. *J. Zool. Bot. Gard.* **2021**, *2*, 17. [CrossRef]
108. Shaw, M.N.; Mcleod, E.M.; Borrie, W.T.; Miller, K.K. Human Positioning in Close-Encounter Photographs and the Effect on Public Perceptions of Zoo Animals. *Animals* **2022**, *12*, 11. [CrossRef] [PubMed]
109. Litchfield, C.A.; Lowry, R.; Dorrian, J. Recycling 115,369 Mobile Phones for Gorilla Conservation over a Six-Year Period (2009–2014) at Zoos Victoria: A Case Study of 'Points of Influence' and Mobile Phone Donations. *PLoS ONE* **2018**, *13*, e0206890. [CrossRef] [PubMed]
110. Hosey, G.; Melfi, V.; Ward, S.J. Problematic Animals in the Zoo: The Issue of Charismatic Megafauna. In *Problematic Wildlife II: New Conservation and Management Challenges in the Human-Wildlife Interactions*, 2nd ed.; Francesco Maria Angelici, L.R., Ed.; Springer: Berlin/Heidelberg, Germany, 2020; pp. 485–508.
111. WTO. Sustainable Tourism Development. *World Tourism Organization.* Available online: https://www.unwto.org/sustainable-development (accessed on 1 February 2022).
112. Moorhouse, T.; D'Cruze, N.C.; Macdonald, D.W. Unethical Use of Wildlife in Tourism: What's the Problem, Who Is Responsible, and What Can Be Done? *J. Sustain. Tour.* **2017**, *25*, 505–516. [CrossRef]
113. Miller, B.; Conway, W.; Reading, R.P.; Wemmer, C.; Wildt, D.; Kleiman, D.; Monfort, S.; Rabinowitz, A.; Armstrong, B.; Hutchins, M. Evaluating the conservation mission of zoos, aquariums, botanical gardens, and natural history museums. *Conserv. Biol.* **2004**, *18*, 86–93. [CrossRef]
114. Nygren, N.V.; Ojalammi, S. Conservation Education in Zoos—A Literature Review. *TRACE* **2018**, *4*, 62–76. [CrossRef]
115. Fennell, D. *Tourism and Animal Ethics*; Routledge: Abingdon, UK, 2011.
116. Burns, G.L. Animals as Tourism Objects: Ethically Refocusing Relationships Between Tourists and Wildlife. *Anim. Tour. Underst. Divers. Relatsh.* **2015**, *67*, 44–59. [CrossRef]
117. Winter, C. A Review of Animal Ethics in Tourism: Launching the Annals of Tourism Research Curated Collection on Animal Ethics in Tourism. *Ann. Tour. Res.* **2020**, *84*, 102989. [CrossRef]
118. Rickly, J.M.; Kline, C. (Eds.) Introduction: Working for the (Hu)Man in the Tourism Industry. In *Exploring Non-Human Work in Tourism*; De Gruyter: Boston, MA, USA, 2021; pp. 1–21.
119. Caton, K. Taking the Moral Turn. *Ann. Tour. Res.* **2012**, *39*, 1906–1928. [CrossRef]
120. Crutzen, P.J.; Stoermer, E.F. The "Anthropocene". *IGPB Glob. Chang. Newsl.* **2000**, *41*, 17–18.
121. Duckworth, J.W.; Timmins, R.J.; Choudhury, A.; Chutipong, W.; Willcox, D.H.A.; Mudappa, D.; Rahman, H.; Widmann, P.; Wilting, A.; Xu, W. Paradoxurus hermaphroditus. *IUCN Red List. Threat. Species* **2016**, e.T41693A45217835. [CrossRef]

animals

MDPI

Commentary

Wicked Problems, Novel Solutions: Nepalese Elephant Tourism and Conservation

Michelle Szydlowski

St. Luke's Campus, University of Exeter, Exeter EX12LU, UK; michelle@szyd.me

Simple Summary: The conservation of endangered Asian elephants (especially when driven by those from non-range countries) is made more complex in cultures where 'owning' individuals for tourism practice is widely accepted. How can a wide variety of stakeholders find common ground upon which to build conservation plans amidst shifting social and environmental pressures?

Abstract: Endangered Asian elephants (*Elephas maximus*) find themselves at the center of debates involving politics, land use, human–wildlife conflict, and environmental justice. The intensity of such debates has led scholars to label conservation challenges as wicked problems with profound implications on local and global practice. In elephant range states such as Nepal, these debates are made more complex through human 'ownership' of endangered individuals for use in tourism, worship, or co-work. Human perspectives on the ethics of using animals for tourism are changing, even in areas heavily reliant on the tourism industry for survival. These debates become inflamed when non-residents take on adversarial positions despite an acceptance of the 'ownership' or 'use' of endangered individuals among local communities. Novel approaches are needed if there is any hope of establishing a common ground upon which to build relationships which may benefit community members, international interests, and endangered individuals.

Keywords: anthrozoology; Asian elephant; captive–wild interface; conservation; multispecies ethnography; Nepal; tourism

1. Introduction

During research involving Greater-one Horned rhinoceros (*Rhinoceros unicornis*) conservation in Nepal, I found myself spending large periods of time within the elephant stables surrounding Chitwan National Park. Through chance interactions with the residents of these hattisars (hatti is elephant in Nepali and hattisar is elephant stable), I developed a working relationship with both Asian elephants (*Elephas maximus*) and humans (*Homo sapiens*) (Using scientific nomenclature for both species is important, as it problematizes the human-centered habit of listing the Latin names of only so called 'other than human' species [1]). I noticed vast differences in the health, welfare, treatment, and behavior of individual elephants who transported researchers into protected areas, served as mounts for anti-poaching patrols, worked with government employees on conservation efforts and forest management, and those who ferried tourists through the national park's protected areas. Depending upon their ownership status and physical location, each elephant faced drastically different challenges as well as levels of legislative protection [2,3]; for many, being 'owned' by different organizations impacted their physical and mental health as well as their expected life span. Likewise, mahouts experienced difficulties arising from their physical location, with some facing inadequate health care or housing at stable facilities. In addition, mahouts face marginalization from community members' preconceived notions regarding their choices of profession, caste, or their socioeconomic status [4–6].

I found myself spending increasing amounts of time examining the reactions of tourists, community members, researchers, and staff from international non-governmental agencies (INGOS) as they visited these elephant–human working pairs, especially those involved in tourism practice (in fact, I ended up revising my doctoral plan and received ethical approval to study human–elephant–NGO relations in Nepal). Both local community members and visitors from other parts of Nepal initially expressed little concern with current elephant-use practices. However, tourists from the Global North, especially those from North America, described themselves as having conflicted or negative perceptions of practices which 'used' elephants solely for entertainment purposes [7]. I was initially hesitant to engage in research involving these conflicting attitudes and the challenges faced by tourism elephants in Nepal. This reluctance was due in part to my education as an anthrozoologist (someone who studies human–other-than-human animal relationships), which was steeped in anthropological theory. As such, I understood the need to be reflexive regarding the potential for perpetuating neocolonial attitudes (ideas or beliefs which assume indigenous knowledge is somehow less valuable than that of so-called 'more-developed' societies [8]) and was concerned about how to represent elephant–human coworking conditions adequately and fairly. I felt uncomfortable discussing the ownership and treatment of elephants in Nepal through a 'western' welfare lens, if, as some alleged, local attitudes were generally supportive of the practice. I admit to experiencing discomfort as feelings of cultural and moral relativism swirled through my head, and I wrestled with conflicting information from elephant owners, mahouts, veterinary staff, biologists, elephant advocates, and other stakeholders. I wrestled with (erroneous) claims by Nepalese interlocuters that elephants had 'always' been used in both forest management and tourism [9]. They explained that thanks to such 'traditional' practice, the use of elephants for any human purpose was justifiable. This information convinced me, temporarily, that due to millennia of embodied knowledge shared with pachyderms, Nepalese mahouts and elephant owners must know better than anyone how to properly use and care for elephants (see Figure 1). It is important to note that the author is neither endorsing nor vilifying the keeping of captive wildlife, a debate beyond the scope of this short paper.

However, I remained bothered by the number of owners who treated both elephants and mahouts with indifference, neglect, or as 'property'. In addition, not all mahouts treated their elephant co-workers with kindness, and some neglected or abused elephants with alarming regularity. Of course, many 'western' bosses also fail to treat their employees as beings worthy of respect, and many co-workers fail to approach colleagues with appreciation and respect. Likewise, animals who share their lives with humans in the US can be mistreated; captive wildlife in zoos or aquaria have their agency removed to fulfil public desire for entertainment and scientific demand for breeding stock. Humans regularly fail to care for, or directly abuse, the companion animals that share their homes and communities or work alongside law enforcement and military personnel [10,11]. However, these failures to treat all species equitably do not automatically mean that all guardianship of companion animals should be outlawed (I prefer this term over 'ownership'). I was, at the time of my initial research, misinformed regarding the problematic and complex history surrounding elephant–human relationships in Nepal, and misguided in my comparison between domestic companion animals and captive wildlife. My initial instinct to 'avoid getting involved' was both naïve and possibly irresponsible. Using a fear of potential accusations of neocolonialism to talk oneself out of advocating is an ethical slippery slope. But as readers will discover throughout this paper, such naivety turned out to be a beneficial characteristic while researching elephant and human relations within the complex conservation and tourist landscape of Nepal; it allowed the use of novel and creative approaches to wicked conservation problems. According to Game et al. (2013) [12], such novel approaches are key to establishing a common ground upon which to build working relationships which may benefit elephants, advocates, conservationists, and other stakeholders in the area around Chitwan National Park (henceforth CNP).

Figure 1. Mahout and elephant co-workers returning home after cutting grass for fodder. Sauraha, Nepal. Photo by Ram Krishna Mahato for this project.

During fieldwork in 2019, I noticed a sign offering 'ethical' elephant activities—a new idea for Sauraha. Out of curiosity and a sense of hopefulness regarding the living situations of both elephants and mahouts, I reached out to the sign's creators. Within hours, the focus of my research veered directly into the path of the largest animal in Nepal. As it turned out, numerous local community organizations, including the nature guide association, local businesses, and school groups, were advocating to cease elephant-backed safari. In addition, several international NGOs had entered the area. Due to these changing attitudes, it seemed the time had come for a study of the ethics and care of captive elephants in Sauraha, and I redirected my research into Nepal's hattisars. However, undertaking this type of study meant that I would have to spend time observing elephants and mahouts as they were mistreated and marginalized. It also meant attempting to incorporate views from a wide variety of individuals: indigenous and international experts, local and foreign veterinary staff, advocates, and welfare specialists from a variety of cultures. It meant trying to equally represent each stakeholder's truth of value laden words such as 'care', 'welfare', and 'conservation'. In other words, it required novel solutions of the kind Game et al. [12] described above. This paper discusses the complex views of stakeholders in Nepal and the ongoing struggles between researchers, owners, advocates, elephants, and governments. It also describes the wicked problems faced by those purporting to practice endangered species conservation. Lastly, it discusses novel approaches to unifying efforts towards improving elephant health.

Information in the following article was derived, in part, from participant observations and interviews with mahouts, owners, I/NGO staff, community members and advocates in ongoing dialogues between 2014 and 2022 in person, via email and messaging apps. These interviews are simply noted as PC (personal communication).

2. Wicked Problem One: Conserving Wild Elephants

Game et al. (2013: 271) describe conservation as 'far more complex' than rocket science. Due to entangled concerns which include politics, caste, land-use, poverty, wildlife conflict, and the long-held belief that humankind inherently has the right to use wildlife and wild spaces, conservation has become a 'wicked problem' [13] with few clear solutions. These issues arise not only from the complex relationships between stakeholders but also from often polarizing 'social, ecological, and economic elements' which lurk beneath the surface of conservation dialogues [14]. As a global biodiversity hot spot (see Figures 2 and 3), the small country of Nepal faces struggles with a growing number of these wicked conservation problems [15].

Figure 2. Hoopoe (*Upupa epops*). Chitwan National Park (henceforth CNP), Nepal. Photos by Nepal Dynamic Eco Tours, used with permission.

Figure 3. Male Himalayan Paradise Flycatcher (*Terpsiphone paradisi*). Chitwan National Park (henceforth CNP), Nepal. Photos by Nepal Dynamic Eco Tours, used with permission.

For example, Nepal has designated more than a third of the country as protected area and has shifted its focus from single-species conservation to landscape-level efforts [16,17]. Nepal has also shifted from centralized governmental control of forestry and conservation programs to more locally representative forest users' groups [16]. To combat illegal wildlife trade, Nepal has implemented greater interagency collaboration, increased enforcement of legislation, and formed anti-poaching teams (Acharya, 2016). Thanks to a combination of these efforts, Nepal has achieved several zero-poaching years for three of its charismatic flagship species: tigers (*Panthera tigris tigris*), elephants, and rhinos [18,19] (see Figures 4 and 5). Nepal has also experienced increases in the numbers of wild rhinos and tigers, with rhino populations rising from fewer than 100 in 1966 to 649 in 2018 [20]. Likewise, tiger populations have more than doubled [17]. However, rebounding populations of these species cannot simply be described as conservation 'successes'; instead, their overall effect on biodiversity preservation and impacts on larger multispecies communities should be considered.

Nepal's conservation efforts have been laudable, and protected areas now represent a large portion of the country (as mentioned above) [17]. However, less than 2% of this protected land is contiguous, the rest is composed of fragmented habitats interrupted by human settlements [21]. Wildlife residing within these protected areas are still at risk from anthropogenic (human-derived) forces, including disease transmission, climate change, and competition for resources [17]. Furthermore, while wild individuals face fewer threats from poaching, they remain vulnerable to potentially harmful tourism practices [22,23]. For example, due to disturbances by tourist jeep safaris, many of Nepal's protected species have become habituated to human noise, litter, and activity [24,25]. This habituation allows for jeeps to approach closer than recommended distances for human and other species' safety [26,27] (see Figures 6 and 7). When disturbances reach levels which are no longer tolerable, wildlife deserts traditional habitat, which in turn encourages tourism operators to venture deeper and deeper into protected areas to provide viewing opportunities for their clients [23,28]. Human incursions into these protected areas are extremely profitable, as entry fees provide the largest source of revenue for the national park [17,27]. For this reason, the government has been reluctant to further limit access to the national park despite decades of academic concern for the wild flora and fauna found within.

Figure 4. Wild Tiger, CNP, Nepal. Photo by Nepal Dynamic Eco Tours, used with permission.

Another issue resulting from the 'success' of charismatic species conservation is that of human–wildlife conflict (HWC). Within protected areas and along their borders, human–rhino and human–tiger interactions are on the rise, as are other conflicts [29]. The proximity of agricultural lands and their associated water sources to protected areas means that 'incursions'—a very anthropocentric word, which seems to imply human entitlement to land which previously served as wild animal habitat—by wildlife occur regularly [29]. Humans face financial and physical losses from these incursions, and wildlife are often killed in retaliation (see below). In addition, as wild populations rise, so do rhino deaths from territorial fights (there is not enough space left in the park, see below), falls into human-made wells, and fatalities from zoonotic (transferable to humans) or transmissible disease from livestock [25,30,31]. The increasing number of wild rhino deaths annually has left the government and researchers concerned that despite an estimated carrying capacity of 800–1000 rhinos [32,33], CNP may already have reached its limit.

Figure 5. Rhino in lake, CNP, Nepal. Photo by Nepal Dynamic Eco Tours, used with permission.

A further problematic result of Nepal's refocused conservation efforts was the forced relocation of indigenous populations to villages outside protected areas [17]. These relocations resulted in increased poverty and an inequitable proportion of HWC fatalities among already marginalized populations, such as indigenous Tharu people, Nepal's oldest ethnic culture [34]. The removal of villages from the park has limited community income potential, thanks to loss of land, access to forest products for sale, and an inability to hunt. In fact, 88% of Tharu families stated they could no longer afford gas for cooking and had returned to a reliance upon firewood [34]. As the collection of forest products is now controlled by park management, families are forced to illegally enter protected areas to collect firewood, putting them at greater risk of death from tiger and rhino attacks [17]. Fatalities and financial losses also occur when villages relocated along the boundaries of protected areas fall prey to crop-grazing by wild rhinos and elephants [35,36]. These populations already faced inequitable representation in forest management and tourism operation ownership, and experienced fewer gains from tourism practices, placing them at greater risk of reliance on forest products for survival [34,37]. Increasing losses are intimately linked to the transformation of landscapes surrounding PAs from traditional habitat to highly desirable (for wildlife and humans) agricultural crops as former forest-dwelling human populations resettled outside the PAs [36,38]. Human population growth is further impacting land use practices, as demand for housing and agricultural land removes more and more wildlife habitat from areas surrounding the PAs [39].

Figure 6. Wild rhino mother and calf, CNP, Nepal. Photo by Nepal Dynamic Eco Tours, used with permission.

Figure 7. Tourists in jeep approaching a wild rhino, CNP, Nepal. Photo by Nepal Dynamic Eco Tours, used with permission.

3. Wild Elephants

Like rhinos and tigers, Nepalese wild elephants face complex conservation problems (see Figures 8 and 9). These wild individuals are in extremely short supply, with only 120–200 remaining among four geographically separated herds [3,40]. These herds are unconcerned with political boundaries, passing regularly back and forth across the Nepal/India border. These migration patterns, coupled with a lack of transboundary cooperation, make obtaining an accurate count difficult [41]. Even though so few remain, tourists list 'seeing' wild elephants as a primary motivation for visiting the protected areas of Nepal and the herds serve as a key selling point for both government agencies and tourism operators [35,42].

Figure 8. Wild bull elephant nicknamed 'Govinda', CNP. Photo by Nepal Dynamic Eco Tours, used with permission.

Unfortunately, Nepalese wild elephants likely represent a 'doomed population', defined as one which is unable to rebound in any meaningful way thanks to low numbers, lack of connected habitat, and the seemingly irreversible pace of habitat loss [43,44]. In fact, early documents recommended ignoring Nepal's herds in favor of populations in other countries with a better chance of survival, or simply giving up and removing them from the wild completely [43]. However, more current research indicates that Nepal's wild elephants represent an invaluable and unique source of genetic material, thanks to their continued geographic and social distance from other herds [45]. These elephants may, in fact, be key to both local and global conservation of the species [40,45].

The habitats of these wild elephants continue to be lost and fragmented thanks to human-derived causes, forcing herds into migration routes which place them ever closer to human settlements [40]. Narrowing routes lead to increased interspecies conflict and result in herds being viewed simultaneously as both incursive and endangered, valuable, and costly [7]. Elephant routes vary seasonally, meaning that herds regularly happen across villages and croplands [40,41]. Current crop-grazing (a term is used in place of 'crop-raiding', which is problematically value-laden, as these crops are planted within former elephant ranges) behaviour can be traced to herds' natural movement patterns, rather than being contributable to the purposeful seeking out of cultivated plants (see Sukumar, 1990). Millet and rice, chosen by humans for their taste, nutritional value, and easy digestibility, likely attract passing elephants for the same reasons [41,46]. Crop-grazing behaviour may also be due to the higher sodium content found in cultivated crops—a mineral which is often low in wild elephants' traditional diet [47]. This behaviour is considered an adaptation in response to declining access to, and variety of, plants within protected areas, and the lack of high-quality, high-nutrition food in their traditional habitat. These food sources are being lost to invasive plants (such as *Mikantha* spp., see Murphy, et al., 2013), competition with other wildlife and human agriculture, and livestock grazing [47]. In addition, some human crops simply taste good. Sugar cane, a popular crop in Nepal, is a known treat for elephants, and draws in passing herds. In addition, elephants can smell water kilometers away, and since water is readily available near agricultural lands, this garners the interest of migrating herds. Surprisingly, Pokharel, et al. (2018) found that elephants who engaged in crop-grazing had lower levels of certain stress hormones, a shocking discovery as elephants exposed to human activity, noise, and human-induced stress typically demonstrate higher levels of stress. They hypothesize that the higher caloric, protein, and general nutritional value of human crops, coupled with the relative ease of access to large quantities of food (rather than having to travel long distances for adequate plant material), instead increases the health and fitness of wild elephants, thus lowering their stress [47].

Figure 9. Wild bull elephant nicknamed 'Renaldo'. Renaldo's tusks have been cut for human protection. Photo by Ram Krishna Mahato for this project.

The ongoing expansion of villages into natural elephant habitat and habitat fragmentation and degradation have long been major contributors to crop-grazing behaviours and will remain a problem if human populations continue to grow at their current rate of 1.7%

annually [48]. Crop-grazing contributes to human–elephant conflict, which has become one of the largest conservation issues in Asia and affects all 13 range states, costing millions of dollars annually [39,40]. These conflicts result in increasing numbers of injuries and fatalities to humans, and elephants face retaliatory killings over crop destruction, injury, or property loss [39,44].

The Nepalese government's stated focus on preserving wild elephants has officials worried about the potential for escalating conflict as human populations increase. Between 2009 and 2020, elephants were responsible for 13% of all human–wildlife conflict in the Chitwan area (with 67 separate incidents of human–elephant conflict), and elephant–human conflicts demonstrated the highest potential for human fatalities [38]. Efforts such as unpalatable crop planting, trenching, noise makers, and fencing have been initiated in attempts to reduce the impacts of conflicts on humans [38,49]. The government has also increased financial compensation for loss in attempts to reduce HWC. However, in some areas, villages are deserting land due to extensive economic losses and property damage caused by elephants, and some families refuse to allow marriages into households which fall within areas of high elephant conflict [3]. As previously mentioned, forcibly removing indigenous people from protected areas has resulted in populations feeling they are less valued than wildlife, and these persons now face increasing risk from elephant–human conflict [50]. Relocating either elephant or human populations is a risky and unrealistic prospect, and unlikely to succeed given the ever-increasing numbers of humans and the declining availability of 'natural' habitat for translocating elephants (Menon and Tawari, 2019). Instead, elephant and human shared spaces need to be more equitably addressed in landscape-level plans which meet the needs of members from all socioeconomic tiers and species [39]. Novel approaches to solving this problem might include offering equal consideration to stakeholders from all involved species [51,52]. Viewing conservation simultaneously through the lenses of sustainability and social justice might also offer more balanced approaches.

4. Wicked Problem Two: Captive Elephants

Of course, conserving wild species may take a backseat to other, more pressing (at least to humans) needs. For example, Nepal's human population faces ongoing social and financial challenges, with 40% of the population surviving on less than $1160 annually [53]. Six percent of the population is reliant upon the tourism industry for survival, and tourism represents ~7.5% of Nepal's GDP [54]. Thanks to the purported (yet controversial) benefits offered by tourism practice (see Bookbinder, et al., 1998; Nyaupane and Poudel, 2011; Puri, 2019), the Nepalese government decided to link poverty reduction with conservation efforts and continues to heavily market nature-based tourism [9]. This marketing works, with 60% of international tourists (430,000 people in 2018) visiting the fragile and biodiverse protected areas of Nepal annually [42]. However, this mass tourism makes it difficult to balance the purported goals of protecting wildlife and wild spaces with their value as commodities [16,55].

Chitwan National Park (CNP) is the busiest protected area in Nepal and has been described as 'the last surviving example of the natural ecosystems of the Terai' [17]. CNP is also home to several wild bull elephants and one of Nepal's four remaining matriarchal herds. It is in the tourist-laden towns surrounding the park where one finds another kind of endangered elephant—those used for tourist safari transportation. These elephant-backed safari rides bring in the second highest amount of revenue for the area [27], and the elephants serving as the vehicles for these adventures live a very different life than those of their governmentally held or wild cousins. While they are members of the same endangered species, privately-owned elephants face increased challenges to their health and welfare [7,30,56]. Unlike governmentally held elephants, privately-owned individuals lack any legal protections regarding their treatment, nutrition, or usage (see Figure 10).

Figure 10. Hatti Chowk (elephant statue public gathering spot). Photo by Ram Krishna Mahato for this project. Used with permission.

Before the COVID-19 pandemic, 52 of these privately owned individuals worked in the Chitwan area. These elephants carry up to six tourists at a time on a howdah (riding platform). Each ride lasts approximately 90 min, and each elephant may perform up to nine rides a day through the jungle [57]. These elephants are massively expensive, costing approximately 90 times more than the average annual worker's salary, meaning that only those of higher socioeconomic standing can obtain them [58]. They represent a huge financial opportunity, and a huge financial burden thanks to their upkeep costs, which may run up to $19,000 USD annually (PC, 2021) [57]. These tourism elephants are caught in a liminal space—seen in Buddhist and Hindu literature and society as divine creatures, as literal vehicles for tourism practice, as symbols of power and wealth, but also as disposable commodities which can be sent off when no longer 'useful' [7,59]. They serve as symbols for international organizations hoping to bring awareness to issues facing elephants in captivity, and as representatives of Nepalese history and culture [60,61]. Working in the hattisars of Nepal is challenging; thanks to my interest in elephant owners and traditions, I initially faced alienation by advocacy organizations hesitant to share their goals lest the owners limit access to elephants. Then, when I began researching advocates and NGOs, elephant owners became hesitant to share their stories, worried that I was a spy for those who might want to see elephant-based tourism end.

Not only are researchers caught between these groups, but so are those who spend the most time with elephants, their mahouts. While elephant caregivers have long been members of marginalized communities [4], at least in the past their choice of profession might have resulted in stable employment with the government or allowed them the use of elephants for farming and logging [62,63]. These elephant caregivers have traditionally come from indigenous tribes and/or low-caste families; they are uneducated, poorly paid, and overworked [64]. They are alternatingly viewed by the community as the keepers of traditional knowledge and ostracized as violent men with gambling or drinking problems [4,34,64].

Like many southeast Asian countries, the mahout–elephant relationship within Nepal has undergone changes over the past five decades [5,65]. For example, the long-term bonds which mahouts once shared with their elephants have, in many cases, been replaced with unstable or temporary working relationships [5,56,66]. In many cases, these relationships have become damaged or neglected as inexperienced mahouts with little training have replaced older mahouts who grew up in hattisars (see also Szydlowski, 2024). The position of mahout as a familial job is no longer the norm (with some exceptions, see Mumby 2019), and it is now young men with little embodied knowledge of elephants who are left to care for these captive endangered beings. The changing skill level of caregivers, the loss of strong mahout–elephant bonds, and the loss of traditional and embodied knowledge places mahouts and elephants at greater risk of neglect, injury, or death [64,67,68].

Within Nepal's hattisars, I have witnessed horrific abuse, malnutrition, and neglect, but I have also seen deep and tender love. These seemingly contradictory human treatments of elephants, and the fact that numerous international organizations attempt to influence the management of elephants in Nepal, are major reasons that the 'elephant situation' within the country is so complex [7] (see Figures 11–13 for an example of Nepalese stables). The influence of both local and international organizations, and the interplay of this influence with changing societal perspectives on the practice of captive-elephant management in Nepal, is beyond the scope of this article but is an area of ongoing study for the author [7,30].

Figure 11. Typical elephant *hattisar* (stable) in Sauraha, Nepal, showing chaining methods. Photos by Ram Krishna Mahato for this project, used with permission.

Figure 12. Another typical elephant *hattisar* (stables) in Sauraha, Nepal, showing chaining by front and back legs. Photos by Ram Krishna Mahato for this project, used with permission.

Figure 13. The only adult male elephant in Sauraha who is privately owned. Kumroj, Nepal. Photo by Ram Krishna Mahato for this project.

5. Additional Challenges

Captive elephants throughout South Asia suffer from a lack of veterinary care as well as low standards for their management, handling, and husbandry, and Nepal is no different [39]. However, Nepal's elephants face vastly different situations than those in the villages of Laos, the camps of Thailand or Myanmar, or the temples and streets of India. There have been studies within other Asian countries demonstrating a range of successful human–elephant relationships, a historical acceptance of elephants as community members, and even positive health and welfare implications of living in certain camp situations (for example, see Bansiddhi, et al., 2020; Lainé, 2019; Lehnhardt and Galloway, 2008). However, given the small number of both wild and captive elephants residing in Nepal, until recently, consideration of their welfare has largely escaped the interest of the international research community (see Szydlowski, 2024). In addition, the housing, nutrition, and management styles used within Nepal vary greatly compared to those seen elsewhere in Asia, making assessing elephant welfare even more difficult.

Another challenge arises from the use of elephant breaking rituals in Nepal. While these rituals, such as the 'crush', pha jaan, or hattiko talim [62,69], have purportedly fallen out of favor in several other Asian countries thanks to international pressures, they remain the standard way of creating 'usable' elephants within Nepal [57,62,70]. This 'desensitization' process, which involves both gentle care (such as singing and rubbing) and violence (fire, abuse, and stabbing), often results in long-lasting trauma to both mother and calf [66,71]. This type of breaking process may also result in injuries or fatalities for mahouts, when formerly 'broken' elephants later attempt to dominate their handlers [71,72].

Rather than living in camp situations as they might in other parts of Asia, elephants in Nepal live singly in small stables behind businesses. These elephants lack any tactile, olfactory, or visual contact with conspecifics, despite the wealth of data documenting the importance of these social connections (see EAZA, 2020; Poole and Granli, 2008; Prado-Oviedo, et al., 2016). In addition, the food provided to captive elephants in Nepal differs greatly from that of wild herds or captives in other Asian countries. As browsers, wild elephants typically spend up to 60% of each day foraging on a large variety of plant species, and camp elephants in other countries often graze throughout their day [73]. Captive elephants in Nepal, however, rely upon domesticated crops or provisioned foods (such as rice, which is not a natural food source for wild elephants in Nepal) (PC, 2017, 2019). These provisioned foods are fed at limited sessions several times a day, and often require the use of threats to get elephants to consume them (PC, 2019, 2022). As one interlocuter stated, 'They are given food they don't like and then beaten when they won't eat it'. In addition, this provisioned food does not allow for typical behaviour during consumption, so while it may fulfil physical needs, it likely does not fulfil mental welfare needs [74]. For example, wild elephants prepare their food in several ways prior to consuming it, such as dropping, breaking, and bending it [75], allowing for further mental and physical exercise opportunities as they manipulate food items. Limiting time for captive elephants to manipulate and consume their food has been shown to cause increases in stereotypic behaviours, changes in body condition scores, decreased welfare, and negatively impacts digestive health [5,72].

Studies of health and welfare among Nepalese captive elephants found that the majority suffered a variety of physical ailments ranging from wounds and abscesses to malnutrition, inappropriate sleeping conditions, broken nails, joint issues, vaginal prolapse, tuberculosis, and other diseases [7,56,76,77]. In addition, these elephants exhibited mental health issues such as maladaptive passivity, stress, stereotypies, loneliness, and other behaviours which may be linked to early separation from their maternal herd or ongoing social isolation.

For example, when not on safari (or during the off season) elephants spend most of their day chained in one position and this inability to move around for exercise or to avoid soiled areas can be detrimental to both physical and mental health [77–79]. Studies have linked bacterial causes of arthritis to unclean conditions or unsuitable substrates, indicating that poor husbandry decreases foot and joint health [80]. The stress caused by resultant arthritis may further impact foot health, creating circular issues [72,80]. In addition, the current chaining method (one front and one back foot using short chains) changes the standing position of these elephants, which further negatively impacts their joint health [57,74,76].

Welfare and health issues are a point of contention between various stakeholders and external agencies involved in captive elephant management within Nepal. At this juncture I would like to note that these privately-owned, captive elephants are (with a few exceptions) illegally obtained. Nepal has been a signatory to CITES since the 1970s, and as such is prohibited from protected animal trade across national borders for commercial purposes [81]. However, removing elephants from the wild has also been prohibited in Nepal since the 1970s, meaning that private owners wishing to continue garnering income from safari must import individuals from India [2] (PC, 2019). These individuals were walked or trucked across the border, and while owners openly acknowledged their illegal activities, officials denied any knowledge of these transfers [7,82]. Even with protective regulations in place, government documents have been quick to point out exactly which Indian cattle fairs potential investors should visit if they would like to purchase elephants for use in the tourist trade [3].

In the fall of 2021, Nepal's supreme court committed to the enforcement of CITES (The Convention for International Trade in Endangered Species) regulations and a cessation of elephant passage across the India–Nepal border. In addition, the government required the registration of all existing privately held elephants (this process is still underway). Unregistered elephants were to be confiscated, although it remains unclear who is responsible for exercising this option; the government, the elephant owners' association, and the NTNC all denied responsibility. Despite issues with implementation, the government's commitment to stopping illegal trade and keeping current elephant residents within the community is a positive step towards affirming the important connections between Nepali culture, history, and transspecies relationships.

In the past six months, mahouts and elephants have been turned back as they attempted to cross into India [83]. However, this new enforcement of legislations is not without danger for elephants. One female, illegally sold by a Nepalese businessman to an Indian buyer, was being transported across the border when reported to authorities by a Nepalese animal advocacy organization. This elephant, Kajol Kali (female elephants are surnamed 'Kali' and males 'Gaj' in honor of Hindu gods), was returned to Sauraha, but without a local owner or stable she had few options. She was severely malnourished, having been traveling for multiple days and kept hidden without adequate food as her owners tried to 'sneak' her across the border. The NTNC (see introduction) and national park staff claimed they were uncomfortable getting involved in disputes involving private owners. However, it is the NTNC's veterinary staff which is responsible for the health care of all privately held elephants in the Chitwan area, thanks to a long-standing agreement with the owners' cooperative [7].

Despite the supreme court requirement that illegally sold elephants be confiscated, no entity reportedly felt they had the authority to enforce the new ruling, and Kajol continued to starve. Local and NTNC veterinary staff were aware of the situation (having been invited to visit the facility by several stakeholders) and refused to examine the elephant. The NTNC veterinary staff did finally offer slight supportive care, after six months. NGOs paid for the necessary supplies and placed pressure on the NTNC. Another local elephant owner offered the use of his stable, and various I/NGOs stepped in to assist with costs associated with feeding and care. One sent staff daily to help ease the burden of the single mahout caring for Kajol (most elephants are assigned two mahouts). Despite the mobilization of NGOs, individuals, and financial support, Kajol did not recover from her malnutrition and fell numerous times, perishing after 8 months of struggle. This case highlights the liminal nature of privately-owned captive elephants in Nepal. With little oversight and no legislation regarding their care, captive elephants remain liminal beings.

They are caught between human desires: the desire to see or be near these embodiments of gods, to use them as transport for viewing their wild cousins, to maintain 'traditional' multispecies relationships, and to garner income at levels inaccessible for most other community members. Yet, as will be discussed next, the health and welfare of these captive individuals is likely key to the survival of wild populations.

6. Wicked Problem Three: The Captive–Wild Interface

Captive individuals impact human safety by providing anti-poaching patrols and can help run off wild elephants in populated areas [17]. In addition, they provide employment for a marginalized population, mahouts, and (purportedly, see previous sections) bring in important tourism revenue for the towns surrounding CNP. However, these activities also increase the risk to populations of wild elephants and other wildlife. The health and welfare of captive individuals is therefore important for reasons that go beyond concerns for the maintenance of the tourism industry, their right to be treated as intrinsically valuable living beings, or the desire of welfare organizations to see an improvement in conditions.

For example, captive elephants residing in private hattisars share their spaces with domesticated livestock as well as local wildlife such as wild boar (*Sus scrofa*, see Figure 14, four species of deer, rhinos, and many smaller species. These elephants travel through protected areas on the way to, and during, tourist safaris each day. Approximately 23% of these captive individuals carry tuberculosis, and others carry elephant endotheliotropic herpes viruses [84]. Other diseases such as rabies, foot and mouth disease, respiratory infections, tetanus, salmonellosis, and internal or external parasites have all been reported in Nepalese elephants [2,76]. Concerns regarding the passage of these diseases into wild populations have become a reality, with fatalities among wild elephants and greater one-horned rhinos increasing [31,85].

Figure 14. Wild boar CNP, Nepal. Photo by Nepal Dynamic Eco Tours, used with permission.

The management of these captive populations of elephants, therefore, is vitally important to the health and welfare of wild elephants [80,85]. Interactions between wild bulls and captive females during mating and overlapping natural foraging areas, for example, create opportunities for the passage of disease into wild populations, puts a strain on forest resources, and may impact the survival of wild populations [2,64]. There is a pressing need for further research which considers the health and welfare of captive individuals as an important part of overall conservation efforts and one world, one health type initiatives.

Thanks to Nepal's reliance on tourism for the survival of many marginalized communities, and the income generated for owners by elephant-backed safari, it is unlikely that elephant tourism will simply disappear from Sauraha. In fact, owners and veterinary staff expressed concern that if elephant safari ended, tourism would follow, and thus are very interested in seeing it continue. What is needed, then, are alternatives which consider the needs of humans, elephants, and other species living in or around the area. One such approach is to identify ways in which disparate entities, such as owners, advocates, I/NGOs, and local communities can work together to ensure continued income from tourism while ensuring elephants are well cared for and healthy.

7. A Novel Approach to Conflict Resolution

Allow me to return to the introduction to this presentation, and my comments on my own naivety. As I undertook my survey of the stables of Nepal, I began to seek out local, national, and international organizations who purported a desire to 'help' captive elephants. These organizations ranged from the quasi-governmental NTNC and the Chitwan Nature Guide Association, to the Elephant Owners' Cooperative and several international interests. Some of these I/NGOs have a permanent presence in Nepal, while the majority rely upon locals to care for their elephants and distribute funds.

Each of these organizations purports to 'help' elephants and profess a sense of responsibility to the captive elephants near Chitwan. The problem with undertaking the 'rescue' of these elephants lies in the lack of knowledge of local elephant–human relationships, Nepal, elephant health, mahout culture, and so on. Most of the people who became involved in these NGOs did so after being moved by an experience with an individual elephant (see Lorimer, 2009; 2010). While some went on to involve experienced staff or local people in their projects, others simply started to fundraise, post on social media, and hope that the future would bring elephants to their facility. Still others had no facility, no plans, and no relationship with elephants beyond their desire to help. Rather than communicating with other NGOs, elephant owners, or the government, they simply began to fundraise using photos of elephants in adverse conditions [7].

During my time in Nepal, I realized that while these agencies were aware of one another, few had any type of contact with one another. In addition, most INGOs attempting to improve elephant welfare were avoiding the owners of said elephants (for a variety of reasons; some viewed owners as 'mafiosos' or 'evil men', some practiced simple avoidance to avoid conflict), and ignoring the NTNC's veterinary staff—the very people who knew mahouts and their elephants best. As I bemoaned this lack of communication during a meeting with the NTNC's veterinarian, he somewhat sarcastically proposed an unusual idea. Would it not make a great addition to my research, he suggested, if I could bring all these entities together for a meeting? While he was joking, the idea continued to nag at me. I decided that thanks to my decades of work in Nepal, and my long efforts to gain the trust of elephant owners, NGOs, grassroots organizations, and community members, I was in a unique position to extend an invitation to what I jokingly entitled the first 'Sanctuary Summit of Nepal'. Sauraha is a small town, and I hypothesized that by getting individuals acquainted to the point where they might say hello when inevitably passing on the street, these groups might find common ground in their efforts to improve elephant welfare. I composed and sent an email inviting nature guides, INGOS, local community leaders, welfare groups, owners, ex-pats, and veterinary staff to coffee at a local shop. The initial response was encouraging, with owners and activists alike agreeing to come. . .if I kept them informed as to who else was invited.

The morning of the summit, I sent reminder emails and made calls to encourage attendance. When NTNC staff, members of an INGO, a visiting veterinarian, and a local hotelier arrived, I was thrilled. While these guests were aware of each other, few were formally acquainted. Introductions were made, and they began to chat about their hopes, needs, and expectations. INGOs and veterinary staff agreed to work together on communication with owners and one another. As we continued to wait, they informed me that it was unlikely any elephant owners would arrive. Rather than refusing my invitation directly and risk disappointing me, they explained that it is more culturally acceptable to simply not show up. In the end, my own cultural expectations of attendance at agreed events set me up for potential disappointment. But only for a few moments, as I soon realized that supporting any opportunities for connecting entities active within Nepal had the potential to ripple into larger collaborations.

Following the summit, other invited organizations began an email chain to summit invitees. These emails continue sporadically, offering occasional updates on organizations and encouraging dialogue with the larger group of elephant advocates. In addition, these email exchanges proved useful in clarifying each organization's definition of 'sanctuary', and 'care', and providing insight into the ethical norms and best practices of each. While the initial number of physical participants was low, the summit proved helpful in facilitating communication among stakeholders in Nepal and continues to provide a link between formerly disparate entities. While I did not realize it at the time, I was using wicked-problem-solving skills. As Mason et al. (2018) suggested, improving interactions among stakeholders may allow for more effective conservation conflict management. Among the suggestions for improving conservation outcomes are decentralizing decision making, bringing in diverse sources of knowledge, being transparent about failures, and focusing on outcomes rather than strategies [14]. Though the summit may have arisen out of a naïve attempt to get others to view elephant care through a more inclusive lens, it was successful in opening a dialogue between interested parties which then encouraged further communication.

Other organizations, however, continue to work alone, often without the permission of elephant owners. These organizations trespass upon private land to make improvements to stables or offer supplemental food for elephants and mahouts. Several focus simply on raising funds for generic elephant 'support' (without financial reports, oversight, boards of directors, etc., there is little hope of tracking where exactly this support goes). Some of these efforts have, according to mahouts and organizational staff, undoubtedly improved the welfare of individual elephants. However, the lack of organizational cooperation, or cooperation with hattisar owners, is problematic. It promotes adversarial relationships between those agencies including elephant owners in the conversation, potential funders of projects, mahouts, and those respecting the ownership rights of property (stables) [86,87]. It also perpetuates neocolonial attitudes, that only outsiders could possibly know 'what is best' for elephants [87,88] (see also Liu and Leung, 2019). It also places mahouts, an already marginalized community, in a position between advocates and owners. While they want to improve their stables to improve the health of their elephant co-workers, they also rely heavily on their jobs to survive a difficult economy. Many report that being a mahout is one of the last remaining jobs for undereducated communities, and worry that without these jobs, their families would not survive.

Lastly, the involvement of INGOs who do not include local advocates, governmentally appointed agencies, or local organizations in the planning process thus limits the ability of these local groups to promote societal change [86,87]. Rather than natural shifts in perspective on the use of animals arising from within the community, owners reported feeling that they are being pressured from exclusively outside interests. Owners report that they are responding to this pressure by disallowing outsider involvement with 'their' elephants, stables, or employees, and doubling down on their commitment to offer elephant-backed safari numerous times a day [7].

8. Conclusions

Rather than seeking a single answer based on conventional approaches, Game et al. (2013) suggest that conservationists should instead incorporate innovative, flexible, adaptive, creative, and diverse methods in their practices. Anthropological, and especially anthrozoological, methodologies and practices may offer novel and interdisciplinary approaches to the wicked problems of conservation.

One such novel approach is to ensure that all stakeholders, including those of multiple species, are offered equal consideration in conservation efforts. To that end, Nepal has shifted from single-species efforts to landscape conservation programs. What is needed next is a shift to include consideration of the agency and the desires of undomesticated species held in captivity, such as endangered elephants. Additionally, both governmental and nongovernmental organizations should focus on supporting mahouts and other marginalized communities which are reliant upon both tourism and conservation efforts, and upon whom captive species rely (see Figure 15). Rather than vilifying owners or typecasting mahouts, I/NGOs and advocates interested in improving elephant living conditions should continue to promote interagency relationships and those with elephant owners and co-workers. As the recognized owners and managers of these elephants, the decision to improve elephant management practices will ultimately fall to them. Creating opportunities for alternatives to safari income might demonstrate that ceasing elephant-backed safari is available, and NGOs might be well served to focus on solutions which equally support mahouts, elephants, and owners. It is also important to consider that while positive bonds between elephants and humans should be supported and celebrated in the hopes of improving the daily lives of both species, perhaps reconsideration is needed of the ethical issues surrounding the keeping of social and intelligent species in captivity.

Figure 15. The author meeting with mahouts near Kumroj, Nepal. Photo by Stand Up 4 Elephants, used with permission.

Multiple other issues surround elephants in Nepal: the passage of disease between wild and captive populations, the reliance on wild bulls for captive reproduction (meaning wild males regularly enter captive stables), and the danger to both humans and elephants which arises from co-work and co-living. These issues require further examination to identify ways in which traditional practices, beliefs, and income sources can be maintained, while adjusting to changing human perspectives on the use of animals for exclusively entertainment purposes such as tourism. Anthrozoology, thanks to its interdisciplinary nature, may prove an ideal discipline for such an examination.

Likewise, the issues facing wild populations, such as distance between matriarchal herds and wild males, too few individuals to 'save the species' in the wild, and increasing HWC, need to be addressed through the lens of environmental justice. Perhaps a novel solution which embraces the needs of both captive and wild populations, as well as humans, can be found. Allowing captive elephants to experience a semi-free-ranging lifestyle has allowed for improvements in elephant health and welfare, along with fewer mahout injuries in other countries, and might be worth further examination in Nepal. Likewise, a semi-free-ranging lifestyle would increase opportunities for expression of agency, choice of mates and subsequent increase in reproduction (although the ethical implications of breeding more elephants for captivity is fraught and requires further reflexivity), ability to access a wider variety of plant material with less work by mahouts and would allow the opportunity for elephants to engage in social interactions with conspecifics. This semi-free-ranging lifestyle may also eventually allow for the eventual release of captive individuals into wild herds, which was attempted in the past and has been successful in other countries (see Kharel, 2002; Varma and Ganhuly, 2011). One elephant owner has been experimenting with this method near CNP, see Figure 16 (a discussion of the experiment can be found in Szydlowski, 2023b [89]). Release efforts, however, can only succeed if captive individuals are free of communicable disease, are in optimal health, have built herd relationships, and have experience with non-provisioned fodder. This solution would require extensive cooperation and planning involving all stakeholders. Therefore, it may be beyond reach in the foreseeable future, and focus should be instead placed on providing legislation which outlines welfare standards for captive individuals, and the placement of females in small, social groups in chain-free facilities (See previous section).

Another novel approach might include embracing Scheper-Hughes (1995) concept of advocacy, in which researchers are encouraged to become involved. As individuals who 'know' their research subjects and are 'morally engaged' in their lives, researchers are in a unique position to offer insight into the perspectives of multiple stakeholders [90]. Choosing not to be involved is a morally indefensible position, according to Scheper-Hughes (1995), as it sets the researcher above their participants and their associated life events. Researcher advocacy of course requires a great deal of reflexivity to ensure that neocolonial attitudes are not being perpetuated and that power relations do not favor the researcher but rather offer equal support to all participants. Approaching the conservation of wild and captive individuals from an advocacy standpoint also allows for an inclusion of environmental and social justice perspectives.

It is likely that the novel solutions needed for the wicked conservation problems outlined here lies in a combination of ideas. What is clear is that singularly focused approaches have been unsuccessful, and there is a pressing need to consider not only the wide variety of human stakeholders involved in conservation efforts, but also members of those species being 'conserved'.

Figure 16. A privately owned, chain-free matriarchal herd in Kumroj, Nepal. Photo by Ram Krishna Mahato for this project.

Funding: No funding was received for this study.

Institutional Review Board Statement: This study was granted ethics approval by the University of Exeter College of Social Science and International Studies Ethics Committee, #201819-054R.

Informed Consent Statement: Informed consent was obtained from all human subjects involved in this study.

Data Availability Statement: Data has not been made publicly accessible due to safety concerns for human and elephant participants. Further inquiries can be directed to the author.

Acknowledgments: Thank you to the mahouts and elephants of Sauraha, Nepal, and the areas surrounding Chitwan National Park. Without you, none of my research would be possible. Thank you to Nepal Dynamic Eco Tours, Ram Krishna Mahato, and Raj Paudel for letting me share your moving and beautiful photographs (and for your many years of assistance in conservation, education, translation, and ethical tourism). Thank you also to Stand Up 4 Elephants, Om Rijal, the National Trust for Nature Conservation, and members of the Chitwan Nature Guide Association. In memory of Chupacabra.

Conflicts of Interest: The author declares there is no conflict of interest.

References

1. Szydlowski, M. *Liminal Beings: On Anthrozoologist and Elephants*; University of Exeter/EASE Working Paper Series; University of Exeter: Exeter, UK, 2023; pp. 269–289. Available online: https://anthrozoologyassymbioticethics.wordpress.com/the-ease-working-paper-series/ (accessed on 15 November 2023).

2. Kharel, F. The challenge of managing domesticated Asian elephants in Nepal. In *Giants on Our Hands: Proceedings of the International Workshop on the Domesticated Asian Elephant*; Baker, I., Kashio, M., Eds.; Food and Agriculture Organization of the United Nations: Bangkok, Thailand, 2002.
3. Government of Nepal. *Elephant Conservation Action Plan for Nepal—2009*; Government of Nepal, Ministry of Forests and Soil Conservation, Department of National Parks and Wildlife Conservation: Kathmandu, Nepal, 2009.
4. Hart, L.; Locke, P. Indian and Nepali Mahouts and their Relationships with Elephants. In *Encyclopedia of Human-Animal Relationships*; MarcBekoff, M.B., Ed.; Greenwood Press: Westport, CT, USA, 2007; Volume 2, pp. 510–515.
5. Kontogeorgopoulos, N. The Role of Tourism in Elephant Welfare in Northern Thailand. *J. Tour.* **2009**, *10*, 1–19.
6. Kontogeorgopoulos, N. Making the Best of a Bad Situation: Elephant Tourism in Northern Thailand. In *Southeast Asia and Environmental Sustainability in Context*; Kukreja, S., Ed.; Rowman and Littlefield: Lanham, MD, USA, 2020.
7. Szydlowski, M. *Elephant Tourism in Nepal: Historical Perspectives, Current Health and Welfare Challenges, and Future Directions*; CAB International: Wallingford, UK, 2024.
8. Watts, M. Neocolonialism. In *International Encyclopedia of Human Geography*; Elsevier: Amsterdam, The Netherlands, 2009; pp. 360–364. [CrossRef]
9. Mishra, H. *The Soul of the Rhino: A Nepali Adventure with Kings and Elephant Drivers, Billionaires and Bureaucrats, Shamans and Scientists and the Indian Rhinoceros*; The Lyons Press: Guilford, CT, USA, 2008.
10. Alleyne, E.; Parfitt, C. Adult-perpetrated animal abuse: A systematic literature review. *Trauma Violence Abus.* **2019**, *20*, 344–357. [CrossRef]
11. Tiplady, C. *Animal Abuse: Helping Animals and People*; CABI: Wallingford, UK, 2013.
12. Game, E.T.; Meijaard, E.; Sheil, D.; McDonald-Madden, E. Conservation in a wicked complex world; challenges and solutions. *Conserv. Lett.* **2014**, *7*, 271–277. [CrossRef]
13. Rittel, H.W.J.; Webber, M.M. Dilemmas in a general theory of planning. *Pol. Sci.* **1973**, *4*, 155–169. [CrossRef]
14. Mason, T.H.; Pollard, C.R.; Chimalakonda, D.; Guerrero, A.M.; Kerr-Smith, C.; Milheiras, S.A.; Roberts, M.; Ngafack, P.R.; Bunnefeld, N. Wicked conflict: Using wicked problem thinking for holistic management of conservation conflict. *Conserv. Lett.* **2018**, *11*, e12460. [CrossRef] [PubMed]
15. Kharal, D.; Dhungana, M. Forest Coverage and Biodiversity in Nepal. In *25 Years of Achievements on Biodiversity Conservation in Nepal*; Ministry of Forests and Environment: Kathmandu, Nepal, 2018; pp. 23–25.
16. Dhakal, M. Major Achievements of 25 Years of Biodiversity Conservation in Nepal. In *25 Years of Achievements on Biodiversity Conservation in Nepal*; Ministry of Forests and Environment: Kathmandu, Nepal, 2018; pp. 17–21.
17. Government of Nepal. *Chitwan National Park and Its Buffer Zone Management Plan*; Ministry of Soil and Forest Conservation, Department of National Parks and Wildlife Conservation: Kathmandu, Nepal, 2015.
18. Acharya, K.P. *A Walk to Zero Poaching for Rhinos in Nepal*; Department of National Parks and Wildlife Conservation: Kathmandu, Nepal, 2016.
19. Lamichhane, S.; Joshi, R.; Poudel, B.; Subedi, P. Role of Community in Leading Conservation: Effectiveness, Success and Challenges of Community-Based Anti-Poaching Unit in Nepal. *Grassroots J. Nat. Resour.* **2020**, *3*, 94–109. [CrossRef]
20. Ellis, S.; Talukdar, B. Rhinoceros unicornis. *IUCN Red List Threat. Species* **2019**, *2019*, e.T19496A18494149. [CrossRef]
21. Wikramanayake, E.; Manandhar, A.; Bajimaya, S.; Nepal, S.; Thapa, G.; Thapa, K. Chapter 10—The Terai Arc Landscape: A Tiger Conservation Success Story in a Human-dominated Landscape. In *Noyes Series in Animal Behavior, Ecology, Conservation, and Management, Tigers of the World: The Science, Politics and Conservation of Panthera Tigris*, 2nd ed.; Tilson, R., Nyhus, P., Eds.; William Andrew Publishing: Norwich, NY, USA, 2010; pp. 163–173.
22. Buckley, R. Tourism and Environment. *Annu. Rev. Environ. Resour.* **2011**, *36*, 397–416. [CrossRef]
23. Doherty, T.S.; Hays, G.C.; Driscoll, D.A. Human disturbance causes widespread disruption of animal movement. *Nat. Ecol. Evol.* **2021**, *5*, 513–519. [CrossRef]
24. Ashraf, M.; Saikia, A.; Sharma, S. Legacy of co-existence between rhino and people in a protected area in India. *Environ. Chall.* **2022**, *9*, 100639. [CrossRef]
25. Szydlowski, M. Renegotiating citizenship: Stories of young rhinos in Nepal. *J. Ecotourism* **2022**, 1–18. [CrossRef]
26. Baral, A.N. (2013) Impacts of Wildlife Tourism on Poaching of Greater One-horned Rhinoceros (Rhinoceros unicornis) in Chitwan National Park, Nepal. Master's Thesis, Lincoln University. Available online: http://www.rhinoresourcecenter.com/pdf_files/136/1365296308.pdf (accessed on 10 January 2020).
27. Subedi, B.S. Biophysical disturbances of elephant safaris in the Royal Chitwan National Park, Nepal. *Banko Janakari* **1999**, *9*, 15–19. [CrossRef]
28. Curry, B.; Moore, W.; Bauer, J.; Cosgriff, K.; Lipscombe, N. Modelling impacts of wildlife tourism on animal communities: A case study from Royal Chitwan National Park, Nepal. *J. Sustain. Tour.* **2001**, *9*, 514–529. [CrossRef]
29. Lamichhane, B.R.; Persoon, G.A.; Leirs, H.; Poudel, S.; Subedi, N.; Pokheral, C.P.; Bhattarai, S.; Thapaliya, B.P.; de Iongh, H.H. Spatio-temporal patterns of attacks on human and economic losses from wildlife in Chitwan National Park, Nepal. *PLoS ONE* **2018**, *13*, e0195373. [CrossRef] [PubMed]
30. Szydlowski, M. Elephants in Nepal: Correlating disease, tourism, and welfare. *J. Appl. Anim. Welf. Sci.* **2022**, *2*, 126–138. [CrossRef] [PubMed]

31. Thapa, J.; Paudel, S.; Vidanta, A.; Shah, Y.; Maharjan, B.; Kaufman, G.E.; McCauley, D.; Gairhe, K.P.; Tsubota, T.; Suzuki, Y.; et al. ycobacterium orygis–associated tuberculosis in free-ranging rhinoceros, Nepal, 2015. *Emerg. Infect. Dis.* **2016**, *22*, 570. [CrossRef] [PubMed]
32. Jhala, H.Y.; Qureshi, Q.; Jhala, Y.V.; Black, S.A. Feasibility of reintroducing grassland megaherbivores, the greater one-horned rhinoceros, and swamp buffalo within their historic global range. *Sci. Rep.* **2021**, *11*, 4469. [CrossRef]
33. Subedi, N.; Lamicchane, B.R.; Amin, R.; Jnawali, S.R.; Jhala, J.V. Demography and viability of the largest population of greater one-horned rhinoceros in Nepal. *Glob. Ecol. Conserv.* **2017**, *12*, 241–252. [CrossRef]
34. Lipton, J.; Bhattarai, U. Park Establishment, Tourism, and Livelihood Changes: A Case Study of the Establishment of Chitwan National Park and the Tharu People of Nepal. *Am. Int. J. Soc. Sci.* **2014**, *3*, 12–23.
35. Dangol, D.; Ghimire, A.; Bhattarai, S.B. Human-elephant conflict in the buffer zone of Chitwan National Park, Nepal. *Nepal. J. Zool.* **2020**, *4*, 36–43. [CrossRef]
36. World Wide Fund for Nature, Nepal (WWF). *A Case Study on Human-Wildlife Conflict in Nepal (With Particular Reference to Human-Elephant Conflict in Eastern and Western Terai Regions*; World Wildlife Fund: Kathmandu, Nepal, 2007; pp. 47–48.
37. Kandel, S.; Harada, K.; Adhikari, S.; Dahal, N.K.; Dhakal, M. Local perceptions of forest rules and interactions between rules, ecotourism, and human-wildlife conflicts: Evidence from Chitwan National Park, Nepal. *Tropics* **2020**, *29*, 25–39. [CrossRef]
38. Kandel, S.; Silwal, T.; Yadav, S.K.; Dhakal, S. Temporal and spatial pattern of wildlife attacks on human in Chitwan National Park, Nepal. *Nepal. J. Zool.* **2023**, *7*, 7–13. [CrossRef]
39. Menon, V.; Tawari, S.K.R. Population status of Asian elephants *Elephas maximus* and key threats. *Int. Zoo Yearb.* **2019**, *53*, 17–30. [CrossRef]
40. Neupane, D.; Baral, S.; Risch, T.S.; Campos-Arceiz, A. Broad scale functional connectivity for Asian elephants in the Nepal-India transboundary region. *J. Environ. Manag.* **2022**, *321*, 115921. [CrossRef]
41. Yadav, B.R.; Dutta, I.C.; Chalise, M.K.; Williams, C. Human-Asian wild elephant conflict and its socio-economic consequences in and around protected areas of Central Terai, Nepal. *Banko Janakari* **2015**, *24*, 7–54. [CrossRef]
42. Government of Nepal. *Nepal Tourism Statistics 2019*; Ministry of Culture, Tourism & Civil Aviation: Kathmandu, Nepal, 2020.
43. Santiapillai, C.; Jackson, P. *The Asian Elephant: An Action Plan for Its Conservation*; IUCN/SSC Asian Elephant Specialist Group: Gland, Switzerland, 1990.
44. Williams, C.; Tiwari, S.K.; Goswami, V.R.; de Silva, S.; Kumar, A.; Baskaran, N.; Yoganand, K.; Menon, V. Elephas maximus. *IUCN Red List. Threat. Species* **2020**, *2020*, e.T7140A45818198. [CrossRef]
45. Flagstad, O.; Pradhan, N.M.B.; Kvernstuen, L.G.; Wegge, P. Conserving small and fragmented populations of large mammals: Non-invasive genetic sampling in an isolated population of Asian elephants in Nepal. *J. Nat. Conserv.* **2012**, *20*, 181–190. [CrossRef]
46. Sukumar, R. Ecology of Asian elephant in south India. II. Feeding habitats and crop raiding patterns. *J. Trop. Ecol.* **1990**, *6*, 33–53. Available online: https://www.jstor.org/stable/2559367 (accessed on 1 May 2023). [CrossRef]
47. Pokharel, S.S.; Singh, B.; Seshagiri, P.B.; Sukumar, R. Lower levels of glucocorticoids in crop-raiders: Diet quality as a potential 'pacifier' against stress in free-ranging Asian elephants in a human-production habitat. *Anim. Conserv.* **2018**, *22*, 177–188. [CrossRef]
48. World Bank. Data: Nepal. World Bank Website. 2023. Available online: https://data.worldbank.org/country/nepal (accessed on 1 November 2023).
49. Government of Nepal. *Nepal National Biodiversity Strategy and Action Plan 2014–2020*; Ministry of Forests and Soil Conservation: Kathmandu, Nepal, 2014.
50. Campbell, B. Resisting the Environmentalist State. In *Resistance and the State: Nepalese Experiences*; Gellner, D., Ed.; Berghahn Books: Oxford, UK, 2007; 383p.
51. Kopnina, H. Wild Animals and Justice: The Case of the Dead Elephant in the Room. *J. Int. Wildl. Law Policy* **2016**, *19*, 219–235. [CrossRef]
52. Puyravaud Jean-Philippe, Davidar Priya, «Justice Spatiale Pour Les Éléphants: Le Cas de la Réserve de Biosphère des Nilgiris (Inde du Sud)» ["Spatial Justice for Elephants: The Case of the Nilgiri Biosphere Reserve (Southern India"], Justice Spatiale | Spatial Justice, n° 17. 2022. Available online: https://www.jssj.org/article/justice-spatiale-pour-les-elephants-le-cas-de-la-reserve-de-bios-phere-des-nilgiris-inde-du-sud/ (accessed on 12 December 2023).
53. International Monetary Fund, Asia and Pacific Group. *Nepal: 2019 Article IV Consultation-Press Release*; Staff Report and Statement by the Executive Director for Nepal; International Monetary Fund: Washington, DC, USA, 2020.
54. World Bank. Project Information Document/Integrated Safeguards Data Sheet. Nepal Sustainable Tourism Value Creation. World Bank Website. 2018. Available online: https://documents1.worldbank.org/curated/en/933281531989529139/pdf/Concept-Project-Information-Document-Integrated-Safeguards-Data-Sheet-Nepal-Sustainable-Tourism-Value-Creation-P165130.pdf (accessed on 1 November 2023).
55. Bajracharya, S.; Dhakal, M. Nature-Based Tourism, Protected Areas and Economic Development in Nepal. In *25 Years of Achievements on Biodiversity Conservation in Nepal*; Ministry of Forests and Environment: Kathmandu, Nepal, 2018; pp. 23–25.
56. deVries, L. *An Elephant is Not a Machine, A Survey into the Welfare of Private Captive Elephants in Sauraha, Chitwan National Park*; Animal Nepal: Kathmandu, Nepal, 2014.

57. Szydlowski, M. Framing Conservation, Colonialism and Care: Captive Endangered Asian Elephants (*Elephas maximus*) in Nepal. Ph.D. Thesis, University of Exeter, Exeter, UK, 2021.
58. Online Khabar. The per Capita Income of Nepalis Is One Thousand 85 US dollars. *Online Khaber*, 13 May 2020. Available online: https://www.semanticscholar.org/paper/A-Forest-Ride-on-Wild-Elephants:-The-Philosophy-of-Ramanathapillai/111498dc80c15be4cdd6306d76189709ea211560 (accessed on 12 December 2023).
59. Ramanathapillai, R. A Forest Ride on Wild Elephants: The Philosophy of Wilderness in Buddhism. *Gajah* **2009**, *30*, 29–33.
60. Barua, M. Circulating elephants: Unpacking the geographies of a cosmopolitan animal. *Trans. Inst. Br. Geogr.* **2014**, *39*, 559–573. [CrossRef]
61. Locke, P. Animals, Persons, Gods: Negotiating Ambivalent Relationships with Captive Elephants in Chitwan, Nepal. In *Conflict, Negotiation, and Coexistence: Rethinking Human-Elephant Relations in South Asia*; Oxford University Press: New York, NY, USA, 2016.
62. Locke, P. Taming and Training in The Human Use of Elephants; The Case of Nepal—Past, Present and Future. In Proceedings of the 'Anthropological and Archaeological Imaginations: Past, Present and Future' the Meeting of the Association of Social Anthropologists, Bristol, UK, 6–9 April 2009. for a panel titled 'Humans and Other Animals'.
63. Sukumar, R. *The Living Elephants: Evolutionary Ecology, Behaviour, and Conservation*; Oxford University Press: New York, NY, USA, 2003.
64. Yadav, B.R. An overview of Elephant Breeding Center at Royal Chitwan National Park, Nepal. *Banko Janakari* **2003**, *13*, 25–33.
65. Hart, L. The elephant-mahout relationship in India and Nepal: A tourist attraction. In *Animals in Person: Cultural Perspectives on Human-Animal Intimacy*; Knight, J., Ed.; BergRoutledge: Oxon, UK, 2005; pp. 163–189.
66. Schmidt-Burbach, J. *Taken for a Ride*; Published Online Only; World Animal Protection: London, UK, 2017.
67. Carlstead, K.; Paris, S.; Brown, J.L. Good keeper-elephant relationships in North American zoos are mutually beneficial to welfare. *Appl. Anim. Behav. Sci.* **2019**, *211*, 103–111. [CrossRef]
68. Lehnhardt, J.; Galloway, M. Carrots and Sticks, People and Elephants. In *Elephants and Ethics: Toward a Morality of Coexistence*; Wemmer, C.E., Christen, C.A., Eds.; Johns Hopkins University Press: Baltimore, MD, USA, 2008.
69. Cohen, E. Young Elephants in Thai Tourism: A Fatal Attraction. In *Animals and Tourism: Understanding Diverse Relationships*; Markwell, K., Ed.; Channel View Publications: Bristol, ON, Canada, 2015.
70. Gautam, M.; Khatiwada, S. *An Assessment of the Welfare Aspect of Elephant Training at the Elephant Breeding Center, Khorsor, Sauraha, Chitwan, Nepal, 2011*; PETA India and Animal Welfare Network Nepal: Norfolk, Nepal, 2011.
71. Rizzolo, J.; Bradshaw, G. Human leisure/elephant breakdown: Impacts of tourism on Asian Elephants. In *Wild Animals and Leisure: Rights and Wellbeing*; Carr, N., Young, J., Eds.; Routledge: New York, NY, USA, 2018; pp. 129–147.
72. Clubb, R.; Mason, G. *A Review of the Welfare of Zoo Elephants in Europe: A Report Commissioned by the RSPCA*; Royal Society for the Prevention of Cruelty to Animals: Southwater, UK, 2002.
73. Sukumar, R. Ecology of Asian elephant in south India. I. Movement and habitat utilization patterns. *J. Trop. Ecol.* **1989**, *5*, 1–18. Available online: https://www.jstor.org/stable/2559605 (accessed on 1 May 2023). [CrossRef]
74. Veasey, J.S. Concepts in the care and welfare of captive elephants. *Int. Zoo Yearb.* **2006**, *40*, 63–79. [CrossRef]
75. Kurt, F.; Garai, M. Food, Tool Use and Sleep. In *The Asian Elephant in Captivity: A Field Study*; Foundation Books: Dehli, India; Cambridge University Press: Cambridge, MS, USA, 2006.
76. Shrestha, S.; Gairhe, K. Nepal. In *Biology, Medicine, and Surgery of Elephants*; Fowler, M.E., Mikota, S.K., Eds.; Blackwell Publishing: Ames, IA, USA, 2006; pp. 253–261.
77. Varma, S.; Ganguly, S. *Captive Elephants in Bardia National Park, Nepal, Investigations into Population, Management, Welfare and a Review of Elephant Training by Working Elephant Programme of Asia, and WWF Finland at Bardia Hattisar, Elephants in Captivity: CUPA/ANCF—Occasional Report. No.18*; Compassion Unlimited Plus Action and Asian Nature Conservation Foundation: Bangalore, India, 2011.
78. Rizzolo, J.; Bradshaw, G. Prevalence and Patterns of Complex PTSD in Asian Elephants (*Elephas maximus*). *Asian Elephants in Culture & Nature*. No Issue Number. 2016. Available online: https://www.researchgate.net/profile/Jessica_Bell_Rizzolo/publication/311511597_Prevalence_and_Patterns_of_Complex_PTSD_in_Asian_Elephants_Elephas_maximus/links/5849aaa608ae5038263d8895.pdf (accessed on 10 December 2022).
79. Vanitha, V.; Thiyagesan, K.; Baskaran, N. Daily routine of captive Asian elephants (*Elephas maximus*) in three management systems of Tamil Nadu, India and its implications for elephant welfare. *J. Sci. Trans. Environ. Technovation* **2010**, *3*, 116–122. [CrossRef]
80. Miller, M.A.; Hogan, J.N.; Meehan, C.L. Housing and Demographic Risk Factors Impacting Foot and Musculoskeletal Health in African Elephants [*Loxodonta africana*] and Asian Elephants [*Elephas maximus*] in North American Zoos. *PLoS ONE* **2016**, *11*, e0155223. [CrossRef]
81. Convention on International Trade in Endangered Species of Wild Fauna and Flora (CITES). *Text of the Convention on International Trade in Endangered Species of Wild Fauna and Flora. Amended 1983*; IUCN: Gland, Switzerland, 1973; Available online: https://cites.org/eng/disc/text.php (accessed on 30 December 2023).
82. Nepal News. Illegal sale and purchase of elephants at the border, regulatory body silent. *Nepal News*, 15 July 2019.
83. Pokhrel, M. Nepal's Illegal Elephant Export. *Nepali Times Online*, 16 September 2021.
84. Gairhe, K.P. Veterinary care and breeding of elephants in Nepal. *Gajah* **2012**, *37*, 27–30.

85. Mikota, S.K.; Gairhe, K.; Kumar, K.; Hamilton, K.; Miller, M.; Paudel, S.; Lyashchenko, K.; Larsen, R.S.; Payeur, J.B.; Waters, W.R.; et al. Tuberculosis surveillance of elephants (*Elephas maximus*) in Nepal at the captive-wild interface. *Eur. J. Wildl. Res.* **2015**, *61*, 221–229. [CrossRef]
86. Gautam, M. One Health, One Welfare in Animal Conservation. In Proceedings of the International Society of Anthrozoology Annual Conference, Online, 4 September 2020.
87. Schuller, M. Gluing Globalization: NGOs as Intermediaries in Haiti. *PoLAR Political Leg. Anthropol. Rev.* **2009**, *32*, 84–104. [CrossRef]
88. Lorimer, J. Elephants as companion species: The lively biogeographies of Asi an elephant conservation in Sri Lanka. *Trans. Inst. Br. Geogr.* **2010**, *35*, 491–506. [CrossRef]
89. Szydlowski, M. One Stable's Novel Approach to Mitigating Human-Elephant Conflict Near Chitwan National Park, Nepal. *Gajah* **2023**, *56*, 23–29. [CrossRef]
90. Scheper-Hughes, N. The primacy of the ethical: Propositions for a militant anthropology. *Curr. Anthropol.* **1995**, *36*, 409–440. [CrossRef]

animals

MDPI

Review

Farm Animal Welfare Is a Field of Interest in China: A Bibliometric Analysis Based on CiteSpace

Lihang Cui [1], Wenjie Tang [1], Xiaoshang Deng [1] and Bing Jiang [1,2,*]

1 College of Economics and Management, Northeast Agricultural University, Harbin 150030, China; b210801002@neau.edu.cn (L.C.); s220801029@neau.edu.cn (W.T.); a08200873@neau.edu.cn (X.D.)

2 Development Research Center of Modern Agriculture, Northeast Agricultural University, Harbin 150030, China

* Correspondence: jiangbing2020@neau.edu.cn; Tel.: +86-132-7450-5764

Simple Summary: Farm animal welfare research conducted in China is relatively less accessed and unknown outside of the country, which might create the impression that farm animal welfare receives limited attention in China. This paper aims to offer a comprehensive review of the existing Chinese farm animal welfare literature. CiteSpace software was employed for the analysis and visualization of various aspects of the literature, including the quantity, species, authors, institutions, journals, and keywords. This process helped reveal the research trends, hotspots, and frontiers in the field. The findings from this bibliometric analysis strongly indicate that farm animal welfare is a subject of considerable interest and research activity in China.

Abstract: Farm animal welfare research conducted in China is not commonly accessed or known outside of China, which may lead to the assumption that farm animal welfare receives relatively little attention in China. Therefore, a bibliometric analysis was conducted on the existing Chinese farm animal welfare literature to provide robust evidence to refute this assumption. A total of 1312 peer-reviewed Chinese studies on farm animal welfare published between March 1992 and June 2023 were retrieved from the Web of Science (WoS) and the China National Knowledge Infrastructure (CNKI) database. CiteSpace software was used to analyze and visualize the number, species, authors, institutions, journals, and keywords of the papers. In China, farm animal welfare research has gone through the processes of an early stage (1992–2001), rapid-growth stage (2002–2007), and mature stage (2008–present), and the scale of research continues to grow. Notably, swine and chickens have received priority attention in this area. A Matthew effect was observed for authors and institutions, with relatively little collaboration among authors and institutions. Most of the papers were published in a small number of journals, with an apparent agglomeration characteristic. The research hotspots, summarized as "feed and diet", "environmental impacts and control", "integrated rearing management", "injury and disease", "behavior and technologies for behavior monitoring", "genetic analysis", "welfare during transport and slaughter", "welfare-friendly animal product consumption", "attitudes toward farm animal welfare", and "healthy breeding". The keywords "computer vision", "recognition", "temperature", "precision livestock farming", "laying hen", and "behavior", represent the major research frontiers in the field, which could indicate potential areas of significant future research. The findings of the present bibliometric analysis confirm the fact that farm animal welfare is a field of interest in China. Farm animal welfare research in China tends to be pragmatic, with a strong emphasis on enhancing growth and production performance, as well as product quality, rather than solely concentrating on improving farm animal welfare. This paper provides insightful references that researchers can use to identify and understand the current status and future direction of the farm animal welfare field in China.

Keywords: farm animal welfare; China; research trends; research hotspots; research frontiers; bibliometric analysis; CiteSpace

1. Introduction

The demand for the most popular farm animal products (i.e., meat, eggs, and milk) in China has expanded from 45.5 kg per capita in 2013 to 60.5 kg per capita by 2021 [1]. In line with this trend, livestock and poultry production in China has experienced significant growth over the past two decades. The production of main farm animal products in China has increased by 8.85% at an average annual rate of 2.43% from 2003 to 2022 [1]. However, to meet the increasing consumption demand, the production growth has primarily stemmed from the rapid growth of intensive livestock and poultry farming practices [2]. Although the intensification of farm animal production significantly boosts productivity and economic benefits, it inevitably leads to farm animal welfare issues and compromises the sustainability of animal husbandry [3].

Farm animal welfare concerns have often been neglected in China due to the Chinese government's prioritization of economic development [4]. This fact is demonstrated by the absence of nationwide laws and official controls pertaining to farm animal welfare [5,6]. In particular, when facing the pressure of the rising demand for farm animal product consumption in the country, it is virtually impossible to reduce intensive livestock and poultry farming practices [4]. For example, the previous version of the Chinese national standard "Construction for Intensive Pig Farms", which was released in 2008, initially stipulated a stocking density of 0.8–1.2 m^2/pig for growing-finishing pigs. However, in the revised version of 2022, the stocking density was adjusted to 0.5–1.0 m^2/pig [7]. In view of China's dominance in the production and exportation of farm animal products, the country is receiving global attention over its farm animal welfare issues [6].

Despite this, there are encouraging signs of positive changes concerning farm animal welfare in China. Several farm animal welfare standards have been established by government departments, social groups, industry associations, and enterprises over the past few years, providing valuable guidance for livestock and poultry production practices. One example is the "Farm Animal Welfare Requirements" released by the International Cooperation Committee of Animal Welfare (ICCAW), which covers various farm animal species, including pigs, beef cattle, mutton sheep, meat-type chickens, laying hens, waterfowl, cashmere goats, and dairy cattle [8]. Moreover, an increasing number of Chinese livestock and poultry enterprises, such as Suzhou OVODAN Foods Co., Ltd. (Suzhou, China), are launching welfare-friendly animal products, particularly eggs and chicken. These products have received certification from organizations such as Humane Farm Animal Care [9]. Furthermore, China is actively, widely, and increasingly participating in the global movement to improve the welfare of farm animals. By the end of 2022, the World Conference on Farm Animal Welfare had been held in China for four consecutive sessions [10]. Driven by food safety considerations, the Chinese public has also shown an increasing level of empathy and concern toward farm animals in recent years [11,12].

Nevertheless, the development of animal welfare in China remains in an early stage, and more practical measures need to be implemented to promote its development [11]. Due to distinctive historical and cultural backgrounds, political systems, basic national conditions, etc., different countries face unique problems and require specific solutions to address farm animal welfare issues. Therefore, farm animal welfare in China is best addressed by using Chinese solutions rather than relying solely on solutions designed by international organizations or dictated by global trends [13]. In many countries with high farm animal welfare standards, reform has been fueled by and based on scientific research in farm animal welfare, particularly in European Union countries [14]. However, the scientific literature on farm animal welfare from China remains relatively scarce when compared to the extensive literature available from Europe, Oceania, and America, where farm animal welfare is highly developed [15]. This has even led to the potentially inaccurate viewpoint that farm animal welfare is of little interest or concern in China [14]. Many worthwhile results and opinions on farm animal welfare in China may also be overlooked by other researchers [15]. For this reason, a systematic review of the existing Chinese farm animal welfare literature is necessary to understand the detailed context of Chinese studies on farm animal welfare.

Very few reviews of Chinese farm animal welfare research have been conducted, with only one systematic review being retrieved. Specifically, a bibliometric analysis was conducted by Sinclair et al. [14] on 854 Chinese-language academic publications related to pig and poultry welfare from Chinese scientific databases between 2008 and 2018. The attention of Chinese academia in the field of farm animal welfare was clarified by identifying the basic characteristics of the publications, including the annual publication count, topic categories, species, age of animals, and production stage [14]. Although several important conclusions were drawn from the study, the current situation and development trend within the field of farm animal welfare research in China remains ambiguous, including aspects such as the publication trends, the contributing authors, institutions and journals, collaborative relationships among authors and institutions, as well as research hotspots and frontiers. Additionally, the study was limited to Chinese-language publications related to pig and chicken species from Chinese scientific databases. Therefore, further research is required to delve deeper into these aspects, address the identified limitations, and provide a more comprehensive understanding of the field.

Therefore, the present study aimed to perform a bibliometric analysis on farm animal welfare research in China. Specifically, publication outputs, the distribution of species, and highly productive authors, institutions, and journals were analyzed. Then, cooperation relationships among authors and institutions were identified. Furthermore, research hotspots and research frontiers were clarified, and future research development directions were predicted. The findings of this study could help readers understand the current status of farm animal welfare research in China and grasp its future direction.

2. Materials and Methods

2.1. Literature Search Strategy

The literature search for this review utilized the databases of the Web of Science (WoS) and China National Knowledge Infrastructure (CNKI). WoS (https://www.webofscience.com, accessed on 12 June 2023) is widely acknowledged as the most commonly used and widely accepted digital database in scientific research, providing researchers with various types of high-quality publications [16]. Therefore, the WoS core collection was selected as the source of non-Chinese-language literature (i.e., Chinese research conducted in China and published in a foreign language), which is consistent with previous studies [17,18]. CNKI (https://www.cnki.net, accessed on 12 June 2023) is the largest and most authoritative continuously updated database for scientific publications in China [19]. It encompasses various academic resources, such as journals, master's and doctoral dissertations, conference proceedings, newspapers, yearbooks, eBooks, patents, standards, etc. [20]. Among these types of publications, the literature published in journals were found to be relatively more continuous, sensitive, and directly related to the academic field [21]. Therefore, the Chinese-language literature (i.e., Chinese research conducted in China and published in the Chinese language) for this review was obtained from academic journals within the CNKI database, which is in accordance with previous studies [14,22]. Furthermore, to ensure the quality of the literature, the literature search was conducted on peer-reviewed journals only [23]. Thus, the Science Citation Index Expanded (SCI) and Social Science Citation Index (SSCI) databases within the WoS core collection, as well as the Chinese Core Journals of Peking University, Chinese Social Sciences Citation Index (CSSCI), and Chinese Science Citation Database (CSCD) within the CNKI academic journals database were selected as literature source categories.

The search term for retrieving literature was not restricted to "farm animal welfare" alone since this could have led to potential errors of data omission. This is because "farm animal" is a generic term for livestock and poultry. When referring to certain specific types of livestock or poultry, scholars often prefer to use the name of specific farm animals directly, rather than the term "farm animal". In the current study, swine, beef cattle, dairy cattle, sheep, goats, broiler chickens, laying hens, ducks, and geese were selected as the major specific species of livestock and poultry for the following reasons: (1) Swine, beef cattle, dairy cattle, sheep, goats, broiler chickens, laying hens, ducks, and geese are considered the most important farm animals in China, representing the largest proportion of the farm animal population and the highest number of animals slaughtered [24]. (2) The exportation and domestic consumption of pork, beef, veal, milk, and lamb, as well as the meat and eggs of chickens, ducks, and geese, ranked highest among farm animal products in China. The welfare of these farm animals is critical since it can have potential impacts not only in China but also in other regions across the world [25]. (3) Welfare-friendly animal products, such as pork, chicken meat, chicken eggs, and milk, have been increasingly and widely promoted in China [26]. (4) Swine, poultry, cattle, and sheep are the key farm animal species for the implementation of the national standard "Welfare on Killing Animals for Disease Control Purposes", which is the only national standard for farm animal welfare currently enforced in China [27].

Given the above, numerous search experiments were conducted using search terms to ensure the comprehensiveness and accuracy of the literature. Referring to Qiang et al. [28], non-Chinese-language literature was limited to research conducted in China by setting the "Address" to "China". The final search term for non-Chinese-language literature was set as {(TS = "farm animal welfare" AND AD = "China") OR [TS = ("swine" OR "pig" OR "sow" OR "boar" OR "piglet" OR "fattening pig", "cattle" OR "cow" OR "bovine" OR "bull" OR "beef cattle" OR "dairy cattle" OR "dairy cow" OR "calves", "sheep" OR "goat" OR "lamb", "chicken" OR "hen" OR "rooster" OR "laying hen" OR "broiler", "duck", "goose", "livestock" OR "poultry" OR "husbandry") AND TS = ("farm animal welfare" OR "animal welfare" OR "welfare") AND AD = "China"]}, including plurals. The final search term for Chinese-language literature was set as {主题 = ("动物福利" OR "农场动物福利") OR 篇关摘 = ("动物福利" OR "农场动物福利") OR [主题 = ("猪" OR "母猪" OR "公猪" OR "仔猪" OR "育肥猪", "牛" OR "母牛" OR "公牛" OR "肉牛" OR "奶牛", "羊" OR "母羊" OR "公羊" OR "绵羊" OR "山羊", "鸡" OR "母鸡" OR "公鸡" OR "蛋鸡" OR "肉鸡" OR "仔鸡", "鸭", "鹅", "畜牧" OR "家禽" OR "畜禽") AND 主题 = ("福利" OR "动物福利")]}. The English translation of the search term for Chinese-language literature is {Subject = ("animal welfare" OR "farm animal welfare") OR Title/Keyword/Abstract = ("animal welfare" OR "farm animal welfare") OR [Subject = ("pig" OR "sow" OR "boar" OR "piglet" OR "fattening pig", "cattle" OR "cow" OR "bull" OR "beef cattle" OR "dairy cattle", "sheep" OR "ewe" OR "ram" OR "ovis aries" OR "carpra hircus", "chicken" OR "hen" OR "rooster" OR "laying hen" OR "broiler" OR "chick", "duck", "goose", "livestock" OR "poultry" OR "livestock and poultry") AND Subject = ("welfare" OR "animal welfare")]}. The effectiveness of the search terms has been confirmed to a certain extent in previous studies [18,20]. Additionally, to obtain literature as widely as possible, no limitations were set for the publication year, language, or document type. The above literature search strategy was conducted through the advanced search function, and the search date was 12 June 2023. Detailed information on the literature search results can be found in the Supplementary Material (Table S1 for the non-Chinese-language literature; Table S2 for the Chinese-language literature).

2.2. Literature Process Strategy

To ensure the rationality and validity of the literature data, each literature record was carefully read and reviewed. Exclusion criteria used for literature selection were as follows: (1) not directly related to farm animals, such as experimental animals (mice, rats, zebrafish, etc.) and wildlife (giant pandas, elephants, tigers, etc.); (2) neither qualitative nor quantitative studies, such as research reports, government documents, conference announcements, or book reviews (for Chinese-language literature only); (3) neither a research article nor a review article, such as proceeding paper, editorial material, or book chapter (for non-Chinese-language literature only); (4) missing essential information, such as authors and institutions; and (5) duplicate literature. After several rounds of manual screening, all the valid literature data were exported in "Refworks" format and saved as plain text files (UTF-8). These files contained basic information about the literature, including the title, authors, institutions, abstract, keywords, journal, document type, published year, volume, issue, and DOI. Details of the valid literature after screening are presented in the Supplementary Material (Table S3 for the non-Chinese-language literature; Table S4 for the Chinese-language literature).

2.3. Literature Analysis Strategy

Bibliometric analysis is a widely used research method that can reveal the research status of a specific academic field by dealing with and analyzing published research papers using mathematical and statistical methods [29]. With the rapid development of computational and information visualization technologies, scientific knowledge mapping has become a practical bibliometric method [30]. However, there is no consensus on which bibliographic software is best [31]. CiteSpace, a Java-based data visualization analysis software developed by Professor Chen Chaomei and his team, can effectively recognize Chinese characters and thus avoid errors or unintelligible codes [21,32]. Therefore, CiteSpace (version 6.2.R2) was used to perform bibliometric analysis and create scientific knowledge maps in the present study.

CiteSpace is essentially an information visualization technique for macro knowledge measurement, which has its own unique measurement indicators and implications. It employs co-citation analysis theory, pathfinding network algorithms, minimum spanning tree algorithms, and other tools to analyze the similarity and measurement of research units (e.g., authors, institutions, and keywords) by measuring specific literature [32]. In addition to the basic statistical analysis, the specific analysis steps undertaken by CiteSpace in the current study were as follows: (1) co-authorship analysis was used to reveal the collaborative relationships among authors and institutions; (2) keyword co-occurrence analysis and keyword cluster analysis were conducted to illustrate research hotspots in the field; (3) citation burst analysis of keywords was performed to explicate research frontiers in the field.

The scientific knowledge maps created by CiteSpace do not use the original matrix consisting of the literature. Instead, they use algorithms to normalize the original matrix and then use the new matrix for network visualization. CiteSpace offers a variety of algorithms for calculating the strength of network connections, such as Cosine, Dice, and Jaccard. In the current study, the default Cosine algorithm was used [33]. The specific formula is as follows:

$$Cosine(C_{ij}, S_{ij}, S_j = \frac{C_{ij}}{\sqrt{S_i S_j}})$$

The range of *Cosine* is 0 to 1, where C_{ij} represents the number of co-occurrences of i and j, S_i represents the occurrence of i, and S_j represents the occurrence of j.

The scientific knowledge maps generated by CiteSpace consist of nodes and lines. For the co-authorship analysis of authors and institutions, each node represents an author or institution, and nodes with a larger size imply that the author or institution has a larger number of publications. The lines between nodes represent collaborative relationships between two authors or institutions. Regarding the co-occurrence analysis of keywords, each node represents a keyword. The larger the node size, the more commonly the keyword appears. The lines between nodes indicate the association of keywords [34,35]. Betweenness centrality is an index that measures the importance of nodes in the network, and CiteSpace uses this index to identify and quantify the status of authors, institutions, and keywords in academic collaboration and information dissemination [36]. The calculation formula is as follows [37]:

$$Centrality(node_i) = \sum_{i \neq j \neq k} \frac{P_{jk}(i)}{P_{jk}}$$

Centrality ranges from 0 to 1, where P_{jk} represents the number of shortest paths between nodes j and k, while $P_{jk}(i)$ indicates the number of paths that pass through $node_i$.

Regarding the cluster analysis of keywords, CiteSpace provides an automatic clustering function based on the spectral clustering algorithm. It also offers three algorithms for extracting cluster topic words from the clustering citing documents: latent semantic index, log-likelihood ratio (*LLR*), and mutual information. *LLR* was selected as the clustering algorithm in the present study, and the calculation formula is as follows [37]:

$$LLR = \log \frac{p(C_j|V_{ij})}{p(\overline{C_j}|V_{ij})}$$

where LLR is the log-likelihood of the keyword W_i for category C_j, while $p(C_j|V_{ij})$ and $p(\overline{C_j}|V_{ij})$ are the density functions in categories C_j and $\overline{C_j}$, respectively. The quality of a cluster is reflected in terms of its silhouette value, which is a measure of cluster homogeneity. Generally speaking, if the silhouette value of a certain cluster is over 0.5, the cluster obtained is considered reasonable. If the silhouette value is over 0.7, the cluster obtained is considered credible and meets the analysis requirements [35,38].

To conduct the aforementioned analysis, the plain text files of the valid literature data were imported into CiteSpace software (version 6.2.R2). The parameters for running the software for statistics and mapping were set as follows: (1) time slicing was set as 1 year per slice, covering the publication period of the literature (2002 to 2023 for non-Chinese-language literature, and 1992 to 2023 for Chinese-language literature); (2) the selection criteria for data extraction was set to the default "g-index, k = 25", which means that the top 25 results were extracted for each time slice; and (3) default settings were all used for other parameters.

As an integrated analysis framework for the current study, the full strategy described above is presented in Figure 1.

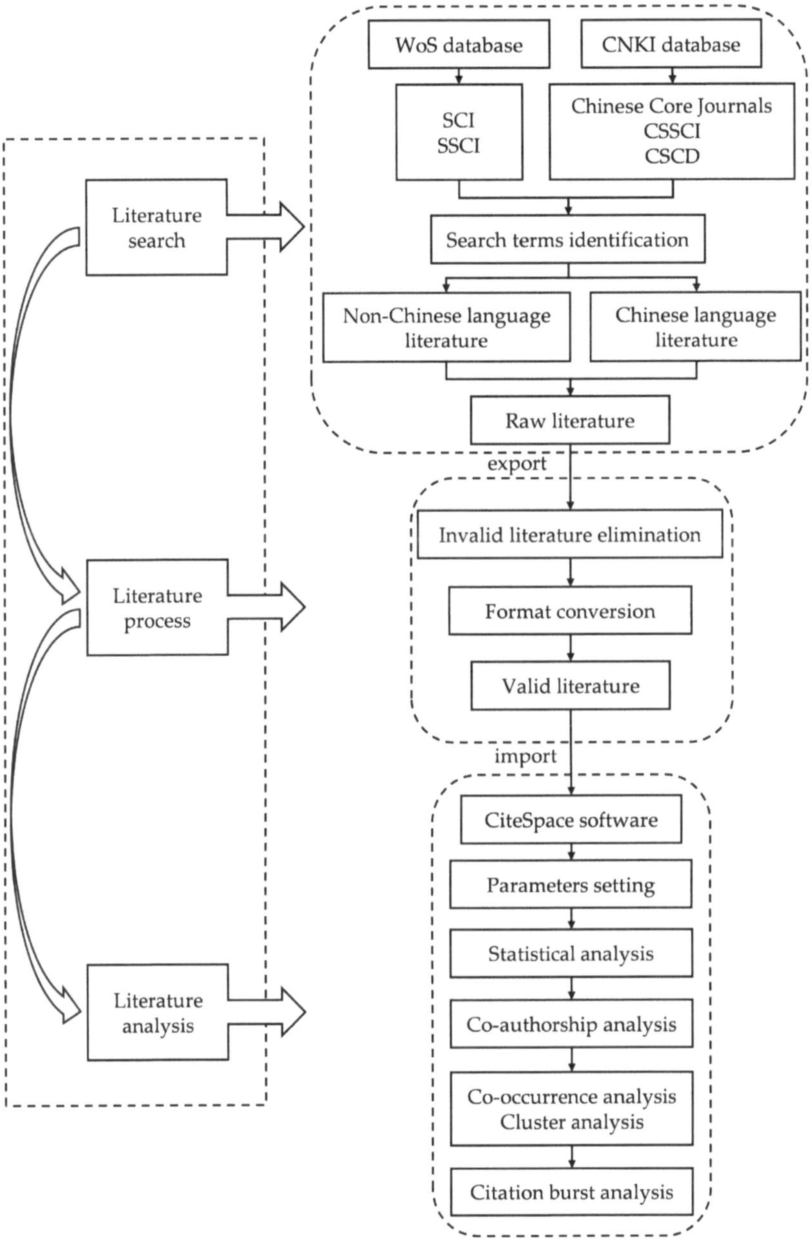

Figure 1. The integrated analysis framework of the study.

3. Results

3.1. Statistical Analysis

3.1.1. Annual Publication Output

A total of 5048 publication records met the search criteria, which were obtained from the WoS and CNKI databases, comprising 2128 non-Chinese-language publication records and 2920 Chinese-language publication records, respectively. After removing all ineligible papers based on the exclusion criteria, a total of 1312 Chinese farm animal welfare articles published between March 1992 and June 2023 were included and used in the final analysis and visualization, including 701 non-Chinese-language articles (from October 2002 to June 2023) and 611 Chinese-language articles (from March 1992 to June 2023). Notably, all of these 701 non-Chinese-language articles were published in English, suggesting that English is the dominant language for Chinese farm animal welfare research published in international journals.

The analysis temporarily excluded articles published in 2023 since the year was not yet finished, and their inclusion may not accurately reflect the overall trends. As shown in Figure 2, a total of 1227 articles were published from 1992 to 2022, averaging nearly 40 articles published per year (mean = 39.58). There was a very low level of prior output in both non-Chinese- and Chinese-language publications, with only four Chinese-language articles published before 2002 and only two non-Chinese-language articles published before 2008. A favorable and continuous fluctuating growth trend in the total number of publications starting from 2002 was observed, with an average annual growth rate of 25.60% from 2002 to 2022, indicating increasing academic attention and interest in the field of farm animal welfare in China. For non-Chinese-language publications, there has been consistent and rapid growth in annual output since 2008, with only one period of decline observed in 2015. Non-Chinese-language publications are growing at a markedly faster rate when compared to Chinese-language publications, surpassing the number of Chinese-language publications for the first time in 2018. It is worth highlighting that the number of non-Chinese-language articles published from 2020 to 2022 represented a substantial proportion (59.19%) of the total number of non-Chinese-language publications. Regarding Chinese-language publications, the annual number of Chinese-language publications has exhibited a consistent growth trend since 2002. The number of Chinese-language publications displayed large fluctuations starting in 2007, yet the average number of Chinese-language articles published per year between 2007 and 2022 consistently exceeded 31 articles (mean = 31.88).

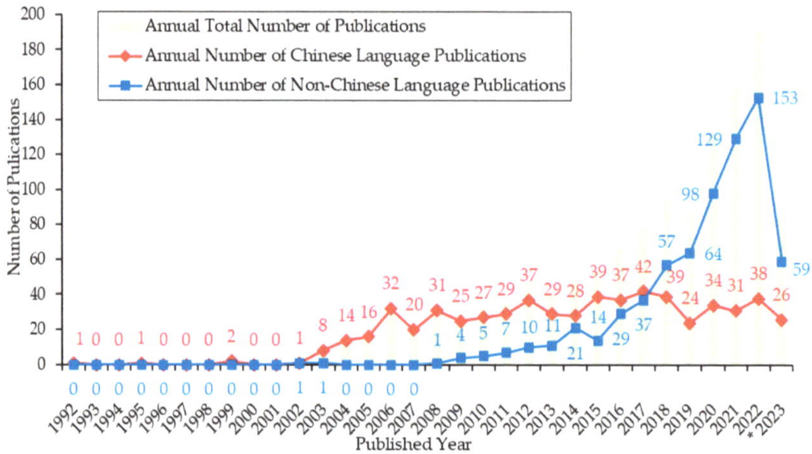

Figure 2. Number of articles published in the field of farm animal welfare in China from 1992 to 2023 June. * The data displayed in the figure for 2023 only includes the number of papers published during the period from January to June 2023.

3.1.2. Distribution of Species

As shown in Figure 3, the main species of interest for Chinese farm animal welfare research, ranked in descending order of priority, were swine (n = 261, 19.89%), chickens (n = 235, 17.91%), cattle (n = 96, 7.32%), sheep (n = 49, 3.73%), ducks (n = 36, 2.74%), and geese (n = 9, 0.69%). Swine are the primary focus in Chinese-language literature (n = 121,

19.80%) and the second focus in non-Chinese-language literature (*n* = 140, 19.97%), highlighting the important role of swine as farm animals in China. Chickens are the most commonly reported species in non-Chinese-language literature (*n* = 152, 21.68%) and the second most commonly reported species in Chinese-language literature (*n* = 83, 13.58%). There is subtle variation in the specific chicken species of interest between the non-Chinese- and Chinese-language literature. More than 60% of the non-Chinese-language literature related to chickens focused on broiler chickens (*n* = 97, 63.82%), while more than half of the Chinese-language literature related to chickens used laying hens (*n* = 46, 55.42%) as the specific species. Regarding cattle, both the non-Chinese- and Chinese-language literature demonstrate a preference for using dairy cows (*n* = 73, 5.56%) as research subjects, rather than beef cattle (*n* = 23, 1.75%). Both the non-Chinese- and Chinese-language literature exhibit a greater focus on sheep (*n* = 32, 2.44%) than goats (*n* = 17, 1.30%). Although ducks and geese received relatively little attention among poultry species, the non-Chinese-language literature showed considerably more interest in these two species than the Chinese-language literature, with nearly four times as many non-Chinese-language publications (*n* = 36, 5.14%) focusing on these species than in the Chinese-language literature (*n* = 9, 1.47%). Limited attention has been given to the welfare of other farm animal species in China, such as rabbits (*n* = 11, 0.84%), horses (*n* = 4, 0.30%), and donkeys (*n* = 4, 0.30%).

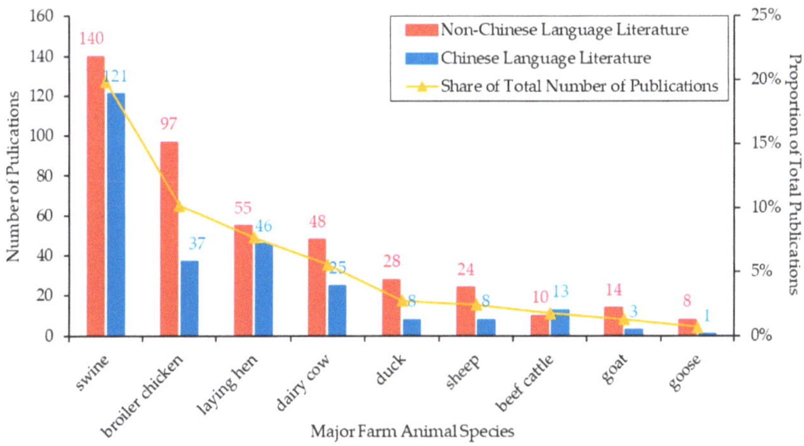

Figure 3. Distribution of species in the field of farm animal welfare in China.

3.1.3. Highly Productive Authors

Table 1 presents the top 10 most productive authors in the non-Chinese- and Chinese-language literature based on the number of publications, which was determined by CiteSpace (g-index, *k* = 25). Bao Jun, a professor at the College of Animal Science and Technology of Northeast Agricultural University and the president of ICCAW, is the most prolific author, ranking first in the non-Chinese-language literature and fourth in the Chinese-language literature. His research career focuses on investigating the behavior and welfare of livestock and poultry. Members of Bao Jun's team, including Li Xiang, Li Jianhong, Liu Honggui, Bi Yanju, Zhang Runxiang, and Wang Chao, also achieved high rankings in the non-Chinese-language literature as co-authors of Bao Jun's publications. Li Baoming, a professor from the College of Water Resources and Civil Engineering of China Agricultural University, emerged as the second most prolific author, ranking first place in the Chinese-language literature and fourth in the non-Chinese-language literature. Shi Zhengxiang, Zheng Weichao, Wang Chaoyuan, and Geng Ailian, who are members of Li Baoming's team, have collaborated on numerous publications with Li Baoming, and they also hold high rankings in both the non-Chinese- and Chinese-language literature. Their main research focus is on environmental control and engineering facilities for welfare-friendly animal production. Gu Xianhong is a researcher at the Institute of Animal Science of the Chinese Academy of Agricultural Sciences that ranked third in the Chinese-language literature. Her main research area relates to the environments, stress, and welfare of livestock and poultry. Shi Binlin is a professor at the College of Animal Science at Inner Mongolia Agricultural University. He ranks fifth in the Chinese-language literature and has devoted himself to the area of nutrition and the environments of livestock and poultry. Yan Huoqi, a professor at the College of Humanities and Social Development at Nanjing Agricultural University, ranks sixth in the Chinese-language literature and has long been engaged in research on the history of animal welfare from the perspective of social sciences and the humanities. Shao Dan and Tong Haibing are researchers from a team at the Poultry Institute of the Chinese Academy of Agricultural Sciences. They are working on the nutrition and welfare of poultry, ranking seventh and eighth in the Chinese-language literature, respectively. Sun Shimin ranks ninth in the Chinese-language literature, is a professor at the College of Economics and Management of Shandong Agricultural University, and has long been concerned with farmers' willingness and behavior to improve farm animal welfare. The centrality of the top 10 most productive authors in the non-Chinese-language literature (mean = 0.03) is slightly higher than that in the Chinese-language literature (mean = 0.00) overall, which indicates that the academic influence and collaboration of highly productive authors in the Chinese-language literature on farm animal welfare require further improvement.

Table 1. Top 10 most productive authors in the field of farm animal welfare in China.

Ranking	Non-Chinese-Language Literature				Chinese-Language Literature			
	Authors	Count	Centrality	Year *	Authors	Count	Centrality	Year
1	Bao J	47	0.01	2012	Li BM	17	0.00	2005
2	Li X	33	0.09	2013	Shi ZX	11	0.00	2005
3	Li JH	26	0.00	2013	Gu XH	9	0.00	1995
4	Li BM	23	0.05	2010	Bao J	8	0.00	2006
5	Liu HG	21	0.05	2014	Shi BL	7	0.00	2010
6	Bi YJ	13	0.00	2018	Yan HQ	7	0.00	2013
7	Zhang RX	13	0.00	2020	Shao D	6	0.00	2014
8	Zheng WC	8	0.00	2018	Tong HB	6	0.00	2014
9	Wang CY	8	0.08	2020	Sun SM	6	0.00	2008
10	Wang C	7	0.00	2018	Geng AL	6	0.00	2007

* The "year" in the table refers to the year when the author's literature first appeared in the field of farm animal welfare in China, as identified by CiteSpace (g-index, $k = 25$).

3.1.4. Highly Productive Institutions

The top 10 most productive institutions in the non-Chinese- and Chinese-language literature obtained using CiteSpace (g-index, $k = 25$), which ranked institutions according to their number of publications, are presented in Table 2. Highly productive institutions of farm animal welfare research in China are dominated by agricultural universities, with 10 of the 14 most productive institutions being agricultural universities. Of the other four highly productive institutions, Zhejiang University and Yangzhou University are comprehensive universities, the Chinese Academy of Social Sciences is a comprehensive research institute focusing on social sciences, and the Chinese Academy of Agricultural Sciences is an agricultural research institute. Except for the Chinese Academy of Social Sciences, which focuses on social sciences and humanities research related to farm animal welfare, the remaining 13 institutions primarily conduct research on farm animal welfare from a natural science perspective. An important commonality of these 13 institutions is the possession of a key discipline in animal science. These 14 highly productive institutions are located in North China, Central China, Northeast China, and Northwest China, where animal husbandry is relatively well developed. The vast livestock and poultry resources in these areas provided a convenient foundation for conducting farm animal welfare research. China Agricultural University and the Chinese Academy of Agricultural Sciences are in the top two positions and far ahead in terms of the number of publications and centrality, representing the high academic influence of these two institutions in the field of farm animal welfare in China. The centrality of the top 10 most productive institutions in the non-Chinese-language literature (mean = 0.12) is higher than those in the Chinese-language literature (mean = 0.07) overall, indicating that the academic influence and collaboration of highly productive institutions in the Chinese-language literature on farm animal welfare require further improvement.

Table 2. Top 10 most productive institutions in the field of farm animal welfare in China.

Literature Type	Ranking	Institutions	Count	Centrality	Year *
Non-Chinese-Language Literature	1	China Agricultural University	120	0.36	2003
	2	Chinese Academy of Agricultural Sciences	72	0.20	2011
	3	Northeast Agricultural University	68	0.08	2010
	4	Nanjing Agricultural University	61	0.13	2009
	5	South China Agricultural University	32	0.06	2016
	6	Huazhong Agricultural University	32	0.15	2010
	7	Northwest Sci-Tech University of Agriculture and Forestry	27	0.08	2013
	8	Sichuan Agricultural University	27	0.06	2002
	9	Zhejiang University	21	0.04	2014
	10	Yangzhou University	19	0.00	2014
Chinese-Language Literature	1	China Agricultural University	71	0.21	2005
	2	Chinese Academy of Agricultural Sciences	70	0.14	1995
	3	Nanjing Agricultural University	25	0.05	2006
	4	Inner Mongolia Agricultural University	25	0.05	2006
	5	Northeast Agricultural University	21	0.06	2005
	6	Henan Agricultural University	17	0.06	2008
	7	Shandong Agricultural University	15	0.03	2008
	8	Northwest Sci-Tech University of Agriculture and Forestry	14	0.06	2005
	9	Chinese Academy of Social Sciences	14	0.03	2003
	10	Yangzhou University	12	0.01	2009

* The "year" in the table refers to the year when the institution's literature first appeared in the field of farm animal welfare in China, as identified by CiteSpace (g-index, $k = 25$).

3.1.5. Journals-of-Choice

The top 10 non-Chinese- and Chinese-language journals, ranked by the number of publications on Chinese farm animal welfare, are listed in Table 3. The top two non-Chinese-language journals and the top two Chinese-language journals with the highest number of publications in the field of Chinese farm animal welfare each accounted for over one-fifth of the respective literature, comprising 23.82% (n = 167) of non-Chinese-language literature and 22.75% (n = 139) of Chinese-language literature. The remaining 18 journals have published less than 50 articles per journal, with a proportion of less than 10%. Nearly half of the non-Chinese-language articles (n = 320, 45.64%) were published in the top 10 non-Chinese-language journals with the highest number of publications in the field of Chinese farm animal welfare, while more than half of the Chinese-language articles (n = 331, 54.16%) were published in the top 10 Chinese-language journals by publication count, with an apparent agglomeration characteristic. The main categories of these 20 journals include veterinary science, zoology, and agricultural engineering, reflecting the multidisciplinary intersection attribution of farm animal welfare research in China. Of these 20 journals, Animals (JCR IF2022 = 3.0) has published the highest number of papers on Chinese farm animal welfare, covering a wide range of farm animal species. On the contrary, China Poultry (CNKI IF2022 = 1.066), which ranked first among Chinese-language journals in the top 10 for the highest number of publications in the field of Chinese farm animal welfare, primarily focused on poultry—especially broiler chickens and laying hens. Coincidentally, the second-ranked non-Chinese-language journal with the highest number of publications in the field of Chinese farm animal welfare, Poultry Science (JCR IF2022 = 4.4), also specializes in poultry research. Computers and Electronics in Agriculture (JCR IF2022 = 8.3), Biosystems Engineering (JCR IF2022 = 5.1), and Transactions of the Chinese Society of Agricultural Engineering (CNKI IF2022 = 3.760) are the journals among these 20 journals that focus on research related to the techniques and equipment of welfare-friendly animal production, which indirectly indicates that research on farm animal welfare in China focuses not only on theoretical innovation but also on practical applications. PLoS One (JCR IF2022 = 3.7) is the only comprehensive journal among these 20 journals. Notably, non-Chinese-language journals specifically dedicated to publishing animal welfare research, such as Animal Welfare (JCR IF2022 = 1.2, $n = 9$, 1.28%) and the Journal of Applied Animal Welfare Science (JCR IF2022 = 1.5, $n = 3$, 0.43%), do not rank among the top 10 non-Chinese-language journals with the highest number of publications in the field of Chinese farm animal welfare.

Table 3. Journals with the top 10 highest number of publications in the field of farm animal welfare in China.

Literature Type	Ranking	Journals	Count	Proportion
Non-Chinese-Language Literature	1	Animals	110	15.69%
	2	Poultry Science	57	8.13%
	3	Computers and Electronics in Agriculture	47	6.70%
	4	Frontiers in Veterinary Science	20	2.85%
	5	Animal	18	2.57%
	6	Biosystems Engineering	18	2.57%
	7	Journal of Animal Science	14	2.00%
	8	PLoS One	13	1.85%
	9	Applied Animal Behaviour Science	12	1.71%
	10	Agriculture	11	1.57%
Chinese-Language Literature	1	China Poultry	71	11.62%
	2	Heilongjiang Animal Science and Veterinary	68	11.13%
	3	Chinese Journal of Animal Science	47	7.69%
	4	Journal of Domestic Animal Ecology	38	6.22%
	5	Transactions of the Chinese Society of Agricultural Engineering	23	3.76%
	6	Animal Husbandry and Veterinary Medicine	22	3.60%
	7	China Animal Husbandry and Veterinary Medicine	21	3.44%
	8	Chinese Journal of Animal Nutrition	15	2.45%
	9	Chinese Journal of Animal and Veterinary Sciences	14	2.29%
	10	Feed Industry Magazine	12	1.96%

3.2. Co-Authorship Analysis

3.2.1. Co-Authorship Analysis of Authors

An author collaboration network map was generated using CiteSpace to illustrate collaborative relationships between the authors of non-Chinese-language articles (Figure 4a) and Chinese-language articles (Figure 4b). The current collaboration among authors in the field of farm animal welfare in China was generally characterized by overall dispersion and partial concentration. Although collaboration was the main form of publication in the field, the scope of cooperation was relatively small and fixed. Most of the research was conducted by mentors and their graduate students, who worked or studied at the same institution, while less collaborative research was performed by researchers from different institutions. According to Figure 4, two relatively large collaborative teams were observed, with Li Baoming and Baǒ Jun as the core of the teams. The team with Li Baoming as the core comes from China Agricultural University and seems to have the most team members while also being strongly interconnected in both the non-Chinese- and Chinese-language literature. They have demonstrated great interest in the study of environmental control and engineering facilities for welfare-friendly animal production. The team with Bao Jun as the core comes from Northeast Agricultural University and has more collaborative relationships in the non-Chinese-language literature, with relatively fewer in the Chinese-language literature. This team is dedicated to studying the behavior and welfare of livestock and poultry. In addition to these two major collaborative teams, the remaining collaborative relationships among authors involved a relatively small number of researchers, with looser interconnections to each other.

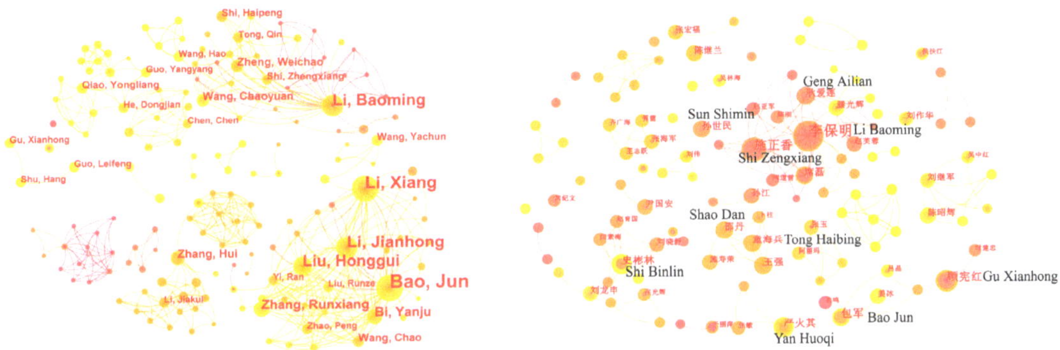

Figure 4. Author collaboration network map of the field of farm animal welfare in China. (**a**) Author collaboration network map of the non-Chinese-language literature; (**b**) author collaboration network map of the Chinese-language literature.

3.2.2. Co-Authorship Analysis of Institutions

Collaborative relationships among institutions associated with the non-Chinese -language literature (Figure 5a) and Chinese-language literature (Figure 5b) were illustrated by an institution collaboration network map created using CiteSpace. Although institutes and universities were the mainstays of farm animal welfare research in China, there is no shortage of participation from livestock enterprises and government departments. This indicates that farm animal welfare has likely become a widespread problem in Chinese livestock and poultry production practices, which required the joint participation of these institutions in researching and finding solutions. This also reflects that research on farm animal welfare had important prospects for application and promotion in China. Based on Figure 5, China Agricultural University and the Chinese Academy of Agricultural Sciences could be identified as the two most influential academic institutions, whose collaborative relationships were far more than those of the others. A normal, but important fact is that most of the collaboration occurred among institutions located in the same city or neighboring cities. However, some collaboration also occurred between Chinese institutions and foreign institutions such as Purdue University, Iowa State University, and the United States Department of Agriculture.

(**a**) (**b**)

Figure 5. Institution collaboration network map of the field of farm animal welfare in China. (**a**) Institution collaboration network map of the non-Chinese-language literature; (**b**) Institution collaboration network map of the Chinese-language literature.

3.3. *Keyword Analysis*

3.3.1. Keywords with High Occurrence

A keyword co-occurrence map of the non-Chinese-language literature (Figure 6a) and Chinese-language literature (Figure 6b) was obtained through keyword co-occurrence analysis using CiteSpace. High-occurring keywords consist of specific species and specific research topics, reflecting popular species and research hotspots in the field of farm animal welfare in China to an extent. In the non-Chinese-language literature, "welfare" is the keyword with the highest occurrence, followed by "behavior", "performance", "animal welfare", and "growth performance". In the Chinese-language literature, "animal welfare" is the keyword with the highest occurrence, followed by "welfare", "laying hen", "healthy breeding", and "behavior". Figure 6 illustrates that the keywords with high occurrence in both the non-Chinese- and Chinese-language literature share similarities in terms of species and research topics, such as "laying hen", "welfare", "behavior", and "animal welfare", indicating that there are common research hotspots in both the non-Chinese- and Chinese-language literature on farm animal welfare research in China.

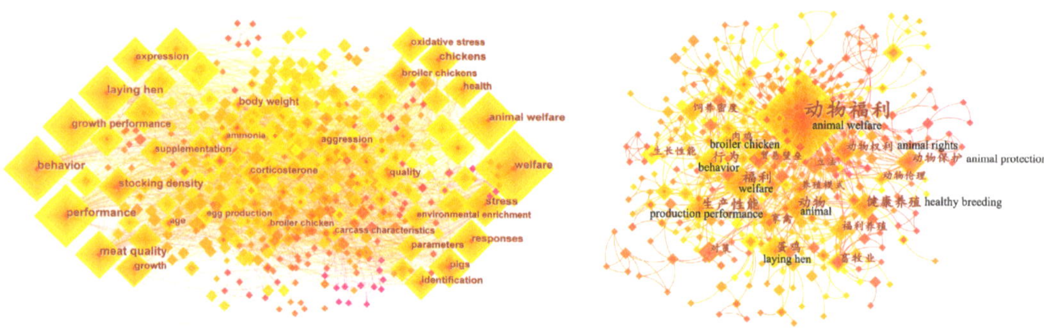

(**a**) (**b**)

Figure 6. Keyword co-occurrence network map of the field of farm animal welfare in China. (**a**) Keyword co-occurrence network map of the non-Chinese-language literature; (**b**) keyword co-occurrence network map of the Chinese-language literature.

To obtain a clearer picture of the specifics of the keywords, 10 keywords with the highest occurrence in the non-Chinese- and Chinese-language literature, which were counted by CiteSpace (g-index, $k = 25$), were sorted and listed in Table 4. Among these keywords, "animal welfare" had the highest centrality in both the non-Chinese- and Chinese-language literature, which illustrates that "animal welfare" occupied a key position in the farm animal welfare research in China and was closely related to other keywords. Other keywords with less centrality, including "behavior", "performance", "growth performance", "meat quality", "expression", "responses", "stress", "healthy breeding", and "production performance", revealed the most popular research content in the field to a certain extent. Keywords related to species, including "laying hen" and "broiler chicken" reflect the most popular species in the field to some extent. Additionally, as variations or synonyms of "animal welfare" and keywords closely related to "farm animal welfare", the centrality of keywords, including "welfare", "animal", "animal protection", and "animal rights", were also relatively low, but still larger than 0.

Table 4. Keywords with the top 10 highest occurrence in the field of farm animal welfare in China.

Literature Type	Ranking	Keywords	Count	Centrality	Year *
Non-Chinese-Language Literature	1	Welfare	145	0.11	2012
	2	Behavior	99	0.12	2010
	3	Performance	98	0.17	2009
	4	Animal welfare	89	0.23	2009
	5	Growth performance	76	0.05	2014
	6	Meat quality	61	0.15	2009
	7	Laying hen	55	0.09	2012
	8	Expression	52	0.13	2011
	9	Responses	50	0.06	2012
	10	Stress	48	0.09	2011
Chinese-Language Literature	1	Animal welfare	202	0.89	2002
	2	Welfare	37	0.09	2006
	3	Laying hen	24	0.05	2008
	4	Healthy breeding	21	0.04	2006
	5	Behavior	21	0.05	2011
	6	Production performance	20	0.09	2009
	7	Animal protection	16	0.02	2003
	8	Animal	15	0.06	2008
	9	Broiler chicken	14	0.03	2013
	10	Animal rights	14	0.01	2002

* The "year" in the table refers to the year when the keyword first appeared in the field of farm animal welfare in China, as identified by CiteSpace (g-index, $k = 25$).

3.3.2. Keywords Clustering Labels with High Occurrence

The keywords were analyzed by cluster analysis using CiteSpace, which grouped keywords from the non-Chinese-language literature (Figure 7a) and Chinese-language literature (Figure 7b) into 9 and 12 clustering labels, respectively. The sequence number of clustering labels was inversely proportional to the size of the clusters. In the non-Chinese-language literature, the largest cluster was labeled as "#0 deep learning", followed by "#1 stereotypic behavior", "#2 heat stress", "#3 quantitative trait loci", "#4 animal welfare", "#5 carcass yield", "#6 meat quality", "#7 tibial dyschondroplasia", "#8 stocking density", and "#9 piglet welfare". In the Chinese-language literature, the largest cluster was labeled as "#0 animal welfare", followed by "#1 laying hen", "#2 behavior", "#3 production performance", "#4 animal", "#5 animal protection", "#6 animal husbandry", "#7 pork", "#8 swine welfare", "#9 international trade", "#10 animal rights", "#11 environmental control", and "#12 food safety".

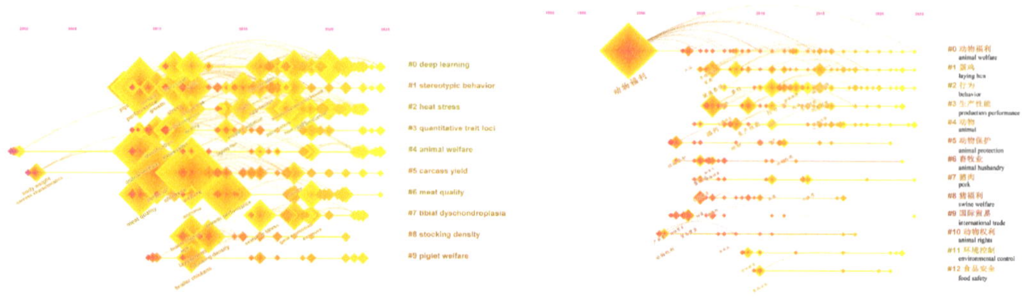

(**a**) (**b**)

Figure 7. Keyword cluster map of the field of farm animal welfare in China. (**a**) Keyword cluster map of the non-Chinese-language literature; (**b**) eyword cluster map of the Chinese-language literature.

Basic information on the top five keywords covered by clusters derived from the LLR algorithm is presented in Table 5. The silhouette values of these clusters were all larger than 0.8, indicating that the clusters were generally considered reasonable and met the analysis requirements. The specific composition of the top five clustering labels in the non-Chinese-language literature is presented as follows. The largest cluster (#0 deep learning) was formed by the keywords of 67 articles published from 2010 to 2023. The second-largest cluster (#1 stereotypic behavior) was composed of the keywords of 60 articles published between 2007 and 2023. Cluster "#2 heat stress" was the third-largest cluster, which was made up of keywords from 53 articles published from 2011 to 2022. The size of cluster "#3 quantitative trait loci" was the fourth largest and consisted of the keywords of 51 articles published from 2009 to 2023. The cluster that ranked fifth was "#4 animal welfare", which contained the keywords of 50 articles published between 2002 and 2023. The specific composition of the top five clustering labels in the Chinese-language literature is presented as follows. The largest cluster, labeled as "#0 animal welfare", comprised keywords from 79 articles published between 1997 and 2023. The second-largest cluster, labeled as "#1 laying hen", consisted of keywords from 46 articles published between 2005 and 2023. The third-largest cluster, labeled as "#2 behavior", included keywords from 42 articles published from 2007 to 2023. The fourth-largest cluster, labeled as "#3 production performance", consisted of keywords from 39 articles published between 2005 and 2022. The fifth-largest cluster, labeled as "#4 animal", contained keywords from 27 articles published between 2004 and 2023.

3.3.3. Keywords with the Strongest Citation Bursts

The top 20 keywords with the strongest citation bursts in the non-Chinese- and Chinese-language literature are listed in Table 6, which highlights the research frontiers in the field of farm animal welfare in China. In the non-Chinese-language literature, the keywords with high burst strength, including "broiler chickens", "performance", and "temperature", have been studied extensively since 2002. The keywords of "computer vision", "recognition", "temperature", and "precision livestock farming" are still experiencing ongoing bursts in 2023, reflecting the present research frontiers and possible future development direction of the non-Chinese-language literature. In the Chinese-language literature, the keywords with the highest burst strength from 2003 were "behavior", "trade barrier", and "welfare-friendly animal production". The keywords "laying hen" and "behavior" continued their prominence in 2023 and represent frontiers of research in recent years, potentially indicating the future development direction of Chinese-language literature.

Table 5. Top five clusters of keywords in the field of farm animal welfare in China.

Literature Type	Cluster	Label (LLR, *p*-Value)	Size	Silhouette Value	Mean (Year) *
Non-Chinese-Language Literature	#0 deep learning	① Deep learning (69.47, <0.001); ② Computer vision (40.61, <0.001); ③ Precision livestock farming (18.40, <0.001); ④ Lameness detection (17.09, <0.001); ⑤ Dairy cattle (14.96, 0.001).	67	0.832	2019
	#1 stereotypic behavior	① Stereotypic behavior (14.44, 0.001); ② Physiology (14.44, 0.001); ③ Growing pigs (10.82, 0.005); ④ Enriched environment (10.82, 0.005); ⑤ Pig (9.50, 0.005).	60	0.802	2014
	#2 heat stress	① Heat stress (14.42, 0.001); ② Gut microbiota (12.49, 0.001); ③ Dairy cow (12.16, 0.001); ④ Rumen fermentation (11.51, 0.001); ⑤ Buffalo (8.10, 0.005).	53	0.814	2019
	#3 quantitative trait loci	① Quantitative trait loci (20.46, <0.001); ② Swine (15.33, <0.001); ③ Candidate gene (15.33, <0.001); ④ Single-nucleotide polymorphism (10.21, 0.005); ⑤ Perch (10.21, 0.005).	51	0.858	2014
	#4 animal welfare	① Animal welfare (29.46, <0.001); ② Willingness to pay (26.24, <0.001); ③ Choice experiment (20.97, <0.001); ④ China (11.36, 0.001); ⑤ Pork (10.46, 0.005).	50	0.824	2014
Chinese-Language Literature	#0 animal welfare	① Animal welfare (41.56, <0.001); ② Legislation (14.03, 0.001); ③ Welfare (9.04, 0.005); ④ Production performance (7.52, 0.010); ⑤ Willingness to pay (6.99, 0.010).	79	0.979	2011
	#1 laying hen	① Laying hen (26.52, <0.001); ② Welfare-friendly breeding (26.52, <0.001); ③ Healthy breeding (26.52, <0.001); ④ Poultry (17.60, <0.001); ⑤ Image processing (13.17, 0.001).	46	0.899	2015
	#2 behavior	① Behavior (37.90, <0.001); ② Growth performance (18.76, <0.001); ③ Beef cattle (11.51, 0.001); ④ Stocking density (9.73, 0.005); ⑤ Straw (9.33, 0.005).	42	0.870	2015
	#3 production performance	① Production performance (34.19, <0.001); ② Welfare (27.62, <0.001); ③ Sow (22.27, <0.001); ④ Light intensity (13.30, 0.001); ⑤ Animal welfare (13.24, 0.001).	39	0.960	2012
	#4 animal	① Animal (23.63, <0.001); ② Dairy cow (17.23, <0.001); ③ Sensor (11.44, 0.001); ④ Heat stress (7.74, 0.010); ⑤ Information monitoring (5.70, 0.050).	27	0.930	2014

* The "mean (year)" in the table refers to the average year when the keywords in the cluster first appeared in the field of farm animal welfare in China, as identified by CiteSpace (g-index, $k = 25$).

Table 6. Top 20 keywords with the strongest citation burst in the field of farm animal welfare in China.

Literature Type	Keywords	Year [1]	Strength	Begin [2]	End [3]	Temporal Distribution [4]
Non-Chinese-Language Literature	Meat quality	2009	3.59	2012	2015	
	Group size	2012	2.60	2012	2017	
	Carcass yield	2012	2.55	2012	2018	
	Stocking density	2013	4.64	2013	2015	
	Performance	2009	5.13	2016	2017	
	Intensity	2016	3.10	2016	2018	
	Identification	2010	2.60	2016	2017	
	Broiler chickens	2012	5.18	2017	2019	
	Systems	2017	2.62	2017	2018	
	Mortality	2018	3.55	2018	2020	
	Tibial dyschondroplasia	2018	3.53	2018	2019	
	Piglets	2008	3.23	2018	2020	
	Indicators	2019	2.85	2019	2020	
	Exposure	2019	2.85	2019	2020	
	Milk	2019	2.69	2019	2020	
	Computer vision	2020	3.69	2020	2023	
	Recognition	2020	3.69	2020	2023	
	Dairy cattle	2020	2.61	2020	2021	
	Temperature	2019	4.27	2021	2023	
	Precision livestock farming	2021	3.71	2021	2023	
Chinese-Language Literature	International trade	2003	2.19	2003	2010	
	Trade barrier	2004	4.13	2004	2010	
	Livestock products	2004	1.95	2004	2008	
	Countermeasure	2005	3.77	2005	2008	
	Animal husbandry	2005	2.30	2005	2009	
	Legislation	2004	2.14	2008	2010	
	Supply chain	2008	1.97	2008	2011	
	Animal protection	2005	2.00	2010	2011	
	Healthy breeding	2006	2.70	2012	2015	
	Broiler chicken	2013	2.43	2013	2017	
	Welfare-friendly animal production	2015	4.06	2015	2019	
	Stocking density	2015	3.65	2015	2019	
	Laying hen	2008	3.23	2015	2023	
	Animal	2008	2.62	2015	2018	
	Sow	2011	2.05	2016	2018	
	Influencing factor	2009	2.38	2017	2021	
	Production performance	2009	2.37	2017	2020	
	Behavior	2011	4.16	2018	2023	
	Stress	2011	2.46	2018	2019	
	Dairy cow	2011	3.72	2019	2021	

[1] The "year" in the table refers to the year when the keyword first appeared in the field of farm animal welfare in China, as identified by CiteSpace (g-index, $k = 25$). [2] The "begin" in the table refers to the year when the keyword started to appear with high occurrence in the field of farm animal welfare in China, as identified by CiteSpace (g-index, $k = 25$). [3] The "end" in the table refers to the year when the keyword stopped appearing with high occurrence in the field of farm animal welfare in China, as identified by CiteSpace (g-index, $k = 25$). [4] The "temporal distribution" for non-Chinese-language literature spans from 2002 to 2023, and the "temporal distribution" for Chinese-language literature spans from 1992 to 2023. The light blue bar indicates that the corresponding term has not yet appeared, while the dark blue bar indicates that the corresponding term has started to appear but is being used relatively infrequently, and the red line represents the burst time when the corresponding terms appeared more prevalently during this period.

4. Discussion

A comprehensive bibliometric analysis of China's farm animal welfare research from the WoS and CNKI databases was conducted using CiteSpace software (version 6.2.R2).

This bibliometric analysis statistically and visually outlines the current status, research hotspots, and research frontiers of the field of farm animal welfare in China, providing the researchers with valuable information on the general landscape and future trends in this field.

4.1. Number of Publications

Within this review, 1312 non-Chinese and Chinese-language publications on farm animal welfare in China published since 1992 were discovered. This validates previous findings that farm animal welfare has indeed become a research area of interest to Chinese scholars, contrary to the perception that farm animal welfare is of little interest or remains unexplored in China [14]. The results revealed that the number of publications increased from 1992 to 2022, indicating that farm animal welfare has emerged as an important topic in China, which has received increasing academic attention and achieved certain progress, despite the growing number of publications in all fields of science internationally [39]. Surprisingly, the number of articles on farm animal welfare published in China appears to surpass that of countries from Latin America. Gallo et al. [40] only collected 663 articles related to the welfare and behavior of farm animals from the WoS database and Centre for Agriculture and Bioscience Abstracts database between 1992 and 2021 in Latin America, of which 287 articles were published in Brazil and 167 articles were published in Mexico. Notably, the present study identified 489 non-Chinese-language publications related to farm animal welfare in the same period in China from the WoS database, far exceeding the number of publications on farm animal welfare in these two countries in Latin America with the highest productivity. It can be speculated that China seems highly likely to make significant progress and take a prominent position in the global field of farm animal welfare research in the near future. For example, Uyanga et al. [41] found that China has emerged as the most prolific institution in the global research on heat stress in poultry. However, research in the field of farm animal welfare in China, which requires more attention and development, remains limited when compared to that of countries in North America and the European Union [15].

The findings showed that the quantity and growth rate of non-Chinese-language literature have surpassed those of Chinese-language literature since 2018. This suggests that the recent expansion in farm animal welfare scientific output in China is primarily driven by non-Chinese-language publications in international journals, and Chinese farm animal welfare research is gaining greater visibility among readers and researchers outside of China [42]. Research has pointed out that the continuous increase in non-Chinese-language publications conducted by Chinese researchers is largely due to the policy of encouraging more international publications in Chinese universities and research institutes [43]. In spite of specific promotion criteria varying across different institutions, including monetary bonus schemes and career-related incentives, a greater emphasis has been placed on SCI and SSCI indexed journal publications than domestic publications [44]. Under such an incentive structure, rational Chinese researchers naturally tend to shift their efforts toward more valued international publishing in pursuit of considerable financial reward.

4.2. Development Process

Based on the changing trend of annual publication outputs (Figure 2), combined with the history of animal husbandry in China, the evolution of farm animal welfare research in China can be categorized into three specific development stages: the early stage (1992–2001), the rapid-growth stage (2002–2007), and the mature stage (2008–present).

In the early stage (1992–2001), only four Chinese-language articles were published, accounting for less than 1% of the total literature (0.30%), while no non-Chinese-language articles were published (Figure 2). The science of animal welfare originated in Western countries, and the concept of animal welfare was introduced to mainland China in the early 1990s [4]. However, in the late twentieth century, livestock and poultry production in China was mainly conducted by rural households, and intensive livestock farming was rare due to limited funding at that time [45]. Farm animal welfare, as a product of intensive livestock farming, initially received little attention from scholars in China [46].

In the rapid-growth stage (2002–2007), a total of 93 articles were published, including 91 Chinese-language articles published consistently and 2 non-Chinese-language articles published sporadically, representing 7.09% of the total literature (Figure 2). With the expansion of livestock and poultry production, intensive livestock and poultry farming was becoming increasingly prevalent in China to further improve production efficiency and economic benefits [47]. Meanwhile, since China entered the World Trade Organization in November 2001, the exportation of Chinese livestock products was widely restricted due to farm animal welfare issues, which resulted in massive economic losses [48]. Therefore, an increasing number of Chinese scholars have begun to focus on farm animal welfare issues and conduct relevant research. However, there was limited substantial progress during this period, with only a few preliminary exploratory experiments conducted, such as the influence of summer ambient temperature on the behavior of small-tailed cold sheep and their postural choices [49], the relationship between shading and environmental enrichment on feather pecking and welfare in laying hens [50], and the effect of transportation on the weight of weaned piglets [51]. The majority of studies in this period focused on qualitative analyses, such as the impact of animal welfare standards in international trade policy on China's farm animal product exports [52,53], strategies and measures for improving farm animal welfare in livestock and poultry production [54,55], and proposals for farm animal welfare legislation [56,57].

In the mature stage (2008–present), 1215 articles were published, consisting of 516 Chinese-language articles and 699 non-Chinese-language articles, which accounted for 92.61% of the total literature (Figure 2). Currently, the priority of animal husbandry development in China has shifted from increasing the volume of livestock and poultry products to quality improvement, while the highlights of the policies have also adjusted from promoting the development of animal husbandry to emphasizing both industry development and environmental protection [58]. In this context, farm animal welfare issues caused by intensive livestock farming have received extensive attention from scholars in China, and Chinese farm animal welfare research in this stage has realized the transformation from the original ethical appeal to comprehensive scientific exploration [59]. Especially in recent years, experimental research from the natural sciences perspective and empirical research from the social sciences perspective have emerged in an endless stream. There is now a broad emerging consensus in the Chinese academic community that improving farm animal welfare has a significant promoting effect on animal husbandry quality [60,61].

4.3. Species Differences

According to the results, the primary focus was placed on swine and chickens within China's farm animal welfare research, with more attention being paid to swine than chickens. This is consistent with the findings of Sinclair et al. [14], which suggested that such a focus is likely to be determined by the value of the pork industry in international trade or the perceived complexity of providing improved welfare for pigs. In contrast to the extensive research conducted on swine and poultry, there has been less farm animal welfare research focused on cattle/cows and sheep/goats in China. This diverges from the research trends observed in other countries and globally. Gautret et al. [62] found that over 30% of studies on livestock and poultry welfare and health in Europe from 2003 to 2014 concentrated on ruminants (34.62%), followed by poultry (17.40%) and pigs (13.13%). Freire and Nicol [15] also revealed that cattle/cows (13.89%) received greater research attention in global animal welfare science than pigs (13.74%) and laying hens (8.46%).

The distribution of species in Chinese farm animal welfare research may be closely related to the welfare status of farm animals in livestock and poultry production. For example, in the United Kingdom, sheep and beef cattle are typically raised using extensive management practices, and their welfare is therefore more likely to be met, with more research focused on addressing welfare concerns for pigs, poultry and dairy cattle [63]. Species distribution is also likely to be influenced by the structure of farm animal production in China. A previous bibliometric review revealed that more research on European organic livestock farming has been conducted on cattle, sheep, and poultry due to their relatively higher proportion among organic farm animals [23]. The ruminant sector, represented by beef cattle, dairy cattle, sheep, and goats, has become one of the rapidly expanding segments within China's livestock and poultry industry, with great growth in both production and consumption of its products [1]. Nonetheless, animal welfare concerns resulting from intensive farming practices remain a major challenge to the sustainable development of the industry [64,65]. Therefore, it is necessary for Chinese researchers to conduct more studies on the welfare of ruminants in the future.

4.4. Major Contributors

A Matthew effect was identified among major contributors in the field of farm animal welfare in China [66]. This implies that most of the highly productive authors in the field are from the same teams which are typically led by scholars with academic influence (Table 1), and most of the highly productive institutions in the field are well-known universities or institutes in China (Table 2). This can be explained by the fact that these highly productive teams and institutions may possess stronger research resources and invest more manpower, financial, and material resources into the implementation of farm animal welfare research to increase scientific output [67,68]. This is unfavorable for newer and less established authors, but similar situations also exist in other research areas [69,70]. Moreover, the low centrality of both authors (Table 1) and institutions (Table 2), particularly in the Chinese-language literature, implies that there is still room for enhancement in terms of the collaboration among highly productive authors and institutions in the field of farm animal welfare in China, as well as the academic impact of their work [71].

Regarding journals, the agglomeration characteristic observed in Chinese farm animal welfare research may be influenced by editorial preferences and the scope of journals. As the field of farm animal welfare research expands and becomes more multidisciplinary, research on this topic may be published in a wider range of journals, potentially resulting in a weaker agglomeration characteristic. Additionally, different from global animal welfare science, non-Chinese-language journals specifically dedicated to publishing animal welfare research, such as Animal Welfare and the Journal of Applied Animal Welfare Science, are not dominate in Chinese farm animal welfare publications [15]. Although there is no Chinese-language journal specifically dedicated to publishing animal welfare research, such positive growth in China's farm animal welfare science is likely to fuel the launch of new scientific journals that specifically focus on animal welfare.

4.5. Collaborative Relationships

A sound collaborative relationship among scholars can establish academic networks to share innovative ideas, concepts, and theories, generate new knowledge, and ultimately reduce the wasting of academic resources and improve the productivity of research [72]. However, it was determined that collaboration in the field of farm animal welfare in China is relatively weak among both authors (Figure 4) and institutions (Figure 5). Although the non-Chinese-language literature demonstrates a slightly higher level of collaboration when compared to the Chinese-language literature, with a greater number of collaborating authors and stronger interconnections, especially within teams, the collaborative relationships in this field could be regarded as relatively fixed on a small scale. The relatively scattered authors and institutions are unfavorable for the exchange of academic knowledge and ideas in the field, which further hinders the expansion of research content. This scenario is likely to be detrimental to the long-term sustainability of Chinese farm animal welfare research. One possible reason for this is the lack of effective academic platforms in this field, such as specialized academic conferences, which makes it difficult for scholars to communicate with and learn from each other [73].

4.6. Research Hotspots

Keywords are domain terms that can briefly and concisely summarize the core theme and reflect the core content of an article at a high level [74]. The results of the keywords co-occurrence analysis (Table 4 and Figure 6) are highly similar to those of international animal welfare science [15], suggesting that the research hotspots in the field of farm animal welfare in China are similar to those of international animal welfare science. This observation was also confirmed in a recent Chinese-language study in China [75]. In addition, the clustering labels of the non-Chinese-language literature tend to be more detailed and specific, while the clustering labels of the Chinese-language literature tend to be less detailed and broader (Table 5 and Figure 7). This phenomenon could be attributed to the difference in linguistic or writing conventions between English language and Chinese language [76].

Combined with a thorough review of the literature searched from databases, the results of keyword co-occurrence analysis (Table 4 and Figure 6) and keyword clustering analysis (Table 5 and Figure 7) indicate that research hotspots in the field of farm animal welfare in China can be summarized in 10 categories: "feed and diet", "environmental impacts and control", "integrated rearing management", "injury and disease", "behavior and technologies for behavior monitoring", "genetic analysis", "welfare during transport and slaughter", "welfare-friendly animal product consumption", "attitudes toward farm animal welfare", and "healthy breeding". As pointed out by Sinclair et al. [14], the Five Freedoms are still the central primary categories for hotspots of Chinese farm animal welfare research, although the publication of the Five Freedoms was around 30 years ago [77].

Despite the specificity of the search terms, it is noteworthy that the research hotspots identified often appear to be only peripherally related to farm animal welfare, rather than directly focused on farm animal welfare. This is mainly reflected in the large number of studies that have been conducted on the effects of factors influencing farm animal welfare on growth and production performance, as well as product quality. These factors are of economic importance, which also at times have positive impacts for farm animal welfare. Sinclair et al. [14] attributed this phenomenon to the pragmatic orientation of farm animal welfare research in China. That is, despite having the ultimate aim of improving farm animal welfare, the studies that are focused on animal husbandry practices seem to always be accompanied by the aim of increasing productivity and improving product quality. This is to be expected given the supply pressure on Chinese farm animal products. One important reason might be potential divergences in the understanding of animal welfare between China and other regions, such as European countries. This is quite understandable given the vast differences in historical and cultural backgrounds, political systems, and basic national conditions. In fact, there still remains confusion about the concepts of "animal welfare" and "animal rights" among a majority of Chinese, and welfare of animals is often wrongly construed to equate with people's welfare in China, such as housing, nutrition, and medical care [78]. Some researchers feel compelled to avoid the term directly for fear of alienating ordinary Chinese people or the government [79]. There are various understandings of animal welfare across countries, but there is a growing consensus that farm animal welfare has to be safeguarded and enhanced.

4.7. Research Frontiers

According to the citation burst analysis of keywords, the current study found that "computer vision", "recognition", "temperature", "precision livestock farming", "laying hen", and "behavior" were the keywords with the strongest citation bursts that are still experiencing ongoing bursts in 2023 (Table 6). Therefore, these keywords can be identified as research frontiers in the field of farm animal welfare in China, and future research may continue to focus on these frontiers.

With increasing capital pouring into the animal husbandry industry, the Chinese Government is encouraging and promoting the adoption of large-scale farming and modern technologies [80]. Precision livestock farming technology, which aims to achieve the real-time monitoring of farm animal health and welfare both automatically and continuously, is being widely adopted in large-scale livestock and poultry production [81]. As a result, research on precision livestock farming, such as computer vision and automatic recognition, is expected to emerge in large numbers.

Climate change is increasingly affecting livestock and poultry production with an overwhelmingly negative effect on the welfare of farm animals [82]. To avoid the serious consequences of the stress response, research on temperature control and management of farm animals in extreme weather conditions may become a potential area of future research.

Daily behavioral patterns are an essential manifestation of the health and welfare status of livestock and poultry [83]. Courses related to animal behavior, such as ethology, have been introduced in most agricultural universities in China since the early 21st century. The growing number of researchers specializing in farm animal behavior have contributed to the rapid development of farm animal behavior research in China [48].

In addition, the welfare of laying hens in China is a particular concern because some practices used for laying hens that have been phased out in other nations for animal welfare reasons remain common in China, such as non-enriched cage systems for egg production [84]. Therefore, laying hens may be a crucial species for future research.

5. Conclusions and Limitations

Overall, the findings of this paper are presented as follows. (1) At present, farm animal welfare research in China is relatively mature, and the scale of research is still growing. Swine and chickens have received priority attention in this area. (2) The authors and institutions in this field are scattered, with only a small number of collaborative networks formed. There also remains a lack of authors and institutions with academic influence. (3) Research hotspots in this field can be summarized in 10 categories: "feed and diet", "environmental impacts and control", "integrated rearing management", "injury and disease", "behavior and technologies for behavior monitoring", "genetic analysis", "welfare during transport and slaughter", "welfare-friendly animal product consumption", "attitudes toward farm animal welfare", and "healthy breeding". (4) It was determined that "computer vision", "recognition", "temperature", "precision livestock farming", "laying hen", and "behavior" represent the major research frontiers in this field. (5) Farm animal welfare research in China is pragmatic. Enhancing growth and production performance, as well as product quality, appear to be prioritized over improving farm animal welfare.

However, the findings of this study must be understood within the confines of its limitations. First, this review applied the quantitative analysis approach only. Thus, future research can analyze China's farm animal welfare research using both quantitative and qualitative analysis approaches. Second, the literature analyzed in this review was only obtained from peer-reviewed journals to ensure the quality of publications. Therefore, some highly relevant publications from non-peer-reviewed journals may have been overlooked, which may have resulted in an incomplete analysis. Third, the analysis conducted for the present study tended to concentrate on the top-ranking authors, institutions, journals, etc., which may have resulted in some less important literature being overlooked. However, this study provides useful baseline data that can be used to understand and track the current status and future direction in the field of farm animal welfare in China.

Supplementary Materials: The following supporting information can be downloaded at: https://www.mdpi.com/article/10.3390/ani13193143/s1, Table S1: Detailed information on the non-Chinese-language literature search results (WoS database); Table S2: Detailed information on the Chinese-language literature search results (CNKI database); Table S3: Detailed information on the valid non-Chinese-language literature after screening (WoS database); Table S4: Detailed information on the valid Chinese-language literature after screening (CNKI database).

Author Contributions: Conceptualization, L.C. and B.J.; methodology, L.C. and B.J.; software, L.C. and W.T.; validation, B.J.; formal analysis, L.C.; data curation, L.C.; writing—original draft preparation, L.C.; writing—review and editing, L.C., W.T., and X.D.; visualization, L.C. and W.T.; supervision, B.J.; project administration, B.J.; funding acquisition, B.J. All authors have read and agreed to the published version of the manuscript.

Funding: This research was funded by the National Natural Science Foundation of China for Young Scholars (grant number 72203034); the Emergency Project Fund of National Natural Science Foundation of China (grant number 71640017); and the Natural Science Foundation of Heilongjiang Province (grant number LH2021G002).

Institutional Review Board Statement: Not applicable.

Informed Consent Statement: Not applicable.

Data Availability Statement: All data generated or analyzed during this study are included in this published article.

Conflicts of Interest: The authors declare no conflict of interest.

References

1. National Bureau of Statistics of China. Available online: https://data.stats.gov.cn/english/easyquery.htm?cn=C01 (accessed on 8 June 2023).
2. Wang, X.; Wu, X.; Yan, P.; Gao, W.; Chen, Y.; Sui, P. Integrated Analysis on Economic and Environmental Consequences of Livestock Husbandry on Different Scale in China. *J. Clean. Prod.* **2016**, *119*, 1–12. [CrossRef]
3. Tarazona, A.M.; Ceballos, M.C.; Broom, D.M. Human Relationships with Domestic and Other Animals: One Health, One Welfare, One Biology. *Animals* **2020**, *10*, 43. [CrossRef]
4. Platto, S.; Serres, A.; Jingyi, A. Chinese College Students' Attitudes towards Animal Welfare. *Animals* **2022**, *12*, 156. [CrossRef]
5. Carnovale, F.; Jin, X.; Arney, D.; Descovich, K.; Guo, W.; Shi, B.; Phillips, C.J.C. Chinese Public Attitudes towards, and Knowledge of, Animal Welfare. *Animals* **2021**, *11*, 855. [CrossRef] [PubMed]
6. Sun, Y.; Ip, P.S.; Jones, M.; Wang, J.J.; An, Y. Determinants of Animal Welfare Disclosure Practices: Evidence from China. *Sustainability* **2021**, *13*, 2200. [CrossRef]
7. Construction for Intensive Pig Farms. Available online: https://std.samr.gov.cn/search/std?q=GB%2FT17824.1-2022 (accessed on 8 June 2023).
8. International Cooperation Committee of Animal Welfare. Available online: http://www.iccaw.org.cn/iccawbz (accessed on 8 June 2023).
9. Humane Farm Animal Care. Available online: http://certifiedhumanechina.org/display/314455.html (accessed on 8 June 2023).
10. World Conference on Farm Animal Welfare. Available online: http://www.wcfaw.com (accessed on 8 June 2023).
11. You, X.; Li, Y.; Zhang, M.; Yan, H.; Zhao, R. A Survey of Chinese Citizens' Perceptions on Farm Animal Welfare. *PLoS ONE* **2014**, *9*, e109177. [CrossRef]
12. Jiang, B.; Tang, W.; Cui, L.; Wei, Y. Factors Influencing Chinese Public Attitudes toward Farm Animal Welfare. *Front. Psychol.* **2023**, *14*, 1049530. [CrossRef] [PubMed]
13. Sinclair, M.; Phillips, C.J.C. Asian Livestock Industry Leaders' Perceptions of the Importance of, and Solutions for, Animal Welfare Issues. *Animals* **2019**, *9*, 319. [CrossRef]
14. Sinclair, M.; Zhang, Y.; Descovich, K.; Phillips, C.J.C. Farm Animal Welfare Science in China-A Bibliometric Review of Chinese Literature. *Animals* **2020**, *10*, 540. [CrossRef]
15. Freire, R.; Nicol, C.J. A Bibliometric Analysis of Past and Emergent Trends in Animal Welfare Science. *Anim. Welf.* **2019**, *28*, 465–485. [CrossRef]
16. Kemec, A.; Altinay, A.T. Sustainable Energy Research Trend: A Bibliometric Analysis Using VOSviewer, RStudio Bibliometrix, and CiteSpace Software Tools. *Sustainability* **2023**, *15*, 3618. [CrossRef]
17. Li, J.; Xue, E.; Cao, J.; He, Y.; Wu, Y.; Hou, H. Knowledge Mapping of the Rural Teacher Development Policy in China: A Bibliometric Analysis on Web of Science. *Sustainability* **2023**, *15*, 7057. [CrossRef]
18. Liu, Z.; Wang, S.; Zhang, Y.; Feng, Y.; Liu, J.; Zhu, H. Artificial Intelligence in Food Safety: A Decade Review and Bibliometric Analysis. *Foods* **2023**, *12*, 1242. [CrossRef] [PubMed]

19. Zhang, D.; Xu, J.; Zhang, Y.; Wang, J.; He, S.; Zhou, X. Study on Sustainable Urbanization Literature Based on Web of Science, Scopus, and China National Knowledge Infrastructure: A Scientometric Analysis in CiteSpace. *J. Clean. Prod.* **2020**, *264*, 121537. [CrossRef]
20. Tu, S.J.; Jin, C.; Chen, B.T.; Xu, A.Y.; Luo, C.; Wang, X.H. Study on the Fusion of Sports and Medicine in China from 2012 to 2021: A Bibliometric Analysis via CiteSpace. *Front. Public Health* **2022**, *10*, 939557. [CrossRef]
21. Dang, Q.; Luo, Z.; Ouyang, C.; Wang, L. First Systematic Review on Health Communication Using the CiteSpace Software in China: Exploring Its Research Hotspots and Frontiers. *Int. J. Environ. Res. Public Health* **2021**, *18*, 13008. [CrossRef]
22. Li, X.; Hu, S.; Jiang, L.; Han, B.; Li, J.; Wei, X. Bibliometric Analysis of the Research (2000-2020) on Land-Use Carbon Emissions Based on CiteSpace. *Land* **2023**, *12*, 165. [CrossRef]
23. Manuelian, C.L.; Penasa, M.; da Costa, L.; Burbi, S.; Righi, F.; De Marchi, M. Organic Livestock Production: A Bibliometric Review. *Animals* **2020**, *10*, 618. [CrossRef]
24. Li, P.J. Exponential Growth, Animal Welfare, Environmental and Food Safety Impact: The Case of China's Livestock Production. *J. Agric. Environ. Ethics* **2009**, *22*, 217–240. [CrossRef]
25. Li, X.; Zito, S.; Sinclair, M.; Phillips, C.J.C. Perception of Animal Welfare Issues during Chinese Transport and Slaughter of Livestock by a Sample of Stakeholders in the Industry. *PLoS ONE* **2018**, *13*, e0197028. [CrossRef]
26. Liang, Y.; Hua, G.; Cai, W.; Li, G.; Wang, H.; Li, H. Knowledge of Animal Welfare and Consumers' Behavioral Intentions in China: A Moderated Mediation Model of Product Cognition and Empathy. *Animals* **2022**, *12*, 1043. [CrossRef]
27. Welfare on Killing Animals for Disease Control Purposes. Available online: https://std.samr.gov.cn/search/std?tid=&q=GB%2FT%2042071-2022 (accessed on 8 June 2023).
28. Qiang, W.; Xiao, C.; Li, Z.; Yang, L.; Shen, F.; Zeng, L.; Ma, P. Impactful Publications of Critical Care Medicine Research in China: A Bibliometric Analysis. *Front. Med.* **2022**, *9*, 974025. [CrossRef] [PubMed]
29. Feng, L.; Shang, S.; Feng, X.; Kong, Y.; Bai, J. Evolution and Trend Analysis of Research Hotspots in the Field of Pollution-Intensive Industry Transfer-Based on Literature Quantitative Empirical Study of China as World Factory. *Front. Environ. Sci.* **2022**, *10*, 732734. [CrossRef]
30. Niu, Y.; Adam, M.; Hussein, H. Connecting Urban Green Spaces with Children: A Scientometric Analysis Using CiteSpace. *Land* **2022**, *11*, 1259. [CrossRef]
31. Pan, X.; Yan, E.; Cui, M.; Hua, W. Examining the Usage, Citation, and Diffusion Patterns of Bibliometric Mapping Software: A Comparative Study of Three Tools. *J. Informetr.* **2018**, *12*, 481–493. [CrossRef]
32. Chen, C. CiteSpace II: Detecting and Visualizing Emerging Trends and Transient Patterns in Scientific Literature. *J. Am. Soc. Inf. Sci. Technol.* **2006**, *57*, 359–377. [CrossRef]
33. Guo, Y.; Geng, X.; Chen, D.; Chen, Y. Sustainable Building Design Development Knowledge Map: A Visual Analysis Using CiteSpace. *Buildings* **2022**, *12*, 969. [CrossRef]
34. Chen, M.; Zhang, Y.; Dong, L.; Guo, X. Bibliometric Analysis of Stroke and Quality of Life. *Front. Neurol.* **2023**, *14*, 1143713. [CrossRef] [PubMed]
35. Zheng, D.; Wu, Z.; Li, L.; Chen, S.; Chang, J. Research Advances and Trends in the Surgical Treatment of Carpal Tunnel Syndrome from 2003 to 2022: A CiteSpace-Based Bibliometric Analysis. *Front. Neurol.* **2023**, *14*, 1124407. [CrossRef] [PubMed]
36. Li, C.; Sun, M.Y.; Tian, S.L. Research Hotspots and Effectiveness of Transcranial Magnetic Stimulation in Pain: A Bibliometric Analysis. *Front. Hum. Neurosci.* **2022**, *16*, 887246. [CrossRef]
37. Zuo, Z.; Cheng, J.; Guo, H.; Li, Y. Knowledge Mapping of Research on Strategic Mineral Resource Security: A Visual Analysis Using CiteSpace. *Resour. Policy* **2021**, *74*, 102372. [CrossRef]
38. Guo, Y.; Xu, Z.Y.R.; Cai, M.T.; Gong, W.X.; Shen, C.H. Epilepsy with Suicide: A Bibliometrics Study and Visualization Analysis via CiteSpace. *Front. Neurol.* **2022**, *12*, 823474. [CrossRef]
39. Larsen, P.O.; von Ins, M. The Rate of Growth in Scientific Publication and the Decline in Coverage Provided by Science Citation Index. *Scientometrics* **2010**, *84*, 575–603. [CrossRef] [PubMed]
40. Gallo, C.; Vejar, L.; Galindo, F.; Huertas, S.M.; Tadich, T. Animal Welfare in Latin America: Trends and Characteristics of Scientific Publications. *Front. Vet. Sci.* **2022**, *9*, 1030454. [CrossRef]
41. Uyanga, V.A.; Musa, T.H.; Oke, O.E.; Zhao, J.; Wang, X.; Jiao, H.; Onagbesan, O.M.; Lin, H. Global Trends and Research Frontiers on Heat Stress in Poultry from 2000 to 2021: A Bibliometric Analysis. *Front. Physiol.* **2023**, *14*, 1123582. [CrossRef] [PubMed]
42. Wang, L. The Structure and Comparative Advantages of China's Scientific Research: Quantitative and Qualitative Perspectives. *Scientometrics* **2016**, *106*, 435–452. [CrossRef]
43. Liu, W.; Hu, G.; Tang, L.; Wang, Y. China's Global Growth in Social Science Research: Uncovering Evidence from Bibliometric Analyses of SSCI Publications (1978–2013). *J. Informetr.* **2015**, *9*, 555–569. [CrossRef]
44. Xu, X.; Rose, H.; Oancea, A. Incentivising International Publications: Institutional Policymaking in Chinese Higher Education. *Stud. High. Educ.* **2021**, *46*, 1132–1145. [CrossRef]
45. Liu, G.; Luo, Q.F.; Zhang, L.X. The Development of Animal Husbandry For 40 Years of Reform and Opening-Up: Achievements, Challenges and Countermeasures. *Chin. Rural. Econ.* **2018**, *408*, 19–36.
46. Bao, J. The Current State of Development of the Animal Welfare Discipline. *J. Domest. Anim. Ecol.* **1997**, *1*, 57–62.
47. Wang, M.L. China's Livestock Industry Development: Achievements, Experiences and Future Trends. *Issues Agric. Econ.* **2018**, *464*, 60–70. [CrossRef]

48. Bao, J.; Li, Y. China Perspective: Emerging Interest in Animal Behaviour and Welfare Science. In *Animals and Us: 50 Years and More of Applied Ethology*; Brown, J., Seddon, Y., Appleby, M., Eds.; Wageningen Academic Publishers: Wageningen, The Netherlands, 2016; pp. 241–252. [CrossRef]
49. Zang, Q.; Zhao, Y.J.; Li, B.M.; Shi, Z.X.; Han, J. Behavioral Variations among Three Developmental Stages of Small-Tailed Sheep in a Summer Free-Range Environment. *Zool. Res.* **2005**, *26*, 279–286. [CrossRef]
50. Zhao, Y.J.; Zhao, Y.; Shi, Z.X.; Li, B.M. Feather Pecking of Broiler Laying Hens in Light Versus Dark Regions in Shade Battery Cages. *J. China Agric. Univ.* **2006**, *11*, 13–17. [CrossRef]
51. Zhou, D.L.; Li, B.M.; Shi, Z.X. Body Weight Loss of the Weaned Piglets during Transporting. *Chin. J. Anim. Sci.* **2007**, *43*, 56–58. [CrossRef]
52. Tang, L. The Impact on International Trade of Animal's Welfare and Our Countermeasure. *Inq. Into Econ. Issues* **2005**, *7*, 44–47. [CrossRef]
53. Hu, J.Y. Impact of "Animal Welfare" on China's Agricultural Output and Countermeasure. *Ecol. Econ.* **2006**, *2*, 113–116. [CrossRef]
54. Xi, L.; Zhou, D.L.; Shi, Z.X.; Li, B.M. Strategy on the Pig Welfare Problem of Raising Environment and Safe Pork Production. *Chin. J. Anim. Sci.* **2006**, *42*, 26–28. [CrossRef]
55. Li, K.J. A Pilot Study on Farm Animal Welfare Issues and Response Strategies. *Chin. J. Anim. Sci.* **2005**, *41*, 52–54. [CrossRef]
56. Chang, J.W. The Independence of Animal Welfare Law from the Perspective of EU Legislation. *Glob. Law Rev.* **2006**, *28*, 343–354. [CrossRef]
57. Xu, Y.C.; Tian, Y.W.; Zhu, B.J.; Wang, L.Y. On the Necessity and Feasibility of Legislation Against Animal Cruelty in China. *Truth Seek.* **2006**, *27*, 103–104. [CrossRef]
58. Chen, Q.H.; Zhang, Y.Y. Evolution of Carbon Reduction Policies for Animal Husbandry: An Analysis Based on 452 Policies. *J. Huazhong Agric. Univ. (Soc. Sci. Ed.)* **2022**, *157*, 10–23. [CrossRef]
59. Yan, H.Q.; Guo, X. The Fusion of Science and Ethic: The Rise of Animal Welfare Science. *J. Dialectics Nat.* **2017**, *39*, 76–83. [CrossRef]
60. Cui, L.H.; Li, C.X.; Bao, J.; Ma, C.P.; Jiang, B. Consumers' Willingness to Pay for Farm Animal Welfare Products and Its Influencing Factors: Based on Animal Welfare Dairy Products. *Res. Agric. Mod.* **2021**, *42*, 724–734. [CrossRef]
61. Yu, F.W.; Huang, X.; Wang, G.L. High-Quality Development of Animal Husbandry: Theoretical Interpretation and Realization Path. *Chin. Rural. Econ.* **2021**, *436*, 85–99.
62. Gautret, M.; Messori, S.; Jestin, A.; Bagni, M.; Boissy, A. Development of a Semi-Automatic Bibliometric System for Publications on Animal Health and Welfare: A Methodological Study. *Scientometrics* **2017**, *113*, 803–823. [CrossRef]
63. Rioja-Lang, F.C.; Connor, M.; Bacon, H.J.; Lawrence, A.B.; Dwyer, C.M. Prioritization of Farm Animal Welfare Issues Using Expert Consensus. *Front. Vet. Sci.* **2020**, *6*, 495. [CrossRef] [PubMed]
64. Li, X.Z.; Yan, C.G.; Zan, L.S. Current Situation and Future Prospects for Beef Production in China—A Review. *Asian-Australas. J. Anim. Sci.* **2018**, *31*, 984–991. [CrossRef]
65. Shi, R.; Shu, H.; Yu, R.; Wang, Y.; Zhang, Z.; Zhang, J.; Gu, X. Current Attitudes of Chinese Dairy Practitioners to Pain and Its Management in Intensively Raised Dairy Cattle. *Animals* **2022**, *12*, 3140. [CrossRef]
66. Wang, J. Unpacking the Matthew Effect in Citations. *J. Informetr.* **2014**, *8*, 329–339. [CrossRef]
67. Kirchner, M.K.; Kost'al, L.; Bilcik, B.; Winckler, C. Mapping Farm Animal Welfare Research in an Enlarged Europe: International Collaboration, Bibliometric Output, Research Resources and Relation to Economic Indices. *Scientometrics* **2017**, *113*, 909–922. [CrossRef]
68. Liao, C.H. The Matthew Effect and the Halo Effect in Research Funding. *J. Informetr.* **2021**, *15*, 101108. [CrossRef]
69. Mostafa, M.M. A Knowledge Domain Visualization Review of Thirty Years of Halal Food Research: Themes, Trends and Knowledge Structure. *Trends Food Sci. Technol.* **2020**, *99*, 660–677. [CrossRef]
70. Vlegels, J.; Huisman, J. The Emergence of the Higher Education Research Field (1976–2018): Preferential Attachment, Small-worldness and Fragmentation in Its Collaboration Networks. *High. Educ.* **2021**, *81*, 1079–1095. [CrossRef]
71. Jayantha, W.M.; Oladinrin, O.T. Bibliometric Analysis of Hedonic Price Model Using CiteSpace. *Int. J. Hous. Mark. Anal.* **2019**, *13*, 357–371. [CrossRef]
72. Khalagi, K.; Mansourzadeh, M.J.; Aletaha, A.; Yarmohammadi, H.; Atlasi, R.; Banar, S.; Fahimfar, N.; Hajipour, F.; Sanjari, M.; Larijani, B.; et al. Co-Authorship Network Analysis of Iranian Researchers on Osteoporosis. *Arch. Osteoporos.* **2021**, *16*, 74. [CrossRef]
73. Nan, M.; Chen, J. Research Progress, Hotspots and Trends of Land Use under the Background of Ecological Civilization in China: Visual Analysis Based on the CNKI Database. *Sustainability* **2023**, *15*, 249. [CrossRef]
74. Yuan, X.; Lai, Y. Bibliometric and Visualized Analysis of Elite Controllers Based on CiteSpace: Landscapes, Hotspots, and Frontiers. *Front. Cell. Infect. Microbiol.* **2023**, *13*, 1147265. [CrossRef] [PubMed]
75. Lv, J.; Wang, Z.Y.; Chen, X.Q. Analysis of Animal Welfare Research Hotspots Based on CNKI and WoS. *J. Domest. Anim. Ecol.* **2022**, *43*, 75–82. [CrossRef]
76. Mu, C. A Multidimensional Contrastive Analysis of Linguistic Features between International and Local Biology Journal English Research Articles. *Scientometrics* **2021**, *126*, 7901–7916. [CrossRef]

77. Mellor, D.J. Updating Animal Welfare Thinking: Moving beyond the "Five Freedoms" towards "A Life Worth Living". *Animals* **2016**, *6*, 21. [CrossRef] [PubMed]
78. Lu, J.; Bayne, K.; Wang, J. Current Status of Animal Welfare and Animal Rights in China. *Altern. Lab. Anim.* **2013**, *41*, 351–357. [CrossRef] [PubMed]
79. Cao, D. Is the Concept of Animal Welfare Incompatible with Chinese Culture? *Soc. Anim.* **2022**, *30*, 284–296. [CrossRef]
80. Tullo, E.; Fontana, I.; Diana, A.; Norton, T.; Berckmans, D.; Guarino, M. Application Note: Labelling, a Methodology to Develop Reliable Algorithm in PLF. *Comput. Electron. Agric.* **2017**, *142*, 424–428. [CrossRef]
81. Li, N.; Ren, Z.; Li, D.; Zeng, L. Review: Automated Techniques for Monitoring the Behaviour and Welfare of Broilers and Laying Hens: Towards the Goal of Precision Livestock Farming. *Animal* **2020**, *14*, 617–625. [CrossRef] [PubMed]
82. Blanco-Penedo, I.; Cantalapiedra, J.; Llonch, P. Climate Change Impacts on Animal Welfare in Livestock Systems. *Inf. Tec. Econ. Agrar.* **2020**, *116*, 424–443. [CrossRef]
83. Jiang, M.; Rao, Y.; Zhang, J.; Shen, Y. Automatic Behavior Recognition of Group-Housed Goats Using Deep Learning. *Comput. Electron. Agric.* **2020**, *177*, 105706. [CrossRef]
84. He, S.; Lin, J.; Jin, Q.; Ma, X.; Liu, Z.; Chen, H.; Ma, J.; Zhang, H.; Descovich, K.; Phillips, C.J.C.; et al. The Relationship between Animal Welfare and Farm Profitability in Cage and Free-Range Housing Systems for Laying Hens in China. *Animals* **2022**, *12*, 2090. [CrossRef]

animals

MDPI

Article

Perceptions of Farm Animal Sentience and Suffering: Evidence from the BRIC Countries and the United States

Fernando Mata [1,*], Bastian Jaeger [2] and Ivo Domingues [3,4]

1 Centre for Research and Development in Agri-Food Systems and Sustainability, Instituto Politécnico de Viana do Castelo, 4900-367 Viana do Castelo, Portugal
2 Department of Social Psychology, Tilburg University, 5037 AB Tilburg, The Netherlands
3 Instituto de Ciências Sociais, Universidade do Minho, 4710-057 Braga, Portugal
4 Centro de Estudos de Comunicação e Sociedade, Universidade do Minho, 4710-057 Braga, Portugal
* Correspondence: fernandomata@ipvc.pt

Simple Summary: The relations between farm animals and humans vary across countries and cultures. It was the aim of this study to understand the position of the population in the BRIC countries (Brazil, Russia, India, and China) and the USA. It was found that perceptions of farm animal sentience and suffering vary a lot with culture, country, gender, and age. This could have important consequences for the globalized trade of animal products does not find common grounds for standardization, and the risk of countries with more advanced animal welfare legislation imposing trade barriers increases. These trade barriers may be precepted as protectionism by exporting countries.

Abstract: In this study, we examined how beliefs about farm animal sentience and their suffering vary across culture and demographic characteristics. A total of N = 5027) questionnaires were administered in Brazil, Russia, India, China, and the USA. Brazilians showed higher and Chinese lower levels of perceived animal sentience. In Russia and India, the perception of suffering and sentience increases with age, with similar levels to those observed in the USA. In all the countries, more people agreed than disagreed that animals are sentient. Men in India show higher levels of agreement with the relation between eating meat and animal suffering, followed by women in Brazil and China. Lower levels of agreement are observed in Americans and Chinese. Women show higher levels of compassion than men. In Russia, there is a slightly higher level of agreement between men and in the USA younger men agree more. Young American men show higher levels of agreement, while in India and China age has the opposite effect. For fair trading competition, it is important to standardize procedures and respect the demand for both animal protein and its ethical production. Overall, our results showed that perceptions of farm animal sentience and suffering vary substantially across countries and demographic groups. These differences could have important consequences for the perceived ethicality of meat production and consumption, and for global trade in animal products.

Keywords: farm animal sentience; farm animal suffering; BRIC countries; USA; ethics of meat consumption; meat trading standards

1. Introduction

Sustainability is a growing concern today. While aiming for agriculture production systems capable of feeding the world, we need to preserve the environment and use natural resources wisely. Under these circumstances, meat production and consumption patterns must also reach sustainable standards. Meat production is the main cause of greenhouse gas emissions and in the consumption of water. Together with sustainable standards, ethical standards are also demanded by today's informed society.

Farm animal welfare (FAW) is part of those standards and is typically defined as "a potentially measurable quality of a living animal at a particular time" [1]. The first step for FAW is the ability to satisfy a basic need, often referred to as the five freedoms. The term "five freedoms" was coined in the UK by the Farm Animal Welfare Council, is now accepted worldwide, and refers to animals having [2]:

1. "Freedom from hunger and thirst, by ready access to water and a diet to maintain health and vigor.
2. Freedom from discomfort, by providing an appropriate environment.
3. Freedom from pain, injury, and disease, by prevention or rapid diagnosis and treatment.
4. Freedom to express normal behavior, by providing sufficient space, proper facilities, and appropriate company of the animal's own kind.
5. Freedom from fear and distress, by ensuring conditions and treatment, which avoid mental suffering."

From the perspective of the human relationship with the non-human animal, though, FAW as a construct must also include human obligations towards animals, often referred to as animal rights. It has been suggested that the five freedoms concept needs a broader interpretation [3] to include the recognition of animal sentience, defined as an animal "having the awareness and cognitive ability necessary to have feelings" [4]. Despite the existence of historical discourse about animals' feelings, ranging from the classic Greek thinkers Hippocrates, and Pythagoras to Charles Darwin [5], most developments in the legal recognition of animal sentience have been made in recent years. European animals gained this official recognition in 1999 through the EU treaty of Amsterdam, which was later complemented with a protocol on their protection and welfare via the 2007 EU Treaty of Lisbon, adopted in 2009, and entering into force with the Directive 2010/63/EU. This event triggered a global reaction, in Western societies (e.g., Canada, Colombia, New Zealand, Switzerland, Turkey, Ukraine, USA) [6]. More recently, 180 countries adopted the OIE Global FAW Strategy 2017, including the recognition of animal sentience, and up to 46 countries supported the UN's Universal Declaration on Animal Welfare [7].

The recognition that positive affective experiences are a good indicator of FAW is paramount in connecting animal sentience with more practical considerations of FAW [7]. In the present, anthropogenic suffering, or the acknowledgement of animal suffering caused by human actions, is seen as the major step towards an ethical relationship between humans and animals [8]. The role of humans in animal suffering is particularly apparent for farm animals that are kept with the explicit purpose of serving human needs [9].

Humans' strong demand for animal products and widely held beliefs about the justifiability of meat consumption [10] makes complete independence from food of animal origin in the near future, unlikely. However, even in the short term, there is room for improvement in the sustainability of animal agriculture to preserve the environment, protect biodiversity, and feed the world's growing population, while also working towards a more ethical treatment of farm animals, for example, by decreasing negative and promoting positive affective experiences in farm animals [11].

Yet, reducing the suffering of animals by changing the way in which humans treat them in food production may be a challenging task. Most people continue to consume meat even though they care about animals. This observation is often referred to as the meat paradox, defined as the cognitive dissonance or contradiction between the compassion felt by humans towards animals and the suffering imposed on them through the consumption of animal products [12]. Eating meat can become a particularly salient moral concern when people acknowledge that the animals, they are eating are sentient and capable of suffering. In fact, research has shown that people show increased moral concern for FAW when they are reminded that animals, such as humans, have the capacity to feel and think [13]. For example, Leach and colleagues [14] examined to what extent a variety of animal characteristics influence the perceived acceptability of eating the animal. People thought that it was less acceptable to eat animals that have the capacity to experience negative emotions. The relationship between perceived animal sentience and attitudes toward meat consumption seems to be bi-directional. That is, to avoid the negative cognitive consequences of holding dissonant attitudes (i.e., eating meat is permissible, but making animals suffer is not), people may categorize farm animals as less sentient and therefore less morally relevant when compared to companion animals [15]. Other strategies used by humans to deal with the cognitive dissonance resulting from meat consumption include endorsing speciesism, or the belief that humans are superior to other animals [16]; dissociating unpleasant ideas by referring to the bodies of sentient animals as 'meat' [15,17]; endorsing consumption as a social norm [18], and other strategies that are encapsulated in the 4N model: that meat is natural, necessary, normal, and nice [10]. Additionally, some people may acknowledge that animal farming can cause some animal suffering while asserting that farming procedures can be improved to minimize that suffering [12]. In short, beliefs about the sentience and suffering of farm animals are psychologically central to how people think about the ethicality of meat production and consumption.

Scholars have started to examine how perceptions of farm animal sentience and suffering differ across demographic groups. Some studies suggest that beliefs may differ between men and women [19,20]. One study [19] with students of different nationalities found higher levels of perceived animal suffering among women, but no difference with regard to perceived animal sentience. Cultural differences between countries were also identified with the authors reaching a generic conclusion that European and American students show stronger beliefs in animal suffering, than Asian students. Another study [20] reports stronger beliefs in animal sentience among female veterinary students, with similar tendencies across countries and cultures. In both studies individuals agreed on a hierarchy of the "capacity to feel" that places companion animals as more sentient, followed by farm animals and others.

The age effect has also been studied as a demographic variable capable of impacting perceived animal suffering and animal sentience. As well, the results obtained are mixed and dependent on the animal setting [21]. However, the authors cite a study [22] to justify a tendency to a natural shift in human perception as people become older. Older people tend to change priorities toward family needs and animals tend to be perceived as less important and seen more from a utilitarian point of view.

It is the aim of this study to examine beliefs about farm animal sentience and suffering of farm animals in the BRIC countries and the USA, and their variability across ages and gender. These beliefs are crucial for understanding how people think about the ethicality of meat production and consumption, which also has consequences for trade between countries. The global economy and trade are inevitable, but to achieve healthy trading competition it is important to standardize procedures and respect the demand for both animal protein and its ethical production. Nevertheless, religion, cultural differences, age, and gender influence public perceptions and awareness of humans' relation with non-human animals [23,24]. The emerging economies of the countries known by the acronym BRIC (Brazil, Russia, India, and China) represent 40% plus of the world population and more than 50% of the world's gross agricultural production in 2018 [25]. Thus, the present study explores beliefs that are central to how people think about the ethicality of meat production, a sector that is attracting increased attention in discussions about sustainability and climate change due to its important role as a greenhouse gas emitter, and the international trade of animal products.

2. Materials and Methods

2.1. Source of Data

The data were collected by Faunalytics as part of an exploratory study on the attitudes and behaviors towards FAW among people in the BRIC countries and the United States (Anderson, 2018b). The data are available in the Open Science Framework repository. The data were collected by YouGov® in May and June of 2018 from The BRIC countries and the USA. The sample of USA individuals is nationally representative, and it was a direct interview. The samples in Brazil, India, and Russia were collected around urban areas, also via direct interviews. The Chinese sample was collected through the internet. All the interviewees are above 18 years old. The individuals were randomly chosen from a YouGov® panel of interviewee volunteers. The sample is stratified by gender and age.

2.2. Data Collection

Data were collected online in the BRIC countries through questionnaires. These questionnaires were originally written in English and then translated into the local languages using a back-translation process in which the original translation from English to the local language was performed by one individual, while another translator who was blind to the original English version translated it back into English so that discrepancies could be caught. This procedure maximizes equivalence between countries and languages. Recommendations for keeping the wording as simple and direct as possible were also followed, using symmetrical response scales, and using both positively and negatively framed items.

The difficulties arriving from cross-cultural research are well studied. In order to overcome these we paid special attention to differences in interpretation, and in the use of response scales. Relatively to differences in interpretation the questions posed to the interviewees were checked against the good question wording [26]. As such the following was adopted to simplify without loss of meaning: short simple sentences, active voice, nouns instead of pronouns, and use of specific instead of general terminology. The following was avoided: metaphors, subjunctive, possessive form, vague words, and verbs suggesting different actions. An expert consultation phase prior to the finalization of the questionnaire was also included [27]. Relatively to differences in the use of response scales, we followed the recommendations to use symmetrical, bipolar response scales with a clear middle point [28].

2.3. Measures

Data included demographics (age, gender, country) and nine survey items. Two of these are part of the present study. Interviewees were asked several questions related to FAW, animal sentience, and their diet. In the present paper, we will investigate the responses related to the perceived sentience of animals ("animals used for food have approximately the same ability to feel pain and discomfort as humans") and the perceived role of meat consumption in contributing to the suffering of animals ("eating meat directly contributes to the suffering of animals"). Participants responded to both items on a five-point Likert scale ranging from 1 (strongly disagree) to 5 (strongly agree).

2.4. Statistical Analysis

The original dataset was explored for the existence of outliers using Tukey's method, with the production of boxplots. As a result, 28 observations were removed from the analysis. To investigate the agreement/disagreement with the statements listed in the previous section, each response scale was individually entered in a multinomial logistic regression as a function of the demographic variables age, gender, and country. The significance of the models was assessed with the -2 log likelihood chi-square test. The Akaike's Information Criterion (AIC) is also shown for each of the models to be adjusted. The parameters were tested for significance using the Wald chi-square test. The statistical package used was the IBM Corp.® SPSS® Statistics, Armonk, NY, USA. Version: 28.0.1.1 (15). For the graphic construction, we used Microsoft® Excel® for Microsoft 365 MSO (version 2204 Build 16. 0. 15128. 20240) 64-bit.

The multinomial logistic models were fit to the data. The main effects were tested all together with a forward stepwise inclusion of interactions. The calculation of the probabilities (P_i) of a national of a country to fall in a particular score while evaluating each of the statements is performed with the generic equation

$$P_i = \frac{\exp(X_i\beta_i)}{1 + \sum_5^{i=2} \exp(X_i\beta_i)} \tag{1}$$

where P_i is the probability to score each of the "*i*" scores (2, 3, 4, 5). The equations' parameters ($\beta_i X_i$) are arranged in a linear manner adopting the form

$$\beta_m X_m + \beta_1 X_1 + \beta_2 X_2 + \beta_3 X_1 X_m + \beta_4 X_1 X_2 + \beta_5 X_1 X_2 + \beta_6 X_1 X_2 X_m \quad (2)$$

where:

β_m is the parameter associated with "Country" (Brazil, China, India, and Russia), being Xm the dummy associated with this parameter, taking the value one when the respective country is considered in the equation and zero otherwise. β_1 is the parameter associated with the covariate "Age", being X_1 the age. β_2 is the parameter associated with "Gender" (Man, Woman), being X_2 the dummy associated and taking the value of one for men and zero for women. β_3, β_4, and β_4 are the parameters associated with the two-way interaction terms ("Age × Country", "Age × Gender", and "Country × Gender"), and β_6 is the parameter associated with the three-way interaction "Age × Gender × Country".

Score 1 (strongly disagree) is used as reference and therefore for the calculation of P_1 in equation (1) the numerator assumes the value 1.

The logistic coefficients (β) for each predictor variable for each alternative score of the statement are shown in the tables associated with the models representing the respective statements. The coefficients β are the expected amount of change in the logit for each one-unit change in the predictor. The logit is the odds of membership for each of the scores. The closer a logistic coefficient is to zero, the less influence the predictor has in predicting the logit. The exp(β) is the odds ratio associated with each predictor. Predictors increasing the logit have exp(β) > 1, those without effect on the logit have exp(β) = 1 and predictors deceasing the logit have exp(β) < 1. These correspond respectively to β coefficients above, equal, or below zero.

The following sections address each of the adjusted models for the different statements, showing the degree of adjustment, the parameter estimates, and the graphic representation and interpretation.

2.5. Analytical Procedure

After adjusting the models with the five categories of the Likert scale, to facilitate the interpretation of the results, and construct simpler graphs, it was decided to aggregate the categories 1 (strongly disagree) with 2 (disagree) and the categories 4 (agree) with 5 (strongly agree). Three new main categories were therefore created: disagree, neutral, and agree. These are the categories herein represented graphically and subject to result interpretation and discussion.

3. Results

3.1. Descriptive Statistics

A total of N = 5172 individuals were entered in the analysis (Brazil n = 1027, China n = 966, India n =1004, Russia n = 1002, USA n = 1173), including n = 2586 men and n = 2586 women. The age distribution within countries is shown in the boxplots in Figure 1. The age distribution for each gender is shown in boxplots in Figure 2. As can be observed data is well balanced for gender, while for age it is slightly skewed towards younger ages in China and India. When considering all the three independent variables together we can observe a slight skewing towards older ages in Chinese men, while for other countries gender is well balanced (Figure 3).

3.2. Fitted Models

3.2.1. Perceived Animal Sentience

The model examining the perceived sentience of animals was significant ($p < 0.001$), −2 log likelihood 5070, chi-square (859, 40df), and AIC 5158. The parameters found to be significant were (−2 log likelihood, chi-square, df, *p*-value): "country" (5176, 106, 16, $p <$ 0.001), "gender" (5146, 75, 4, $p < 0.001$), and the interaction "age x country" (5141, 71, 20, *p*

< 0.001). The percentage of respondents per score and country are given in Table 1. The description of the significant parameters of the model are given in Table 2. Figure 4 is the graphical representation of the model.

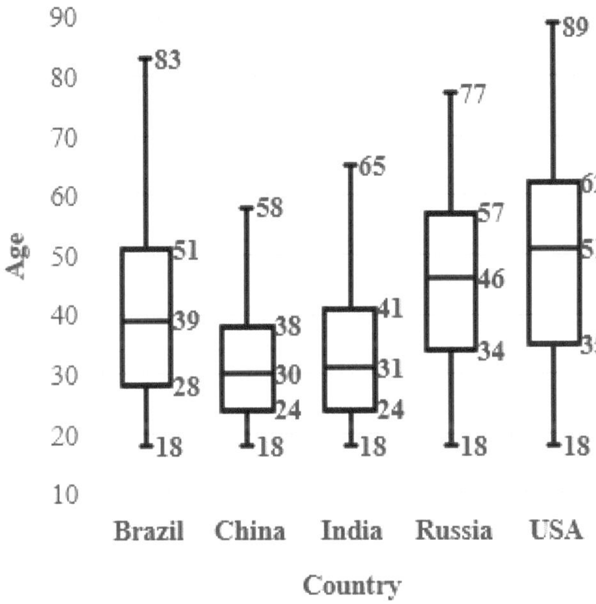

Figure 1. Sampled variables distribution. Ages within countries.

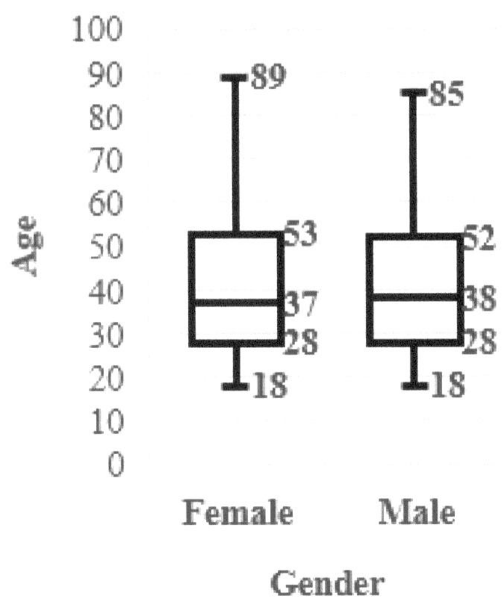

Figure 2. Sampled variables distribution. Ages within gender.

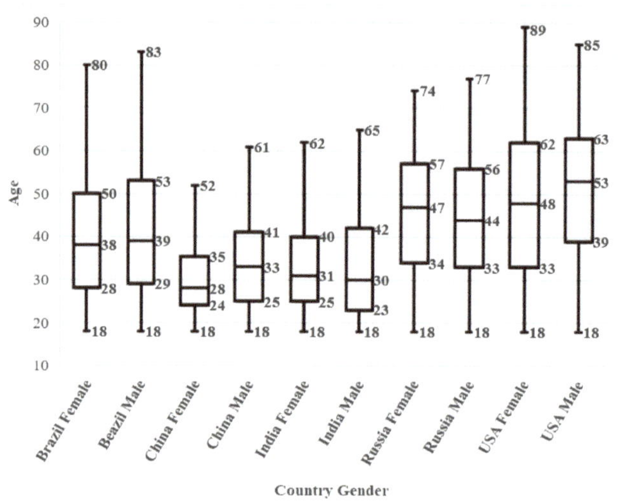

Figure 3. Sampled variables distribution. Ages within gender and countries.

Table 1. Percentage of respondents to the question in each of the countries in each of the categories disagree, neutral and agree.

Question	Brazil	Russia	India	China	USA
Animals used for food have approximately the same ability to feel pain and discomfort as humans					
Disagree (%)	7	14	14	12	10
Neutral (%)	15	24	18	50	28
Agree (%)	79	63	67	37	62

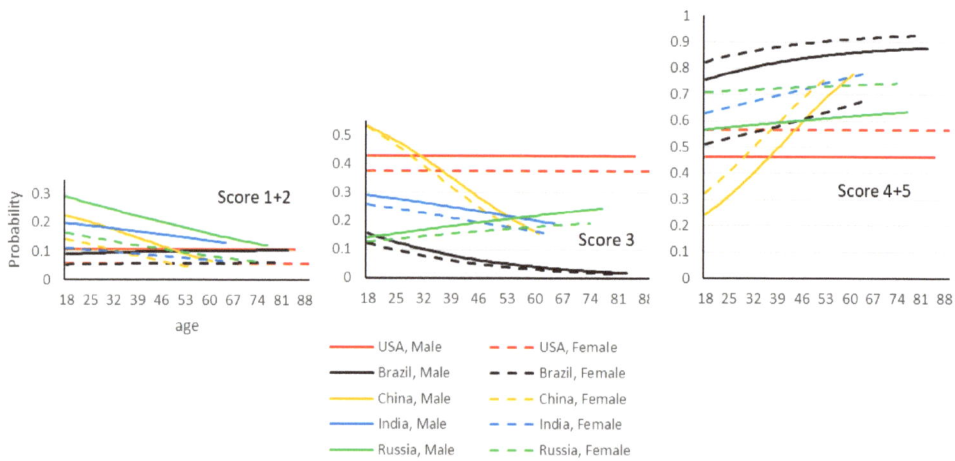

Figure 4. Graphical representation of the multinomial logistic model fitted to the data. The scores 1, 2 and 4, 5 are aggregated. Probabilities associated with disagreement scores (1 + 2), neither agree nor disagree score (3), and agreement scores (4 + 5) given to the statement *"animals used for food have approximately the same ability to feel pain and discomfort as humans"*.

Table 2. Parameters of the multinomial logistic model fitted to the data. The ordinal dependent variable is the statement *"animals used for food have approximately the same ability to feel pain and discomfort as humans,"* modeled as a function of the independent variables "Country", Gender" and "Age" together with the interactions between these. Only significant ($p < 0.05$) parameters are shown. Score 1 is used as reference in the model.

Score	Parameter		β	exp(β)
2	Country	Brazil	2.336 **	10.345
	Gender	Male	−0.407 *	0.665
3	Country	Brazil	3.902 ***	49.518
		India	1.533 ***	4.633
		China	2.016 *	7.509
		USA	2.541 ***	12.692
	Gender	Male	−0.638 ***	0.529
	Country × Age	Brazil, Age	−0.036 *	0.965
		Russia, Age	0.024 *	1.024
4	Country	Brazil	4.432 **	84.058
		Russia	1.285 **	3.614
		India	1.738 **	5.684
		USA	2.123 **	8.359
	Gender	Male	−0.883 ***	0.414
	Country × Age	China, Age	0.062 *	1.064
5	Country	Brazil	4.508 *	90.744
		Russia	1.177 *	3.243
		India	1.276 **	3.581
		USA	2.376 *	8.435
	Gender	Male	−1.050 ***	0.350
	Country × Age	China, Age	0.104 ***	1.110
		India, Age	0.025 *	0.021
		Russia, Age	0.024 *	1.024

Note: * $p < 0.05$, ** $p < 0.01$, *** $p < 0.001$.

In general Brazilian participants showed stronger beliefs in animal sentience than other nationalities in the study. Chinese participants showed the lowest belief in animal sentience, and this was particularly true for Chinese participants below the age of 45. Note that the Chinese sample was on average much younger compared to other countries which may have impacted the overall results. Older participants in Russia and India had indicated stronger beliefs in animal sentience than younger participants in these countries. Both have similar levels of perception to those in the USA. China has the highest levels of neutral responses. Women showed stronger beliefs in animal sentience than men across the different countries. The levels of disagreement are all bellow a probability of 0.3 in the different countries, but men from Russia, India and China show the higher levels. Disagreement decreased with age in China, India, and Russia. All the countries showed a higher prevalence of participants who do (vs. do not) believe in animal sentience.

3.2.2. Perceived Animal Suffering

For this statement, the model is significant ($p < 0.001$), −2 log likelihood 5648, chi-square (1500, 80df), and AIC 5808. The parameters found to be significant were the interactions (−2 log likelihood, chi-square, degrees of freedom, p-value): "country x age" (5885, 238, 40df, $p < 0.001$), and "gender x age x country" (5748, 100, 40df, $p < 0.001$). The percentage of respondents per score and country is given in Table 3. The description of the significant parameters of the model is given in Table 4. Figure 5 is the graphical representation of the model.

Table 3. Percentage of respondents to the question in each of the countries in each of the categories disagree, neutral, and agree.

	Brazil	Russia	India	China	USA
Eating meat directly contributes to the suffering of animals					
Disagree (%)	27	32	21	23	33
Neutral (%)	28	33	28	45	36
Agree (%)	45	35	51	31	30

Table 4. Parameters of the multinomial logistic model fitted to the data. The scores given to the statement *"eating meat directly contributes to the suffering of animals,"* used as the dependent variable, are modeled function of the independent variables "Country", Gender" and "Age" together with the interactions between these. Only significant ($p < 0.05$) parameters are shown. Score 1 is used as a reference in the model.

Score	Parameter		β	exp(β)
2	Country × Gender	China, Male	1.686 *	5.396
		India, Female	1.851 *	6.365
		Russia, Female	2.308 ***	10.050
	Country × Gender × Age	China, Female, Age	0.150 *	1.162
3	Country × Gender	Brazil, Male	1.174 *	3.235
		Brazil, Female	1.693 **	5.435
		China, Male	2.586 ***	13.275
		India, Female	2.379 ***	10.796
		Russia, Male	1200 *	3.322
		Russia, Female	2.191 **	8.947
		USA, Male	2.038 ***	7.673
		USA, Female	1.571 ***	4.809
	Country × Gender × Age	China, Female, Age	0.147 *	1.159
		USA, Male, Age	−0.031 ***	0.970
4	Country × Gender	Brazil, Female	1.698 **	5.464
		India, Female	2.387 ***	10.878
		Russia, Female	1.730 *	5.643
		USA, Male	1.288 **	3.625
	Country × Gender × Age	China, Female, Age	0.166 **	1.181
		USA, Male, Age	−0.025 **	0.975
5	Country × Gender	Brazil, Female	1.193 *	3.298
		China, Male	−2.126 *	0.119
		China, Female	−3.934 *	0.20
		India, Female	−2.057 **	7.823
		USA, Male	1.097 *	2.996
	Country × Gender × Age	China, Male, Age	0.051 *	1.053
		China, Female, Age	0.163 *	1.178
		India, Male, Age	0.047 **	1.048
		USA, Male, Age	−0.032 ***	0.968

Note: * $p < 0.05$, ** $p < 0.01$, *** $p < 0.001$.

Men in India showed stronger beliefs in the relation between eating meat and animal suffering than women, followed by women in Brazil and China. Lower levels of agreement are observed in Americans and Chinese, especially younger Chinese men. Americans and Chinese also indicated neutral responses more often. Women reported stronger beliefs in animal suffering than men, and the difference is especially large in China, Brazil, and among older individuals in India. Stronger beliefs were also observed in Russia among men and in the USA among younger men. The decline with age in the USA stands in opposition to Indian men, and Chinese individuals, where the prevalence of belief in animal suffering increases with age.

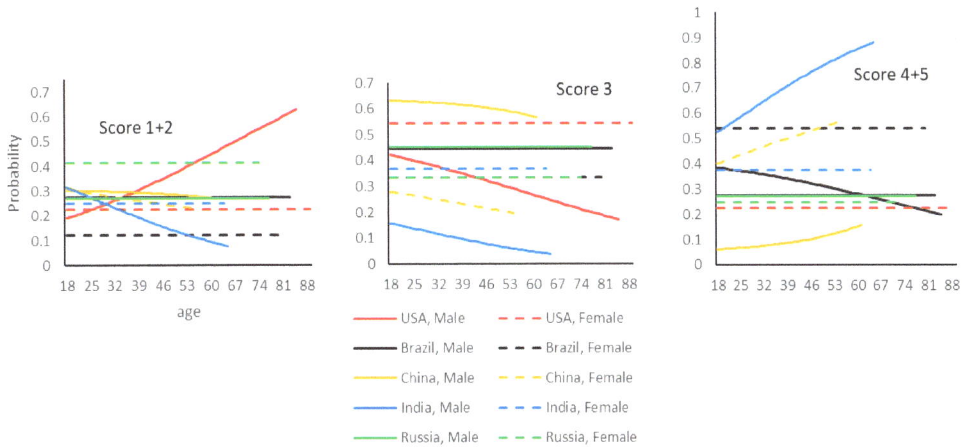

Figure 5. Graphical representation of the multinomial logistic model fitted to the data. The scores 1, 2 and 4, 5 are aggregated. Probabilities associated with disagreement scores (1 + 2), neither agree nor disagree score (3), and agreement scores (4 + 5) given to the statement *"eating meat directly contributes to the suffering of animals"*.

Neutral responses are especially common among Chinese men (decreasing with age) and American women. Neutrality decreases with age in Chinese participants, and in American men and Indian men.

Disagreement with the relationship between eating meat and animal suffering is especially high in older American men and Russian women. It is lower for Brazilian women and older Indian men.

4. Discussion

4.1. Country Effects

China is a global leading producer of meat (poultry, pork, dairy, and wildlife products) [29]. However, the stark increase in production that took place come at the expense of FAW [30,31]. Farm animal welfare science originated in Western societies and was introduced in China only in the 1990s [32]. There is however a growing awareness among the Chinese scientific community and in the population in general, due to the influence of Western culture and the awareness of the relationship between FAW and meat quality [33,34]. Despite this growing awareness, FAW issues are still enormous, including a lack of legislation and poor slaughter procedures [30,35]. When considering the countries in the present study, China has the overall lowest animal protection index.

The major Chinese religions (Buddhism, Confucianism, and Taoism) shape cultural attitudes toward animals in the country, more than any Western philosophy [36]. Buddhism, however, shares similar values with the animal liberation theories of the West, where consideration is given to every living being, along with the right to a life free from suffering [37]. Confucianism and Taoism give the same consideration to animals as to humans [38]. Nevertheless, popular practices of animal trading and consumption still cause a lot of animal suffering [39]. Despite prescribing respect for non-human animals, Confucianism places non-human animals in an inferior status in relation to humans and advocates for rituals and offers involving the consumption of animals [40].

Chinese youth underwent an important shift in market economy, urbanization, and globalization when compared with the previous generation [41]. The globalization of Chinese society brought cultural and ideological conflicts with the Western countries and the younger generations entered into a societal change [33]. On the other hand, the increasing competition pressure is also shaping the younger generations in China, widening the generation gap [41]. A survey of Chinese college students showed that the majority recognize animal sentience [42], with stronger recognition in rural areas.

Considering animal sentience, in this study, Chinese participants showed a response pattern that completely differentiates them from the other nationalities. A positive age effect is very pronounced both in men and women, with the acknowledgement of animal sentience growing exponentially approximately up to the age of 50, reaching extremely prominent levels at this stage. A possible explanation for this may be the higher degree of commitment to traditional religious practices, and the rural origin of much older Chinese, which typically exhibit a stronger endorsement of animal sentience [32]. This can also explain the higher levels of neutral responses, especially in younger ages. Nevertheless, perceptions of animal sentience are lower than in any other countries in the study, which is partly due to the exceptionally low levels of perceived sentience among younger Chinese participants.

With regard to belief in farm animal suffering, most Chinese men took a neutral position, while the agreement was higher among women. Again, a positive effect of age was apparent. Levels of disagreement are similar to other countries in the study.

In India, cruelty against animals has always been under philosophic religious scrutiny due to beliefs in non-violence purity, and reincarnation [43]. Cattle have an important religious role for Hindu people being widely venerated and typically only slaughtered by individuals from lower casts [44].

The cultural differences between Indians and Western cultural attitudes towards animals have been evident from the moment Europeans arrived and settled in India. The condemnation of some native practices prompted the British to pass legislation against animal cruelty introduced in Bengal in 1869. However, the colonial practices of slaughtering cattle to feed armies or hunting expeditions were perceived as contradictory and have not been well understood by the locals [45].

Indians were perceived to be vegetarians, but meat has always been a central part of their diet. However, the colonizers' diet included much more meat, which led to the perception of vegetarianism by comparison [46,47]. Nevertheless, the nationalism that led to Indian independence was followed by legislation prohibiting the slaughtering of the sacred cow [48]. More recently the ban on beef in India raised issues with the Christian, Muslim, and Dalit (lower caste) communities, who accused the government of forgetting the multicultural heritage of India, favoring Hindus [49].

Indian society experienced significant changes in the 1990s, with a new middle class being able to afford an improved diet to include meat [50]. Nowadays about 80% of the population is Hindu and 70% of the population eats meat [51]. Nevertheless, despite recent increases in meat consumption, India remains one of the nations in the world with the lowest meat consumption per capita [52]. Vegetarians are especially motivated by the moral concern related to farming and slaughtering [46].

Indians showed high levels of perceived animal sentience, especially among older participants. Older individuals are more religious and therefore tend to be more aligned with the vegetarianism tradition of the Hindu culture [46]. This age effect is even more pronounced when examining agreement with the statement that animal farming causes suffering. The rate of neutral respondents is similar to other countries.

Brazil is an important world meat producer and holds a tradition of meat consumption, with the livestock sector of the country accounting for 8.7% of the GDP in 2019 [53]. As an emerging economy, Brazil is still experiencing changing diets including the inclusion of more animal protein, and therefore meat consumption is still expected to increase [54].

Brazil is a multicultural nation; however, Christianity is predominant. As such there are no meat-eating restrictions. Certain Afro-Brazilian religions, predominantly Candomblé, gained the status of national tradition, without ethnical or social boundaries, and are even incorporated by the local Christians [55]. These traditions include rituals of animal blood sacrifices offered to the orisha (gods). These sacrifices are taken in festivities which include eating the animals, and may include, goats, chicken, ducks, pigs, sheep, other birds, turtles, cavy, and snails [56].

The Brazilians showed the strongest belief in animal sentience, higher than all other countries that were part of this study, which tallies the results obtained in a study comparing Brazilian and French [57]. Important levels of animal sentience perceptions among sheep farmers in Brazil have also been reported [58]. In the current study, the perceived suffering of animals due to meat consumption was especially acknowledged by Brazilian women.

Meat consumption has also been growing in Russia. Recently during the embargo imposed by the EU in 2014, the government has implemented policies to attain self-sufficiency in meat production and to reach higher levels of animal protein consumption in poorer segments of society [59]. Indigenous populations of Siberia still rely on the wild meat of ungulates, but consumption of other animals such as bears is also common [60]. Russia has an Orthodox Christian heritage which translates to lack of restrictions on eating meat. Russians have a strong preference for meat and meat consumption can be expected to grow if affordable [61].

Russian participants showed similar beliefs in animal sentience and compared to participants of other nationalities in this study. Russian women showed particularly higher levels of disagreement with the statement relating meat consumption and animal suffering.

The USA is among the top consumers of meat in the world [52]. Meat-based diets, especially red and processed meat are common in the USA, which is positively correlated with obesity (Rouhani et al., 2014). The USA tops the list of countries of meat production and consumption per capita [52] and is among the countries with higher obesity rates [62].

Despite the health effects and the effects on the sustainability of the environment, consumerism in the USA is still finding room for increasing consumption. The "growth in advocacy for animal rights and environmental sustainability often cannot match the powerful lobby of meat producers that exerts influence on politicians (e.g., via donations to certain candidates) and the public (e.g., via TV commercials). In a study [63] about the influence of the media on meat consumption in the USA, it was found that individuals more exposed to media content, dissociate more often meat from its animal origin and eat more meat. Examples of dissociation include the use of terms such as drumsticks and buffalo wings to dissociate from chicken; the use of the words "pork" or "beef" to dissociate from pigs and cattle; sculptured made nuggets to appeal to children and dissociate the product from chicken, and processed meat such as cheeseburgers. The livestock industry in the USA is the most productive, per capita, in the world.

USA participants showed weaker beliefs in animal sentience in comparison to the BRIC countries. Only China showed lower levels of perceived animal sentience. While the prevalence of neutral responses was especially high in China, disagreement with the statement that animals can feel pain and discomfort was higher in the USA.

Regarding the acknowledgement of farm animal suffering caused by meat consumption, scores in the USA were comparable to scores in the BRIC countries. However, in older men acknowledgment was particularly low compared to the other countries. In a recent study in the USA, about 30% of the population was reported to acknowledge that farm animal suffering should be given the same consideration as human suffering [63].

4.2. Age Effects

Younger people consume more meat in the countries that were part of this study: USA [64], Brazil [65], China [66], and India [67], but not in Russia [68]. Older individuals consume less meat which correlates positively with a lower need to justify the morals of farming and fishing [69]. Older people more frequently justify the morals of eating meat with the 4 Ns (natural, necessary, normal, or nice) [10]. Older people also score lower on perceived animal sentience [70]. However contradictory results are also found in research relating age to meat eating and its moralities [71].

Perceptions of animal sentience showed a positive correlation with age. Perceptions of animal suffering also increased with age among Chinese individuals and Indian men, while the opposite is observed for men in the USA. The other pairs' country/gender show no age effects.

4.3. Gender Effects

Men and women have been shown to behave differently in relation to meat consumption and its perceived morality. Men consume proportionally more meat than women, and meat consumption is stereotypically linked to power, dominance, and masculinity, the roles traditionally embraced by men [72]. Men also show lower levels of perceived animal sentience, and more frequently deny farm animal suffering [72]. Men also use the 4Ns strategy more often to justify eating meat [10]. Women on the other hand have been shown to be more concerned about animal rights and ethics [73]. Women present stronger critical attitudes towards meat eating, disgust, and repulsion [74]. The suffering of animals is a stronger concern for women who are also more likely to turn vegetarian [75]. Gender attitudes can also influence meat-eating attitudes, and societies with higher levels of hostile sexism and lower levels of women empowerment correlate with greater gender differences in meat-eating attitudes [76].

The consumption of meat by men is higher than that of women in all the countries of the present study: USA [64], Brazil [77], China [78], India [50], and Russia [61]. We found that women reported stronger beliefs in animal sentience across all the countries in this study. Regarding the perceived suffering of farm animals, the gender results are mixed. Women acknowledged farm animal suffering more than men in Brazil, China, and from the age of 80 in the USA, but not in other countries. Particularly Russia has a prominent level of women that do not acknowledge the suffering of farm animals.

4.4. The Ethicalities of Meat Consumption

Studies show us that meat consumption raises a diversity of ethical questions among consumers [79,80]. Namely, meat provides essential nutrients, such as amino acids, energy, and micronutrients, and gives sensorial pleasure due to unique organoleptic properties [10]. However, if consumed in excess is also a cause health problems, such as cardiovascular diseases and colorectal cancer [79]. Meat consumption is legitimated by cultural tradition but also raises sustainability issues due to high levels of water consumption and greenhouse gas emissions [79,80].

Due to this dichotomy between benefits and costs, meat consumption has become a topic of ideological battles and paradoxical tensions. Some scholars argue that meat consumption is "unhealthy, unsustainable, and unethical". These paradoxical tensions can be seen as the "two sides of the same coin" as the need for meat production to satisfy human needs brings the referred issues [81].

Some of the solutions being proposed include the reduction of meat consumption and the diversification of protein sources from the non-animal origin [82]. However, at the same time, it is proposed that meat should not be fully replaced [81]. Therefore, meat consumption is characterized by this systemic paradoxical tension for which, and with the actual food technology state of the art, an optimal solution was still not found.

The actuality shows that most people agree that doing harm to animals is wrong [83] and that sentient beings should be treated morally [84]. In the present study, we have found high levels of agreement with both the suffering and the sentience question, which reveals the ethical consideration that people give to animals. This stands in contrast with the widespread harm still inflicted on animals in society.

The present study demonstrates the diversity of beliefs in farm animal sentience and anthropogenic animal suffering. Although we examined variation across gender, nationality, and age, we acknowledge that other factors are also a source of variation. Other variables not included in this study may impact the perceived animal suffering and animal sentience, and these include personality, religious and/or political stance, and ownership of companion animals [23].

4.5. Limitations and Future Research

The current analysis was quantitative and descriptive in nature, examining relationships between central beliefs about animals (animal sentience and suffering of farm animals) and demographic characteristics (country, gender, and age)). The results obtained to allow a snapshot of the situation, but do not necessarily fully explain the reality. Future research may investigate other type of theoretical models from a cultural perspective. Studies focused on the comparison of societal diversity have shown that culture has an enormous influence on people's behaviors, and therefore social behaviors vary immensely [85].

Culture is defined [86] as "the collective programming of the mind which distinguishes the members of one human group from another". This does not however mean that every member of a particular society shares the same programming; the national culture is only the dominant mental program [86]. National culture naturally varies accordingly to gender and age; being masculinity, femininity, and age determined by national characteristics, therefore subject to analysis [86].

Through the definition of culture, we can immediately infer limitations in this study. First of all, these type of studies are hard to implement; secondly, once they are directed to a specific societal subject, reduces the plausibility of a broader national culture; and third, it involves a society confined to a geography, eventually hiding the impact of variables such as religious differences mainly in nations such as China and India, but also in the others, and the original cultures of immigrants, mainly in the case of the USA and Brazil.

Research-based on the theory of lifestyles may be useful in the phenomenon herein studied and analyzed. This theory is based on habits, which is seen as the principle in the genesis of the portrayal of intrinsic and relational characteristics in a unitary lifestyle; in other words, a homogeneous set of people's choices, beliefs, and practices, allowing the explanation of practices and judgments in a distinct circle [87]. Therefore, the selection or creation of lifestyles is influenced by group and peer pressures and the visibility of their role models, together with the socio-economic variables. As a result, lifestyles have been used in a theoretical model for analysis of reality, and in the support of segmentation strategies to study consumers' behaviors [88–90]. Hence, the adoption of a theoretical framework based on the lifestyle's theory could improve the scientific capacity to analyze the judgements and practices of consumers related to the ontological nature of animals and the consumers' impact on its suffering, as well as perception of animal sentience.

Moreover, the sampling strategy implemented by YouGov® does not fully represent a pure random sample of the vast population in the studied countries. Despites the large sampling effort of around one thousand interviewees per country. Despites all the efforts to implement a valid construct of the questions posed to the interviewees, the existence of cross-cultural differences may impact negatively the evaluation of the sentience and suffering concepts across nations.

4.6. Implications

The results of this study contribute knowledge to stakeholders, policymakers and traders. These may be used by these supply chain players to program and overcome problems related to the global FAW standards. For fair trading competition it is important to standardize procedures and respect the demand for both animal protein and the ethics of its production. If international common grounds for standardization are not found, the risk of countries with more advanced animal welfare legislation imposing trade barriers increases. These trade barriers may be perceived as protectionism by exporting countries.

5. Conclusions

Farm animal welfare is growing in the political agenda in many countries, and the global trade of animal products is a reality. However, FAW, sentience, suffering, and the ethicalities of meat consumption vary between cultures, gender, and across ages.

In Russia and India, perception of FAW suffering and FAW sentience increases with age, with similar levels to those observed for the USA. Women have higher levels of perceived animal sentience and suffering in all the countries included in this study. Also, in all the countries, more people agreed than disagreed that animals are sentient. Men in India show higher levels of agreement with the relation between eating meat and animal suffering, followed by women in Brazil and China. Lower levels of agreement are observed in Americans and Chinese. In Russia there is a slightly higher level of perceived FAW suffering between men and between younger Americans. In India and China age has the opposite effect.

The paradoxicality of meat consumption remains an ethical concern promoting the debate of sustainability.

Author Contributions: Conceptualization, F.M.; Methodology, F.M.; Formal analysis and investigation, F.M.; Writing—original draft preparation, F.M., B.J. and I.D.; Writing—review and editing: F.M., B.J. and I.D.; Funding acquisition: N/A; Resources: F.M.; Supervision: F.M. All authors have read and agreed to the published version of the manuscript.

Funding: This research received no external funding.

Institutional Review Board Statement: Not applicable.

Informed Consent Statement: Informed consent was obtained from all subjects involved in the study.

Data Availability Statement: The data used in this study is available at the Open Science Framework repository: Anderson, J., 2018b. BRIC dataset.sav [data set]. Faunalytics page in the Open Science Framework repository. https://osf.io/stbxc/ (accessed on 12 March 2022).

Acknowledgments: To the Foundation for Science and Technology (FCT, Portugal) for financial support to CISAS UIDB/05937/2020 and UIDP/05937/2020, including the contract with Fernando da Mata. To the Charity Faunalytics (Olympia USA) for sharing the data and encouraging this research.

Conflicts of Interest: The authors declare no conflict of interest.

References

1. Broom, D.M. A history of animal welfare science. *Acta Biotheor.* **2011**, *59*, 121–137. [CrossRef]
2. Farm Animal Welfare Council. Farm Animal Welfare in Great Britain: Past, Present and Future, 2009. Available online: https://assets.publishing.service.gov.uk/government/uploads/system/uploads/attachment_data/file/319292/Farm_Animal_Welfare_in_Great_Britain_-_Past__Present_and_Future.pdf (accessed on 18 March 2022).
3. McCulloch, S.P. A critique of FAWC's five freedoms as a framework for the analysis of animal welfare. *J. Agric. Environ. Ethics* **2013**, *26*, 959–975. [CrossRef]
4. Broom, D.M. The evolution of morality. *Appl. Anim. Behav. Sci.* **2006**, *100*, 20–28. [CrossRef]
5. Preece, R. *Awe for the Tiger, Love for the Lamb: A Chronicle of Sensibility to Animals*, 1st ed.; UCB Press: Vancouver, BC, Canada, 2002; pp. 3–355.
6. Blattner, C.E. The recognition of animal sentience by the law. *J. Anim. Ethics* **2019**, *9*, 121. [CrossRef]
7. Mellor, D.J. Welfare-aligned sentience: Enhanced capacities to experience, interact, anticipate, choose and survive. *Animals* **2019**, *9*, 440. [CrossRef] [PubMed]
8. Kona-Boun, J. Anthropogenic suffering of farmed animals: The other side of zoonoses. *Anim. Sentience* **2020**, *30*, 20. [CrossRef]
9. Farm Animal Welfare Council. *Evidence and the Welfare of Farmed Animals Part 1: The Evidence Base. London: DEFRA, 2014.* Available online: https://assets.publishing.service.gov.uk/government/uploads/system/uploads/attachment_data/file/324480/FAWC_report_on_evidence_and_the_welfare_of_farmed_animals_part_1_the_evidence_base.pdf (accessed on 18 March 2022).
10. Piazza, J.; Ruby, M.B.; Loughnan, S.; Luong, M.; Kulik, J.; Watkins, H.M.; Seigerman, M. Rationalizing meat consumption. The 4Ns. *Appetite* **2015**, *91*, 114–128. [CrossRef]
11. Campbell, D.L.M.; Lee, C. A perspective on strategic enrichment for brain development: Is this the key to animal happiness? *Front. Vet. Sci.* **2021**, *8*, 720422. [CrossRef]
12. Loughnan, S.; Bastian, B.; Haslam, N. The psychology of eating animals. *Curr. Dir. Psychol. Sci.* **2014**, *232*, 104–108. [CrossRef]
13. Bastian, B.; Costello, K.; Loughnan, S.; Hodson, G. When closing the human–animal divide expands moral concern: The importance of framing. *Soc. Psychol. Personal. Sci.* **2011**, *3*, 421–429. [CrossRef]
14. Leach, S.; Sutton, R.M.; Dhont, K.; Douglas, K.M. When is it wrong to eat animals? The relevance of different animal traits and behaviours. *Eur. J. Soc. Psychol.* **2021**, *51*, 113–123. [CrossRef]
15. Bastian, B.; Loughnan, S.; Haslam, N.; Radke, H.R. Don't mind meat? The denial of mind to animals used for human consumption. *Pers. Soc. Psychol. Bull.* **2012**, *38*, 247–256. [CrossRef] [PubMed]
16. Dhont, K.; Hodson, G.; Leite, A.C.; Salmen, A. The Psychology of Speciesism. In *Why We Love and Exploit Animals*; Routledge: Abingdon-on-Thames, UK, 2019; pp. 29–49.

17. Kunst, J.R.; Hohle, S.M. Meat eaters by dissociation: How we present, prepare and talk about meat increases willingness to eat meat by reducing empathy and disgust. *Appetite* **2016**, *105*, 758–774. [CrossRef] [PubMed]
18. Joy, M. *Why We Love Dogs, Eat Pigs, and Wear Cows: An Introduction to Carnism*, 3rd ed.; Red Wheel/Weisser LLC: Newburry Port, MA, USA, 2020; pp. 85–104.
19. Phillips, C.J.C.; McCulloch, S. Student attitudes on animal sentience and use of animals in society. *J. Biol. Educ.* **2005**, *40*, 17–24. [CrossRef]
20. Clarke, N.; Paul, E.S. Veterinary Students' Beliefs About Animal Sentience: What Role Does Gender Play? *Anthrozoös* **2019**, *32*, 581–595. [CrossRef]
21. Knight, S.; Vrij, A.; Cherryman, J.; Nunkoosing, K. Attitudes towards animal use and belief in animal mind. *Anthrozoös* **2004**, *17*, 43–62. [CrossRef]
22. Kellert, S.R.; Berry, J.K. Attitudes, knowledge, and behaviors toward wildlife as affected by gender. *Wildl. Soc. Bull.* **1987**, *15*, 363–371.
23. Taylor, N.; Signal, T.D. Empathy and attitudes to animals. *Anthrozoös* **2005**, *18*, 18–27. [CrossRef]
24. Norwood, F.B.; Lusk, J. Animal Qualia: Investigating Animal Sentience. In *Compassion by the Pound*; Oxford University Press: Oxford, UK, 2011; pp. 67–93.
25. Ren, Y.; Li, Z.; Wang, Y.; Zhang, T. Development and prospect of food security cooperation in the BRICS countries. *Sustainability* **2020**, *12*, 2125. [CrossRef]
26. Brislin, R.W. The Wording and Translation of Research Instruments. In *Field Methods in Cross-Cultural Research*; Lonner, W.J., Berry, J.W., Eds.; Sage: Beverly Hills, CA, USA, 1986; pp. 137–164.
27. Straus, M.A. Phenomenal identity and conceptual equivalence of measurement in cross-national comparative research. *J. Marriage Fam.* **1969**, *81*, 233–239. [CrossRef]
28. Smith, T.W. *Improving Cross-National Survey Research by Measuring the Intensity of Response Categories*; GSS Cross-National Report No. 17; National Opinion Research Center, University of Chicago: Chicago, IL, USA, 1997.
29. Liu, R.; Xing, L.; Zhou, G.; Zhang, W. What is meat in China? *Anim. Front.* **2017**, *7*, 53–56. [CrossRef]
30. Carnovale, F.; Jin, X.; Arney, D.; Descovich, K.; Guo, W.; Shi, B.; Phillips, C.J.C. Chinese public attitudes towards, and knowledge of, animal welfare. *Animals* **2021**, *11*, 855. [CrossRef] [PubMed]
31. Li, P.J. Exponential Growth, Animal Welfare, Environmental and Food Safety Impact: The Case of China's Livestock Production. *J. Agric. Environ. Ethics* **2009**, *22*, 217–240. [CrossRef]
32. Bayne, K.; Ramachandra, G.S.; Rivera, E.A.; Wang, J. The evolution of animal welfare and the 3Rs in Brazil, China, and India. *J. Am. Assoc. Lab. Anim. Sci.* **2015**, *54*, 181–191. [PubMed]
33. Cao, D. Is the concept of animal welfare incompatible with Chinese culture? *Soc. Anim.* **2020**, *30*, 284–296. [CrossRef]
34. Sinclair, M.; Zhang, Y.; Descovich, K.; Phillips, C.J.C. Farm animal welfare science in China-A bibliometric review of Chinese literature. *Animals* **2020**, *10*, 540. [CrossRef] [PubMed]
35. Li, X.; Zito, S.; Sinclair, M.; Phillips, C.J.C. Perception of animal welfare issues during Chinese transport and slaughter of livestock by a sample of stakeholders in the industry. *PLoS ONE* **2018**, *13*, e0197028. [CrossRef]
36. Henry, C.; Chow, V.; Grinberg, N. A Literature Review of the Current Consideration of Animals in China. 2021. Available online: https://osf.io/e9whf/ (accessed on 23 March 2022).
37. Lecso, P.A. To do no harm: A Buddhist view on animal use in research. *J. Relig. Health* **1988**, *27*, 307–312. [CrossRef] [PubMed]
38. Deborah, C. Confucianism and Daoism. In *The Routledge Handbook of Religion and Animal Ethics*; Linzey, A., Linzey, C., Eds.; Routledge: Abingdon-on-Thames, UK, 2018; pp. 56–62.
39. Li, P.J. Animal suffering in China. Anim. *Sentience* **2016**, *7*, 8. [CrossRef]
40. Fan, R. How should we treat animals? A Confucian reflection. *Dao* **2010**, *9*, 79–96. [CrossRef]
41. Li, C. Children of the reform and opening-up: China's new generation and new era of development. *J. Chin. Sociol.* **2020**, *7*, 18. [CrossRef]
42. Platto, S.; Serres, A.; Jingyi, A. Chinese college students' attitudes towards animal welfare. *Animals* **2022**, *12*, 156. [CrossRef] [PubMed]
43. Dave, N.N. Witness: Humans, animals, and the politics of becoming. *Cult. Anthropol.* **2014**, *29*, 433–456. [CrossRef]
44. Kennedy, U.; Sharma, A.; Phillips, C.J.C. The sheltering of unwanted cattle, experiences in India and Implications for cattle industries elsewhere. *Animals* **2018**, *8*, 64. [CrossRef] [PubMed]
45. Chakrabarti, P. Beasts of burden: Animals and laboratory research in colonial India. *Hist. Sci.* **2010**, *48*, 125–151. [CrossRef]
46. Khara, T.; Riedy, C.; Ruby, M.B. A cross cultural meat paradox: A qualitative study of Australia and India. *Appetite* **2021**, *164*, 105227. [CrossRef]
47. Pandian, M.S.S. One step outside modernity: Caste, identity politics and public sphere. *Econ. Polit. Wkly.* **2002**, *37*, 1735–1741. Available online: https://www.jstor.org/stable/4412071 (accessed on 10 March 2022).
48. Burgat, F. Non-violence towards animals in the thinking of Gandhi: The problem of animal husbandry. *J. Agric. Environ. Ethics* **2004**, *17*, 223–248. [CrossRef]
49. Mahendra, S.K. Ban on beef and the politics of food: An intercultural analysis. *New Life Theol. J.* **2016**, *6*, 84–94.
50. Donner, H. New vegetarianism: Food, gender and neo-liberal regimes in bengali middle-class families. *South Asia J. South Asia Stud.* **2008**, *31*, 143–169. [CrossRef]

51. Arora, R.S.; Brent, D.A.; Jaenicke, E.C. Is India ready for alt-meat? Preferences and willingness to pay for meat alternatives. *Sustainability* **2020**, *12*, 4377. [CrossRef]
52. Ritchie, H.; Roser, M. Meat and Dairy Production. 2017. Available online: https://ourworldindata.org/meat-production (accessed on 10 March 2022).
53. Heidemann, M.S.; Taconeli, C.A.; Reis, G.G.; Parisi, G.; Molento, C.F.M. Critical perspective of animal production specialists on cell-based meat in Brazil: From bottleneck to best scenarios. *Animals* **2020**, *10*, 1678. [CrossRef] [PubMed]
54. Vranken, L.; Avermaete, T.; Petalios, D.; Mathijs, E. Curbing global meat consumption: Emerging evidence of a second nutrition transition. *Environ. Sci. Policy* **2014**, *39*, 95–106. [CrossRef]
55. Prandi, R. African gods in contemporary Brazil: A Sociological introduction to Candomblé today. *Int. Sociol.* **2000**, *15*, 641–663. [CrossRef]
56. Tamioso, P.R.; Rucinque, D.S.; Miele, M.; Boissy, A.; Molento, C.F.M. Perception of animal sentience by Brazilian and French citizens: The case of sheep welfare and sentience. *PLoS ONE* **2018**, *13*, e0200425. [CrossRef]
57. Tamioso, P.R.; Guimarães, P.R.B.; Molento, C.F.M. Attitudes of South Brazilian sheep farmers to animal welfare and sentience. *Cienc. Rural.* **2017**, *47*, 1–6. [CrossRef]
58. Gorlov, I.F.; Fedotova, G.V.; Slozhenkina, M.I.; Mosolova, N.I. The Meat Products Supply of Population in Russia. In *Growth Poles of the Global Economy: Emergence, Changes and Future Perspectives*; Popkova, E.G., Ed.; Springer International Publishing: New York, NY, USA, 2020; pp. 311–318.
59. Ozeretskovskaya, N.N.; Mikhailova, L.G.; Sabgaida, T.P.; Dovgalev, A.S. New trends and clinical patterns of human trichinellosis in Russia at the beginning of the XXI century. *Vet. Parasitol.* **2005**, *132*, 167–171. [CrossRef]
60. Honkanen, P. Food preference based segments in Russia. *Food Qual. Prefer.* **2010**, *21*, 65–74. [CrossRef]
61. World Population Review. Most Obese Countries 2022. Available online: https://worldpopulationreview.com/country-rankings/most-obese-countries (accessed on 30 March 2022).
62. Weathers, S.T.; Caviola, L.; Scherer, L.; Pfister, S.; Fischer, B.; Bump, J.B.; Jaacks, L.M. Quantifying the valuation of animal welfare among Americans. *J. Agric. Environ. Ethics* **2020**, *33*, 261–282. [CrossRef]
63. Lewis, K.C. A Meat Paradox: Media's Role in Mitigating the Omnivore's Dilemma. Master's Thesis, University of Southern Mississippi, Hattiesburg, MS, USA, 2018.
64. Frank, S.M.; Jaacks, L.M.; Batis, C.; Vanderlee, L.; Taillie, L.S. Patterns of red and processed meat consumption across North America: A nationally representative cross-sectional comparison of dietary recalls from Canada, Mexico, and the United States. *Int. J. Environ. Res. Public Health* **2021**, *18*, 357. [CrossRef]
65. Bezerra, I.N.; Goldman, J.; Rhodes, D.G.; Hoy, M.K.; de Moura Souza, A.; Chester, D.N.; Martin, C.L.; Sebastian, R.S.; Ahuja, J.K.; Sichieri, R.; et al. Difference in adult food group intake by sex and age groups comparing Brazil and United States nationwide surveys. *Nutr. J.* **2014**, *13*, 74. [CrossRef] [PubMed]
66. Min, S.; Bai, J.-F.; Seale, J.; Wahl, T. Demographics, societal aging, and meat consumption in China. *J. Integr. Agric.* **2015**, *14*, 995–1007. [CrossRef]
67. Khara, T. What are Consumer Attitudes in Urban India like towards Ethical Food Products and What Influences Their Attitudes? Master's Thesis, Curtin University, Perth, Australia, 2015. Espace—Curtin's Institutional Repository. Available online: http://hdl.handle.net/20.500.11937/1656 (accessed on 30 March 2022).
68. Maksimov, S.; Karamnova, N.; Shalnova, S.; Drapkina, O. Sociodemographic and Regional Determinants of Dietary Patterns in Russia. *Int. J. Environ. Res. Pub. Health* **2020**, *17*, 328. [CrossRef] [PubMed]
69. Hartmann, C.; Siegrist, M. Our daily meat: Justification, moral evaluation and willingness to substitute. *Food Qual. Prefer.* **2020**, *80*, 103799. [CrossRef]
70. Peden, R.S.E.; Camerlink, I.; Boyle, L.A.; Loughnan, S.; Akaichi, F.; Turner, S.P. Belief in pigs' capacity to suffer: An assessment of pig farmers, veterinarians, students, and citizens. *Anthrozoös* **2020**, *33*, 21–36. [CrossRef]
71. Gradidge, S.; Zawisza, M.; Harvey, A.J.; McDermott, D.T. A Structured literature review of the meat paradox. *Soc. Psychol. Bull.* **2021**, *16*, 1–26. [CrossRef]
72. Rothgerber, H. Real men don't eat (vegetable) quiche: Masculinity and the justification of meat consumption. *Psychol. Men Masc.* **2013**, *14*, 363–375. [CrossRef]
73. Herzog, H.A. Gender differences in human–animal interactions: A review. *Anthrozoös* **2007**, *20*, 7–21. [CrossRef]
74. Ruby, M.B.; Heine, S.J. Too close to home. Factors predicting meat avoidance. *Appetite* **2012**, *59*, 47–52. [CrossRef]
75. Forestell, C.A.; Nezlek, J.B. Vegetarianism, depression, and the five factor model of personality. *Ecol. Food Nutr.* **2018**, *57*, 246–259. [CrossRef]
76. Allcorn, A.; Ogletree, S.M. Linked oppression: Connecting animal and gender attitudes. *Fem. Psychol.* **2018**, *28*, 457–469. [CrossRef]
77. Filho, R.G.C. Meat Masculinity. *Contexts* **2014**, *13*, 56–58. [CrossRef]
78. Yuan, Y.-Q.; Li, F.; Meng, P.; You, J.; Wu, M.; Li, S.-G.; Chen, B. Gender difference on the association between dietary patterns and obesity in Chinese middle-aged and elderly populations. *Nutrients* **2016**, *8*, 448. [CrossRef] [PubMed]
79. Godfray, H.C.J.; Aveyard, P.; Garnett, T.; Hall, J.W.; Key, T.J.; Lorimer, J.; Pierrehumbert, R.T.; Scarborough, P.; Springmann, M.; Jebb, S.A. Meat consumption, health, and the environment. *Science* **2018**, *361*, eaam5324. [CrossRef]
80. Font-i-Furnols, M.; Guerrero, L. Understanding the future meat consumers. *Meat Sci.* **2022**, *193*, 108941. [CrossRef]

81. Leroy, F.; Fabien, A.; Beal, T.; Dominguez-Salas, P.; Gregorini, P.; Manzano, P.; Rowntree, J.; Van Vliet, S. Animal board invited review: Animal source foods in healthy, sustainable, and ethical diets–An argument against drastic limitation of livestock in the food system. *Animal* **2022**, *16*, 100457. [CrossRef]
82. Siddiqui, S.A.; Bahmid, N.A.; Mahmud, C.M.; Boukid, F.; Lamri, M.; Gagaoua, M. Consumer acceptability of plant-, seaweed-, and insect-based foods as alternatives to meat: A critical compilation of a decade of research. *Crit. Rev. Food Sci. Nutr.* **2022**, *in press*. [CrossRef]
83. Schein, C.; Gray, K. The theory of dyadic morality: Reinventing moral judgment by redefining harm. *Pers. Soc. Psychol. Rev.* **2018**, *22*, 32–70. [CrossRef]
84. Gray, K.; Young, L.; Waytz, A. Mind perception is the essence of morality. *Psychol. Inq.* **2012**, *23*, 101–124. [CrossRef]
85. Henrich, J. Culture and social behavior. *Curr. Opin. Behav. Sci.* **2015**, *3*, 84–89. [CrossRef]
86. Hofstede, G. *Culture's Consequences: International Differences in Work-Related Values*, 2nd ed.; Sage Publications Inc.: Newbury Park, CA, USA, 1984; pp. 13–153.
87. Bourdieu, P. *La Distinction: Critique Sociale du Jugement*; Les edition du Minuit: Paris, France, 2016; pp. 14–672.
88. Baruk, A.I. Co-creation of a food marketing offer by final purchasers in the context of their lifestyles. *Br. Food J.* **2020**, *123*, 1494–1512. [CrossRef]
89. Beekman, V. You are what you eat: Meat, novel protein foods, and consumptive freedom. *J. Agric. Environ. Ethics* **2000**, *12*, 185–196. [CrossRef]
90. Korthals, M.; Komduur, R. Uncertainties of nutrigenomics and their ethical meaning. *J. Agric Environ. Ethics* **2010**, *23*, 435–454. [CrossRef]

animals

MDPI

Article

Ubiquitous Love or Not? Animal Welfare and Animal-Informed Consent in Giant Panda Tourism

David A. Fennell [1,*] and Yulei Guo [2]

[1] Department of Geography & Tourism Studies, Brock University, St. Catharines, ON L2S 3A1, Canada
[2] Chengdu Research Base of Giant Panda Breeding Research Department, Chengdu 610081, China
* Correspondence: dfennell@brocku.ca

Simple Summary: Understanding the relationship that tourists have with the giant panda can be taken-for-granted given this species' iconic status and cuteness. Based on a self-reporting questionnaire, we found that Chinese "fans" of giant pandas generally paid greater attention to the welfare of the captive pandas in comparison to "non-fans". Additionally, fans tend to perceive the welfare of captive pandas more positively, who tend to believe that giant pandas offer more prominent consent to being used as tourist attractions. This study provides new insight into an understanding of the human–panda connection. The ubiquitous love we have for pandas can translate into a better life for these animals down the road.

Abstract: Scholars argue that the ubiquity of the "virtual panda"—the panda people meet in zoos and consume as souvenirs, online memes, or videos—exists in a state of hybridity between wild and domesticated. The species has garnered a significant amount of attention because of their iconic status and because of how cute they are to an adoring crowd. However, given the degree of regard tourists have for the panda, there is a dearth of research on different types of visitors to captive panda venues. In filling this gap, we investigated (1) how deeply Chinese "fans" and "non-fans" consider the welfare of captive giant pandas, and (2) if these groups differ in their assessment of whether giant pandas consent to being used as tourist attractions. In both aims, we apply a recent model on animal welfare and animal consent to giant pandas of the Chengdu Research Base of Giant Panda Breeding.

Keywords: tourism; giant panda; animal-informed consent; animal welfare

1. Introduction

China's giant panda has value on many fronts. The species has political value in the form of panda diplomacy, or the sending of pandas to other countries for diplomatic reasons [1], cultural value as a national icon [2,3], conservation value both inside China and globally [4], and touristic value as a revenue generator [5]. In 2022, the number of captive pandas doubled over the course of the previous decade to 673 individuals [6]. The wild panda population increased to 1864 individuals based on an expanding network (parks) of the panda's natural habitat [7], prompting the IUCN in 2016 to reclassify the giant panda from "endangered" to "vulnerable" [8]. The conservation value of pandas is estimated to be worth between US$ 2.6–6.9 billion per year, some 27× more than the cost to the Chinese government to facilitate the panda's conservation [7]. The giant panda has remained a global celebrity in zoos since the exhibition of the first panda, Su-Lin, outside of China in 1937, renowned as "a sensation" [9]. By 2019, 27 zoos in 21 countries outside of mainland China hosted pandas, with significant economic advantages. Pandas at the National Zoo in Washington D.C., for example, attracted an estimated 12 million people in the 1970s [10], while in Japan, pandas are leased or loaned to other institutions for the purpose of gaining huge profits [11]. In 2019, the Chengdu Research Base of Giant Panda Breeding (entry ticket RMB¥55 ≈ US$7.8) was visited by 9 million visitors [12].

However, in contrast to other Ursidae family members, such as polar bears [13–19], brown bears [20–22], sloth [23,24], and black bears [25,26], that can be viewed in their natural habitats, the giant panda is strictly protected by the Wildlife Conservation Law of China and the newly established Giant Panda National Park system [27]. Restrictions on interacting with pandas in the wild places captive pandas at the center of the panda-visitor industry, despite the push to re-establish panda in protected areas [28–30]. This one-dimensional manner by which to view pandas in captive venues has been discussed by Nicholls [31], who argues that wild pandas have become largely intangible because of their persistent invisibility from human society in wild spaces. Instead, Nicholls [31] suggests the ubiquity of the "virtual panda"—the panda people meet in zoos and consume as souvenirs, online memes, or videos—exists in a state of hybridity between wild and domesticated. We find this state worthy of investigation because in virtual form, as in captivity, society has embraced the panda as one of the most beloved animals of all.

While the captive tourism industry sets the standard for the production and consumption of virtual or hybridized pandas, we argue that this hybrid state has not been sufficiently interrogated [32]. While conservation is a central feature in the presentation of pandas to visitors, little is known about the keeping of pandas, as similar or different, than the practices of captive sites that place individual (e.g., tourists) and organizational (e.g., the breeding facilities interests in profit and growth) interests above the interests of animals [33]. It has been argued, consistently, that the giant panda garners significant attention because of how attractive or "cute" the species is to an adoring crowd [32,34–36]. However, does being attractive or cute provide justification for life as a commodity? To be produced and consumed for profit and pleasure? To justify conservation efforts? Hence, the purpose of this paper is to investigate what it means to be an animal with hybrid and virtual significance in contemporary Chinese society by (1) assessing how deeply Chinese "fans" and "non-fans" consider the welfare of captive giant pandas, and (2) investigating if panda fans and non-fans differ in their assessment of whether giant pandas *consent* to being used in the manner described above. In both aims, we apply a recent model by Fennell [33] on animal welfare and animal consent in the animal-based tourism domain to giant pandas of the Chengdu Research Base of Giant Panda Breeding.

1.1. Literature Review

The review of literature follows two main streams. The first provides an overview of research on animal welfare as a distinct theoretical and applied perspective in a broader realm of animal ethics scholarship. Included in this review is current research on animal-informed consent in tourism. The second stream discusses the literature on the concept of "fandom" and the establishment of core differences between panda fans and non-fans at the Chengdu Research Base of Giant Panda Breeding.

1.2. Animal Welfare and Animal-Informed Consent

The literature on the ethical use of animals in tourism has advanced considerably over the course of the last two decades, with a flurry of activity in recent years [37]. Not only is the theoretical nature of animal ethics being explored through a range of different theories [38], but also how these theories apply in certain practices and for certain species. Examples include animal welfare and elephants [39], animal rights in reference to hunting [40], the balancing of costs and benefits (utilitarianism) in marine wildlife tourism [41], Burns et al., [42] on dingoes at Fraser Island, Australia, and ecofeminism and its application to polar bears in northern Canada [18].

Studies have also sought to provide greater scope and understanding on how personhood, agency, and stakeholder apply to the use of animals in tourism. In the case of personhood, Cavalieri [43] argues that some animals, like the great apes, must be granted life, liberty, and the prohibition of torture in living lives of dignity and respect. Agency has been discussed in the context of sled dogs in Finland, who should be given more freedom, control, and influence in the events of their own lives in striving for responsible sled dog practices [44]. Furthermore, animals are now being discussed regularly in tourism and animal ethics studies as stakeholders [38,45]. Compelling in this narrative is work by Kenehan [46] in the business and organizational culture context who argues, following Evan and Freeman [47], that a stakeholder is anyone who is "vital to the survival and success of the firm" (p. 58). While not referring to animals in tourism as stakeholders, Kenehan [46] argues from an organizational context that animals used in experimentation are actors who are essential in bringing products to market.

A novel advancement in theory and practice for the use of animals in tourism surrounds the concept of animal-informed consent [33,48]. This perspective disrupts the traditional order of human–animal relationships by placing value into the cognitive, emotional, biological, and behavioral state of an animal instead of solely on the needs of tourism operators and tourists. Szydlowski [48] discussed the challenges of receiving consent from elephants in Nepal, while Fennell [33] developed a comprehensive framework on levels of assent and dissent for sled dogs based on the Five Domains model of animal welfare by Mellor et al. [49].

Traditional assessments of animal welfare typically focus on the cognitive and physiological well-being of animals which, if all conditions are met, indicate that an animal is faring well. The Five Freedoms model, for example, focuses on the freedom from hunger and thirst; freedom from discomfort; freedom from pain, injury, and disease; freedom to express normal behavior; and freedom from fear and distress [50]. In contrast, Mellor et al., [49] in their Five Domains model, include cognitive and physiological assessments, but also the subjective experiences of animals. Supplementary Tables S1–S4 provide an overview of these different positive (enhancing welfare) and negative (reducing welfare) conditions according to the following categories: (1) Nutritional conditions (food and water intake); (2) physical environment conditions (fresh air, good ventilation, shelter from adverse weather), (3) health conditions (hygiene, disease, wounds, malnutrition, obesity), (4) behavioral interactions (with the environment, with other animals, and with humans), and (5) the combination of these four sets of conditions which provide an overall assessment of the welfare of animals. An added dimension to Mellor et al.'s [49] system is the ability to grade negative affective impacts due to the range of different ways in which humans use animals. Therefore, for example, aversive training of companion animals could have a range of negative affective impacts from "none" to "very severe".

Fennell [33] adapted the Mellor et al. [49] model for the purpose of establishing an animal-informed consent continuum, ranging from active animal-informed consent (extremely positive and moderately positive affective impact), passive animal-informed consent (negligibly positive and negligibly negative affective impact), and no animal-informed consent (moderately negative and severely negative affective impacts) and how they correspond to the animal welfare indicators discussed above. Fennell [33] also suggested that empirical data could further support the development of animal-informed consent indicators. In the following, we consider the possibility that panda fans and non-fans perform as evaluators of welfare indicators.

1.3. Panda Fans and Consent

The WWF logo of the giant panda has been recognized as one of the most valuable trademarks that has ever been designed [5]. The ability to attract public investment is only one facet of the giant panda's "celebrity status". Blewitt [51] shows that the giant panda becomes an animal celebrity within a complex of "image, conservation status, mystery, politics, bizarre diet, and global cultural merchandising". For the Chinese, Songster [2] suggests that the celebrity status of giant pandas is a part of the story of the rise of the People's Republic of China in 1949. Hence, the giant panda also symbolizes the nation's modernity. The wild panda and the captive panda, waved into a virtual hybridity, allow the animal to gain a distinct advantage in being an animal celebrity.

The celebrity status of captive giant panda in zoos and enclosures almost makes questions about the animals' welfare superfluous compared to other species living in zoos or breeding facilities. Although historical records show that NGOs and international organizations have struggled to improve the welfare of captive giant pandas [11], pandas have now been given unparalleled care and attention by most hosting organizations [52]. For example, Wang Wang and Funi in the Adelaide Zoo live in a more than 3000 square meters enclosure with abundant choice, comfort, and stimuli [53]. In the Netherlands, Xing Ya and Wu Wen enjoy a 3400 square meters Chinese-style enclosure built at the cost of seven million euros [54]. Panda enclosures in the Chengdu Research Base of Giant Panda Breeding are named "Villas", suggestive of a luxury lifestyle befitting animals of such high status and attractiveness. In 2020, Ueno Zoo opened a new enclosure, Panda no Mori, spanning about 6800 square meters to accommodate the panda couple, Ri Ri and Shin Shin [55]. The name Panda no Mori, literarily translated into "panda forest", is in line with efforts to implement a design ideology consistent with the panda's natural habitat.

Just like human celebrities who often have an extravagant lifestyle, the offer of spacious and luxurious enclosures to captive pandas is only one way that humans pay tribute to these animal celebrities. More often, in parallel to fields such as sports, music, and films, enthusiastic communities towards celebrities are formed. We suggest that "panda fan" or "panda fandom" be employed as a term to describe this unique and intense relationship between tourists and captive pandas. Reflecting on Sandvoss' [56] definition that fandom is the "regular, emotionally involved consumption of a given popular narrative or text", we show that panda fandom can be described as a regular and emotional involvement with virtual captive pandas through consumptive experiences such as visitation, souvenir collection, and social media networking. Panda fandom is therefore said to be integral to panda tourism and sustains the production, reproduction, and consumption [57] of these hybrid beings.

The connection to their object of interest is at the core of the differences between fans and non-fans. Duffett [58] notes that precisely because fans develop strong emotional bonds to their objects and fans use the attachments to create relations with their "heroes and with each other", fans are more than consumers. The emotional connection between fans and their chosen object allows for the demonstration of deep knowledge and exceptional expertise about the object [59–64]. The introduction, sharing, and engagement with the knowledge and expertise further contribute to the social connection and capitalization between fans and the fan communities [65]. For example, Hills [63] shows that the fan community of the television show Doctor Who has grown in expertise and knowledge on fandom-specialized wikis that has fundamentally contested the official showrunners' knowledge and experiences. In this study, we place this attachment and knowledge about captive pandas as the key to distinguishing panda fans from non-fans. We show that the invested time, knowledge, and energy in the captive pandas have legitimated panda fans to evaluate and make sense of the relationship through self-identify with objects, and to even express the relationship by labeling the idol as a family member or close friend [66–69].

Fennell [33] explains that humans often fail to capture the expression and significance of animal-expressed indicators even if humans have played numerous roles, such as owner [70–73], trainer [74,75], experimental subject [76,77], and even colleague [78] of animals. We acknowledge the self-identification process of the panda fans as they experience and build intense attachments with these animals by exploring whether they have elevated welfare concerns and acceptance of animal-informed consent over their non-fan counterparts.

2. Methods

2.1. Questionnaire Design

The researchers designed a two-part questionnaire to evaluate panda fans' and non-fans' engagement with giant panda welfare indicators and informed consent. The first part of the survey determined whether participants were fans or non-fans. Studies addressing differences between fans and non-fans have often categorized the two groups based on participants' self-evaluation [79–81]. For example, in their study of celebrity business fans and non-fans, Teng, Liao, and Wei [82] asked respondents to answer the question, "Are you fans of Ashin? Select Yes or No.", for the purpose of helping researchers to distinguish the two groups. Researchers have also differentiated fans from non-fans based on the two groups' engagement with social media [83], with the former group actively engaging and following social media updates.

Working from Sandvoss' [56] definition of fans, above, we differentiate panda fans from non-fans based on self-identification and engagement with panda-related social media. On the one hand, we asked participants to evaluate the extent to which they identify with panda fans on a 6-point Likert scale (1 = Severely negative; 6 = Extremely positive). Second, we asked participants whether they follow panda-themed social media accounts (Yes/No question). Participants who follow panda-themed social media accounts and identified with panda fans positively (including 4 = Negligibly positive, 5 = Moderately positive, and 6 = Extremely positive) were categorized as panda fans. Furthermore, the questionnaire asked participants to evaluate their relationship with the giant panda with the question, "How do you describe your intimate relationship with the panda?" Available answers include "Family members", "Friends", "Idol", "Research object", "Visitor attraction", and "Other". The second part of the survey was built from Fennell's [33] table of animal-informed consent of 5 main indicators (Table 1). The fourth indicator, "behavioral interactions", has four sub-indicators concerning the panda's interactions with the environment, other animals, keepers, and tourists (Table 1).

Table 1. Panda welfare and consent indicators.

Indicator		Description		
1. Nutrition		a. Water supply	b. Food quantity	c. Food quality d. Variety
2. Physical environmental conditions		a. Spacious enclosures	b. Enriched enclosures	c. Enclosures protect pandas against thermal extremes
3. Health conditions		a. The panda does not suffer from acute and chronic injuries	b. Healthcare treatment	c. The panda is fit and attends to exercises
4. Behavioral interaction	4.1 Environment	a. Enclosures reflect the panda's life stage	b. Males and females are separated	c. Multiple choices to move and be active
	4.2 Other animals	a. The panda socializes with a group	b. The panda plays with its peers	c. The panda enjoys community life
	4.3 Keepers	a. Keepers are kind and friendly	b. Keepers are qualified and skillful	
	4.4 Tourists	a. The panda is at ease despite the presence of tourists	b. The panda can respond to my attention	
5. Mental conditions		a. Pandas give consent to being used as tourist attractions.	b. It is possible for pandas to express to humans their consent to being used as tourist attractions	c. Humans can recognize if pandas are expressing their consent to being used as tourist attractions

Based on an experimental test of the questionnaire with tourists, the research team constructed the indicators of conditions. The rationale behind the conditioned question design is participants' reflections on the test that they did not pay equal attention to each indicator in their visiting experiences. For example, participants in the test phase reflected that if they did not think about the panda's mental conditions or physical environment at all, it was impossible to evaluate these two indicators. Given the feedback, the research team designed a conditioned questionnaire with an 8-indicator matrix for participants to evaluate their knowledge or observations of the pandas first. The matrix required participants to evaluate their observations of the pandas through the question, "I pay close attention to the panda's nutrition/physical environmental conditions/health conditions/interactions with the environment/interactions with other animals/interactions with keepers/interactions with tourists/mental conditions". The participants could suggest that they were either concerned with one, a few, or all eight indicators (answer "yes"), or they did not pay sufficient attention (answer "I do not know"). For example, participants offering "yes" to the panda's nutrition were then directed to the evaluation of the panda's water supply, food quality, food quantity, and food variety. With the answer "I do not know", participants did not need to evaluate the details for nutrition. Evaluations of the details were all based on the 6-Likert scale (1 = Severely negative; 6 = Extremely positive). Demographic data, including gender, age, occupation, educational background, and place of residence, were collected at the beginning of the questionnaire.

2.2. Data Collection

Based on long-term fieldwork experiences at the base, the research team collected the questionnaire at the panda base's exit and panda nursery house to ensure the data's quality. The questionnaire was published on Wenjuanxing, a survey data collection website, and issued a QR code, which the research team printed out to allow participants to access it in the field. Wenjuanxing automatically recorded all data and ensured all questionnaires were complete. The research team collected data from 14 November to 20 November 2022, generating a pool of 221 respondents. Four questionnaires had unrealistic age entries and were excluded from the pool, leaving a total of 217 valid samples. On average, a questionnaire took 135.8 s to finish. All participants were Chinese, since the ongoing COVID-19 pandemic still puts international tourists in China on hold. In November, the pandemic and control measures continued to influence the number of visitors at the panda base. All participants contributed to the questionnaire voluntarily with prior consent before they scanned the QR code with their own mobile phones. In exchange, the research team offered panda magazines for the participants' time. The research team recognizes the use of the QR code as a means to access the questionnaire may pose a barrier for tourists with less internet literacy (e.g., tourists over the age of 45 and with a less-than-college degree). Additionally, using panda magazines as an incentive could lead to further differences (e.g., gender, age, and level of education) in the sample. Table 2 profiles participants based on their demographic characteristics.

Table 2. Demographic summary of participants.

Measure	N	%
Age	217	Mean = 28.3; SD = 9.10
Gender		
Female	139	64.1%
Male	78	35.9%
Level of education		
Less than college	24	11.1%
Some college	50	23.0%
Undergraduate	97	44.7%
Postgraduate and above	46	21.2%
Occupation		
Full-time	111	51.2%
Student	48	22.1%
Freelance	32	14.7%
Unemployed	8	3.7%
Retired	5	2.3%
Other	13	6.0%
Place of residence		
Chengdu	81	37.3%
First-tier cities	44	20.3%
Other regions in Sichuan	11	5.1%
Provincial capitals	40	18.4%
Other	41	18.9%

2.3. Data Preparation

The study identifies two groups, panda fans and non-fans, depending on the participants' self-identification with fans and consumption of social media content. The two conditions allowed us to identify 117 panda fans (53.9%) and 100 non-fans (46.1%). We conducted *t*-tests to evaluate panda welfare differences between panda fans and non-panda fans (Figure 1).

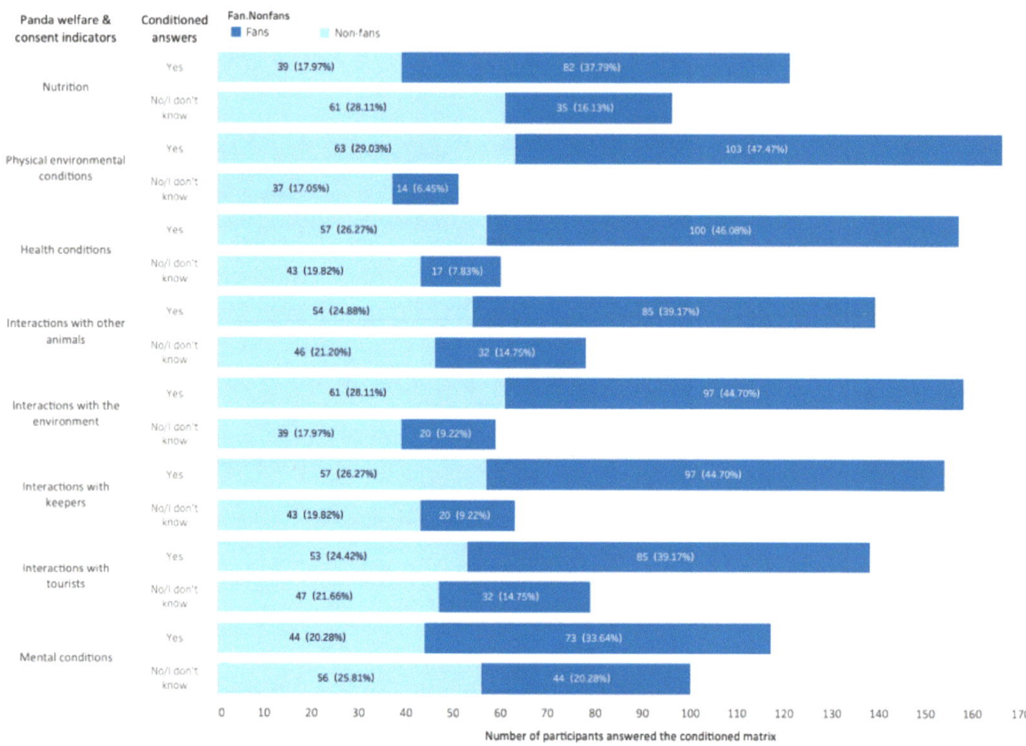

Figure 1. Conditioned answers on eight indicators of fans and non-fans.

3. Results

3.1. Panda Consent Indicators

Figure 2 shows that identified panda fans paid greater attention to all panda consent indicators. In contrast, non-fans were more likely to suggest that they did not know or care about these indicators. This finding demonstrates that panda fans, in general, have invested more time and effort into knowing about pandas. We also note that fans (47.47%) and non-fans (29.03%) tend to carefully investigate the pandas' physical environmental conditions. Fans (33.64%) addressed the mental conditions of the pandas the least, while non-fans (17.79%) paid the least attention to the panda's nutrition.

Regarding participants' relationship with pandas (Figure 2), 195 fans and non-fans answered the question: "how do you describe your intimate relationship with the panda" (22 non-fans skipped this question). Results indicate that most panda fans (47.86%) and non-fans (31.00%) tended to view pandas as their friends. More non-fans (25.00%) than fans (20.51%) considered the panda as mere tourist attractions. Fans (14.53%) also have a stronger tendency to view pandas as family members than non-fans (6.00%). The fans, in general, demonstrated a stronger and more intimate connection with the pandas by including the animal as friends and families.

Figure 2. Participants' relationship with pandas.

3.2. Panda Welfare and Consent Indicators in Detail

Table 3 and Figure 3 compare the performances of panda fans and non-fans on panda welfare and consent indicators. We note that, on average, all indicators were given more than 5 points (between moderately positive to extremely positive) in this study, showcasing the care and attention that giant pandas receive at the panda base. The results further affirm Li's [52] observation that giant pandas have been given unprecedented welfare benefits in China since the 1990s. Table 3 affirms that the giant pandas are, at least from the point of view of visitors, experiencing the best institutional facilities and relationship with humans possible. We note that panda fans scaled all indicators on average higher than non-panda fans, except for enclosure enrichments.

Table 3. *t*-tests on panda welfare and consent indicators and details.

Indicators		Description	Fans			Non-Fans				
			n	*M*	SD	*n*	*M*	SD	*t*	*p*
Nutrition		1a. Water supply	82	5.78	0.52	39	5.54	0.82	1.684	0.098
		1b. Food quantity	82	5.71	0.64	39	5.64	0.87	0.472	0.638
		1c. Food quality	82	5.77	0.50	39	5.54	0.88	1.511	0.137
		1d. Food variety	82	5.76	0.60	39	5.67	0.77	0.697	0.487
Physical environmental conditions		2a. Spacious enclosures	103	5.49	0.92	63	5.52	1.05	1.495	0.137
		2b. Enriched enclosures	103	5.25	1.19	63	5.30	1.09	0.266	0.790
		2c. Enclosures protect pandas against thermal extremes	103	5.50	0.88	63	5.40	0.93	0.683	0.496
Health conditions		3a. The panda does not suffer from acute and chronic injuries	100	5.51	0.89	57	5.35	1.20	0.944	0.347
		3b. Healthcare treatment	100	5.67	0.64	57	5.54	0.83	1.070	0.286
		3c. The panda is fit and attends to exercises	100	5.68	0.68	57	5.54	0.87	1.089	0.287
Behavioral interactions	Environment	4.1a. Enclosures reflect the panda's life stage	97	5.59	0.83	61	5.41	0.99	1.219	0.225
		4.1b. Males and females are separated	97	5.61	0.73	61	5.46	1.01	1.002	0.319
		4.1c. Multiple choices to move and be active	97	5.62	0.77	61	5.44	0.94	1.226	0.223
	Other animals	4.2a. The panda socializes with a group	85	5.49	0.95	54	5.44	1.04	0.290	0.772
		4.2b. The panda plays with its peers	85	5.62	0.72	54	5.44	0.98	1.154	0.252
		4.2c. The panda can enjoy solidarity	85	5.45	0.93	54	5.12	1.30	1.427	0.145
	Keepers	4.3a. Keepers are kind and nice	97	5.80	0.51	57	5.72	0.70	0.862	0.390
		4.3b. Keepers are qualified and skillful	97	5.81	0.53	57	5.72	0.73	0.938	0.350
	Tourists	4.4a. The panda is at ease despite the presence of tourists	85	5.64	0.81	53	5.53	0.85	0.740	0.461
		4.4b. The panda can respond to tourists	85	5.56	0.96	53	5.25	1.11	1.685	0.094
Consent indicators		5a. Pandas give consent to being used as tourist attractions	73	5.27	1.28	44	5.05	1.29	0.930	0.354
		5b. It is possible for pandas to express to humans their consent to being used as tourist attractions	73	5.63	0.72	44	5.09	1.22	2.674	0.010
		8c. It is possible for humans to recognize if pandas are expressing consent	73	5.48	0.96	44	5.30	1.17	0.923	0.358

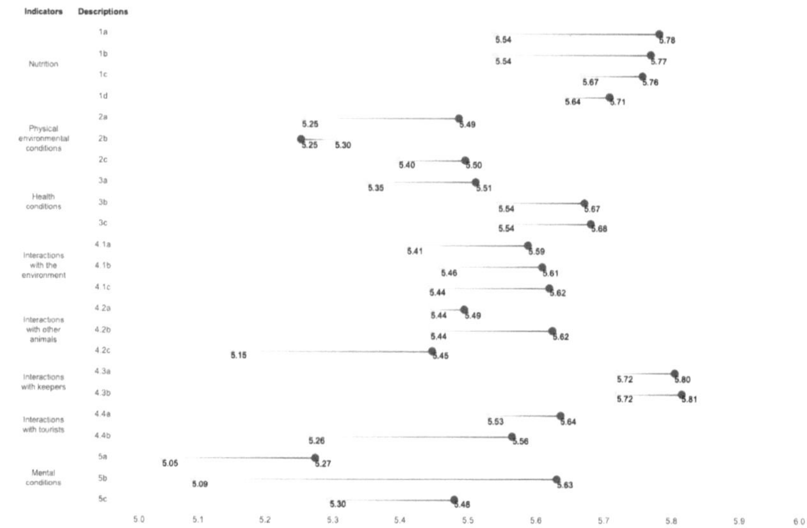

Figure 3. Means of panda welfare and consent indicators by panda fans and non-fans.

The results indicate that fans and non-fans ranked the pandas' interaction with keepers highest (4.3a and 4.3b). Fans believed that keepers are nice and kind (M = 5.80) and skillful and well-trained (M = 5.81). Non-fans had also rated keepers highly. However, we note that no significant difference can be observed in this indicator between fans and non-fans ($p > 0.05$).

Panda fans rated the enclosures' enrichment (2b) lowest (M = 5.25) in all fan-generated scales. Although panda fans and non-fans did not demonstrate a significant difference in evaluating the enclosure enrichments ($p = 0.790$), it is the only time that fans generated a lower scale than non-fans (M = 5.30) on all indicators. In addition, in all evaluations non-fans were least positive in evaluating the panda's consent to being a tourist attraction (5a) (M = 5.05). Additionally, we note that fans (M = 5.27) were more conserved in evaluating 5a.

Panda fans and non-fans only differ significantly ($p = 0.010$) in the possibility that pandas could express to humans their consent to being used as tourist attractions (5b). Fans believed firmly in the pandas' ability to express their feelings to human beings while being a tourist attraction (M = 5.63), whereas non-fans only confirmed the panda's ability moderately.

4. Discussion

In their evaluation of Zoo Atlanta's Giant Panda Conservation Center, Wilson et al. [84] show that 23 staff and 145 zoo visitors evaluated the exhibit favorably in terms of its design, construction, enclosure, working experiences, and visitor experiences. On a 5-point scale, the staff offered a mean rating of 3.64 for the panda enclosures and their own working space, while visitors provided an average rating of 4.50 for their visits. The favored evaluation results justify Zoo Atlanta's goal as the organization strives to provide the best possible environment for the pandas and offer visitors educational opportunities [84]. Barua ([5], p. 9) studied the panda enclosure enrichment projects executed around zoos and organizations to improve the enclosures' biological relevance. Nicholls [31] explains that the enriched enclosures provided for pandas aim at improving animal welfare on the one hand and increasing the species' productivity on the other hand. Our study is confirmation of the above research results. Giant pandas at the Panda Base, from the point of view of the tourists, are well taken care of and enjoy a very comfortable lifestyle [52]. All animal welfare indicators fall between moderately positive to extremely positive. We suggest that the current welfare indicators of giant pandas substantiate and justify the possible efforts and investment that the tourism industry could use to endorse animals in this industry.

A critical finding of this study is that panda fans and non-fans were "invested" at different (and significant) levels in pandas. Panda fans evaluated the details of welfare and consent indicators at a higher level than non-fans, who tended to be less concerned about the pandas. However, our study shows that panda fans' evaluation of the panda's welfare does not differ from non-fans substantially in all indicators, except for the panda's expression of consent to being used as a tourist attraction (5b). Panda fans believed that pandas willingly play the role of a tourist attraction and would be able to express this willingness. The study also shows that fans more passionately build intimate relationships with the pandas as they identified the animals as their friends and family members. However, this intimate relationship between panda fans and pandas does not support the hypothesis that panda fans know or understand pandas better than non-fans. Instead, this panda fandom could mean that fans deeply anthropomorphize pandas [85], as fans believe they could experience the panda's feelings and subjectivity. The tourist–panda relationship, building on the notion of fandom, may be a starting point for constructing more ethical contact between the two species. However, the fan's love and passion towards the animal celebrity, it seems, can also be blind without proper education programs. More studies need to determine whether the panda fandom could further support animal welfare indicators evaluation, especially when international tourism is allowed in China and at the Panda Base.

Keepers, rather than pandas, received the highest credit from panda fans and non-fans in our study. Keepers are believed to be professionals who are kind and nice to the animals. In contrast, neither fans nor non-fans merited the panda's subjectivity highly, as both had the least concerns with the panda's mental conditions. Fennell [33] writes that animals can speak for themselves, but humans have missed these "voices" because of ignorance and self-interest. The results in this study support Fennell's statement, showing that panda fans might also have, wittingly or unwittingly, prioritized human self-interest over the interests of their fanned animal. The professional care, kindness, and attention giant pandas have received mirror the idealized harmony and companionship between the animal world and human beings, but these human efforts and idealizations can also suppress giant panda voices.

The findings of our study indicate that animal welfare evaluations of the nature conducted at the panda base were a novel experience for the Chinese public. Even though the researchers carefully translated the indicators into lay words that participants could understand, the researchers still received questions from participants that some questions/statements were completely new to these participants. We selected panda fans as the target research group because we believed that fans were more concerned about panda welfare and could reflect more thoughtfully on the indicators. While animal protection and welfare are becoming more important in China, Li [52] suggests that the nation is "almost 200 years behind the most developed countries in terms of animal protection legislation". As such, we, as researchers, felt a deep need to enhance the public's literacy on animal welfare. A good starting point for this discussion and action would be participants such as panda fans, who have invested in these animals to a greater extent than non-fans. Whether the role of animal fans can become a springboard for further animal protection and welfare enhancement demands more intensive study.

We further contend that captive animal sites, like the Panda Base, generate passionate, emotional responses to pandas, while the pandas themselves are being used in ways that are mainly focused on pleasure and profit, even though conservation is a desired end. As such, we need to move beyond the self in turning our attention to animals not simply as objects to satisfy our curiosity, or indeed political and economic ends, but rather as subjects that exist independent of these ends—or at least alongside such ends. Time will tell how successful breeding sites will be at repopulating the panda's wild habitat, and the fate of the panda at such time. Until then, the panda exists in a virtual and hybrid state where the few are required to be "flag bearers" for the success of the species and the amusement of followers.

5. Conclusions

The purpose of this paper was to investigate how deeply Chinese "fans" and "non-fans" consider the welfare of captive giant pandas, and to better understand if panda fans and non-fans differ in their assessment of whether giant pandas *consent* to being used as tourist attractions. Our general conclusion is that panda fans paid greater attention to all panda consent indicators over their non-fan counterparts. Our approach, especially the latter question, provides new inroads into the relationship that people have with charismatic animals like pandas. We view our research as a benchmark from which to build more of an understanding of the human–panda connection, and if the ubiquitous love we have for pandas translates into a better life for these animals down the road. Furthermore, we argue that cross-cultural research will provide a better foundation of knowledge in this area (as tourism to China and the Panda base reopens in the future), as well as efforts to refine the constructs investigated in our study.

Supplementary Materials: The following supporting information can be downloaded at: https://www.mdpi.com/article/10.3390/ani13040718/s1, Table S1: Domain 1: Nutrition and affects; Table S2: Domain 2: Physical environmental conditions and affects; Table S3: Domain 3: Health conditions and affects; Table S4: Domain 4: Behavioural interactions and affects.

Author Contributions: Conceptualization, D.A.F. and Y.G.; methodology, Y.G.; software, Y.G.; validation, D.A.F. and Y.G.; formal analysis, D.A.F. and Y.G.; investigation, Y.G.; resources, D.A.F. and Y.G.; data curation, Y.G.; writing—original draft preparation, D.A.F. and Y.G.; writing—review and editing, D.A.F.; visualization, Y.G.; supervision, D.A.F.; project administra-tion, D.A.F.; funding acquisition, D.A.F. and Y.G. All authors have read and agreed to the published version of the manuscript.

Funding: This research and the APC were funded by the Central Guidance on Local Science and Technology Development Fund of Sichuan Province grant number [2021ZYD0156].

Institutional Review Board Statement: Ethical review and approval were waived for this study because data are properly anonymized and informed consent was obtained at the time of original data collection.

Informed Consent Statement: Informed consent was obtained from all subjects involved in the study.

Data Availability Statement: Data generating results of the paper are available upon request.

Conflicts of Interest: The authors declare no conflict of interest.

References

1. Anderlini, J. Panda Politics: The Hard Truth about China's Cuddliest Diplomat. Available online: https://www.ft.com/content/8a04a532-be92-11e7-9836-b25f8adaa111 (accessed on 3 November 2017).
2. Songster, E.E. *Panda Nation: The Construction and Conservation of China's Modern Icon*; Oxford University Press: Oxford, UK, 2018.
3. Wang, A. The Significance of the Panda. Available online: https://thedailychina.org/the-significance-of-the-panda/ (accessed on 25 May 2017).
4. WWF. Why Should We Save the Giant Panda? 2020. Available online: https://wwf.panda.org/discover/knowledge_hub/endangered_species/giant_panda/panda/why_we_save_the_giant_panda/ (accessed on 10 January 2023).
5. Barua, M. Affective economies, pandas, and the atmospheric politics of lively capital. *Trans. Inst. Br. Geogr.* **2020**, *45*, 678–692. [CrossRef]
6. Hu, L.; Yu, L. The Number of Captive Pandas in the World Has Doubled in the Last Decade. Available online: http://www.news.cn/mrdx/2022-10/12/c_1310669406.htm (accessed on 12 October 2022).
7. Wei, F.; Costanza, R.; Dai, Q.; Stoeckl, N.; Gu, X.; Farber, S.; Nie, Y.; Kubiszewski, I.; Hu, Y.; Swaisgood, R.; et al. The Value of Ecosystem Services from Giant Panda Reserves. *Curr. Biol.* **2018**, *28*, 2174–2180. [CrossRef] [PubMed]
8. Swaisgood, R.R.; Wang, D.; Wei, F. Panda downlisted but not out of the woods. *Conserv. Lett.* **2018**, *11*, e12355. [CrossRef]
9. Brookfield Zoo. Milestones. 2022. Available online: https://www.czs.org/Brookfield-ZOO/About/History.aspx (accessed on 27 October 2022).
10. Chinoy, M. Pandamonium, or everything you always wanted to know about pandas. In *New China*; Spring, 1975; pp. 15–17.
11. Schaller, G.B. *The Last Panda*; University of Chicago Press: Chicago, IL, USA, 1994.

12. Huang, Z. Panda Base Caps Number of Visitors during Holiday. Available online: https://www.chinadaily.com.cn/a/201909/24/WS5d89535da310cf3e3556d067.html (accessed on 4 July 2021).
13. Bueddefeld, J.; Benbow, M. 12 The greening of polar bears: Lively commodities in a climate change economy. In *Exploring Non-Human Work in Tourism*; De Gruyter Oldenbourg: Berlin, Germany, 2021; pp. 207–224.
14. D'Souza, J.; Dawson, J.; Groulx, M. Last chance tourism: A decade review of a case study on Churchill, Manitoba's polar bear viewing industry. *J. Sustain. Tour.* **2021**, *31*, 14–31. [CrossRef]
15. Dybsand, H.N.H. In the absence of a main attraction—Perspectives from polar bear watching tourism participants. *Tour. Manag.* **2020**, *79*, 104097. [CrossRef]
16. Lemelin, R.H. The gawk, the glance, and the gaze: Ocular consumption and polar bear tourism in Churchill, Manitoba, Canada. *Curr. Issues Tour.* **2006**, *9*, 516–534. [CrossRef]
17. Shi, Q.; Gao, Y. *Sustainable Development of Rural Household Economy: Transition of Ten Villages in Zhejiang, China, 1986–2002*; Shanghai Jiao Tong University Press Springer Nature: Shanghai, China, 2020.
18. Yudina, O.; Grimwood, B.S. Situating the wildlife spectacle: Ecofeminism, representation, and polar bear tourism. *J. Sustain. Tour.* **2016**, *24*, 715–734. [CrossRef]
19. Yudina, O.; Grimwood, B.S.; Berbary, L.A.; Mair, H. The gendered natures of polar bear tourism. *Tour. Cult. Commun.* **2018**, *18*, 51–66. [CrossRef]
20. Kojola, I.; Heikkinen, S. Problem brown bears Ursus arctos in Finland in relation to bear feeding for tourism purposes and the density of bears and humans. *Wildl. Biol.* **2012**, *18*, 258–263. [CrossRef]
21. Penteriani, V.; Lamamy, C.; Kojola, I.; Heikkinen, S.; Bombieri, G.; del Mar Delgado, M. Does artificial feeding affect large carnivore behaviours? The case study of brown bears in a hunted and tourist exploited subpopulation. *Biol. Conserv.* **2021**, *254*, 108949. [CrossRef]
22. Penteriani, V.; López-Bao, J.V.; Bettega, C.; Dalerum, F.; del Mar Delgado, M.; Jerina, K.; Kojola, I.; Krofel, M.; Ordiz, A. Consequences of brown bear viewing tourism: A review. *Biol. Conserv.* **2017**, *206*, 169–180. [CrossRef]
23. Carder, G.; Plese, T.; Machado, F.C.; Paterson, S.; Matthews, N.; McAnea, L.; D'Cruze, N. The impact of 'selfie tourism' on the behaviour and welfare of brown-throated three-toed sloths. *Animals* **2018**, *8*, 216. [CrossRef] [PubMed]
24. Page, A. "This Baby Sloth Will Inspire You to Keep Going": Capital, Labor, and the Affective Power of Cute Animal Videos. In *The Aesthetics and Affects of Cuteness*; Routledge: London, UK, 2016; pp. 85–104.
25. Campbell, M.; Lancaster, B.L. Public attitudes toward black bears (*Ursus americanus*) and cougars (*Puma concolor*) on Vancouver Island. *Soc. Anim.* **2010**, *18*, 40–57. [CrossRef]
26. Massé, S.; Dussault, C.; Dussault, C.; Ibarzabal, J. How artificial feeding for tourism—Watching modifies black bear space use and habitat selection. *J. Wildl. Manag.* **2014**, *78*, 1228–1238. [CrossRef]
27. Huang, Q.; Fei, Y.; Yang, H.; Gu, X.; Songer, M. Giant Panda National Park, a step towards streamlining protected areas and cohesive conservation management in China. *Glob. Ecol. Conserv.* **2020**, *22*, e00947. [CrossRef]
28. Tang, J.; Wang, C.; Zhang, H.; Zhao, J.; Guo, W.; Mishra, S.; Kong, F.; Zeng, B.; Ning, R.; Li, D. Gut microbiota in reintroduction of giant panda. *Ecol. Evol.* **2020**, *10*, 1012–1028. [CrossRef] [PubMed]
29. Wei, F.; Swaisgood, R.; Hu, Y.; Nie, Y.; Yan, L.; Zhang, Z.; Qi, D.; Zhu, L. Progress in the ecology and conservation of giant pandas. *Conserv. Biol.* **2015**, *29*, 1497–1507. [CrossRef]
30. Yang, Z.; Gu, X.; Nie, Y.; Huang, F.; Huang, Y.; Dai, Q.; Hu, Y.; Yang, Y.; Zhou, X.; Zhang, H. Reintroduction of the giant panda into the wild: A good start suggests a bright future. *Biol. Conserv.* **2018**, *217*, 181–186. [CrossRef]
31. Nicholls, H. *The Way of the Panda: The Curious History of China's Political Animal*; Pegasus Books: New York, NY, USA, 2012.
32. Guo, Y. Cute, but get up and work! The biophilia hypothesis in tourists' linguistic interactions with pandas. In *Exploring Non-Human Work in Tourism: From Beasts of Burden to Animal Ambassadors*; De Gruyter Oldenbourg: Berlin, Germany, 2021; p. 225.
33. Fennell, D.A. Animal-informed consent: Sled dog tours as asymmetric agential events. *Tour. Manag.* **2022**, *93*, 104584. [CrossRef]
34. Lee, H.; Yan, L. Selling cute destinations to East Asia. *J. Hosp. Tour. Insights* **2020**, *4*, 282–299. [CrossRef]
35. Lv, X.; Liu, Y.; Luo, J.; Liu, Y.; Li, C. Does a cute artificial intelligence assistant soften the blow? The impact of cuteness on customer tolerance of assistant service failure. *Ann. Tour. Res.* **2021**, *87*, 103114. [CrossRef]
36. Sun, J.; Nazlan, N.H.; Leung, X.Y.; Bai, B. "A cute surprise": Examining the influence of meeting giveaways on word-of-mouth intention. *J. Hosp. Tour. Manag.* **2020**, *45*, 456–463. [CrossRef]
37. Winter, C. A review of research into animal ethics in tourism: Launching the Annals of Tourism Research curated collection on animal ethics in tourism. *Ann. Tour. Res.* **2020**, *84*, 102989. [CrossRef]
38. Fennell, D.A.; Sheppard, V. Tourism, animals and the scales of justice. *J. Sustain. Tourism.* **2021**, *29*, 314–335. [CrossRef]
39. Duffy, R.; Moore, L. Global regulations and local practices: The politics and governance of animal welfare in elephant tourism. *J. Sustain. Tour.* **2011**, *19*, 589–604. [CrossRef]
40. Lovelock, B. An introduction to consumptive wildlife tourism. In *Tourism and the Consumption of Wildlife: Hunting, Shooting and Sport Fishing*; Lovelock, B., Ed.; Routledge: London, UK, 2008; pp. 3–30.
41. Dobson, J. Towards a Utilitarian Ethic for Marine Wildlife Tourism. *Tour. Mar. Environ.* **2011**, *7*, 213–222. [CrossRef]
42. Burns, G.L.; MacBeth, J.; Moore, S. Should dingoes die? Principles for engaging ecocentric ethics in wildlife tourism management. *J. Ecotourism* **2011**, *10*, 179–196. [CrossRef]
43. Cavalieri, P. The meaning of the great ape project. *Politics Anim.* **2015**, *1*, 16–34.

44. Äijälä, M. Knowing through interspecies relationality in tourism? Animal agency in human-sled dog encounters. *Finn. J. Tour.* **2019**, *15*, 45–50. [CrossRef]
45. García-Rosell, J.C.; Tallberg, L. Animals as tourism stakeholders: Huskies, reindeer, and horses working in Lapland. In *Exploring Non-Human Work in Tourism: From Beast of Burden to Animal Ambassadors*; Rickly, J.M., Kline, C., Eds.; De Gruyter: Berlin, Germany, 2021; pp. 103–121.
46. Kenehan, S. The moral status of animal research subjects in industry: A stakeholder analysis. In *Animal Experimentation: Working towards a Paradigm Change*; Herrmann, K., Jayne, K., Eds.; Brill: Boston, MA, USA, 2019; pp. 209–223.
47. Evan, W.M.; Freeman, E. A stakeholder theory of the modern corporation: Kantian Capitalism. In *Ethical Theory and Business*; Prentice-Hall: Hoboken, NJ, USA, 1988; pp. 97–106.
48. Szydlowski, M. Asking consent from pachyderm persons: Facing ethical complexities in multispecies research. In *Animal Life and Human Culture: Anthrozoology Studies*; Frasin, I., Bodi, G., Bulei, S., Vasiliu, C.D., Eds.; Presa Universitara Clujeana: Cluj-Napoca, Romania, 2022.
49. Mellor, D.J.; Beausoleil, N.J.; Littlewood, K.E.; McLean, A.N.; McGreevy, P.D.; Jones, B.; Wilkins, C. The 2020 five domains model: Including human-animal interactions in assessments of animal welfare. *Animals* **2020**, *10*, 1870. [CrossRef] [PubMed]
50. Farm Animal Welfare Committee. Assessment of Farm Animal Welfare—Five Freedoms and a Life Worth Living. 2022. Available online: https://www.gov.uk/government/groups/farm-animal-welfare-committee-fawc#assessment-of-farm-animal-welfare---five-freedoms-and-a-life-worth-living. (accessed on 15 October 2022).
51. Blewitt, J. What's new pussycat? A genealogy of animal celebrity. *Celebr. Stud.* **2013**, *4*, 325–338. [CrossRef]
52. Li, P.J. *Animal Welfare in China: Culture, Politics and Crisis*; Sydney University Press: Sydney, NSW, Australia, 2021.
53. Mackay, S.; Horton, T.; Persons, N.; Hall, A.; Mitchell, E.; Schunke, A.; Wright, J.; Palmer, J.; Reed, A.; Fraser, H.; et al. Adelaide Zoo Giant Panda Forest. 2009. Available online: https://www.archdaily.com/70657/adelaide-zoo-giant-panda-forest-hassell (accessed on 10 December 2022).
54. PhysOrg. Panda Stars Get First Taste of Life in The Netherlands. Available online: https://phys.org/news/2017-05-panda-stars-life-netherlands.html (accessed on 30 May 2017).
55. Greuner, T. Ueno Zoo has opened a new panda enclosure that resembles the bears' habitat. *TimeOut* 11 September 2020. Available online: https://www.timeout.com/tokyo/news/ueno-zoo-has-opened-a-new-panda-enclosure-that-resembles-the-bears-habitat-091120 (accessed on 6 June 2022).
56. Sandvoss, C. *Fans: The Mirror of Consumption*; Polity: New York, NY, USA, 2005.
57. Diamond, J. Evolution, consequences and future of plant and animal domestication. *Nature* **2002**, *418*, 700–707. [CrossRef]
58. Duffett, M. *Understanding Fandom: An Introduction to the Study of Media Fan Culture*; Bloomsbury Publishing: New York, NY, USA, 2013.
59. Brooker, W. *Using the Force: Creativity, Community and Star Wars Fans*; Bloomsbury Publishing: New York, NY, USA, 2002.
60. Busse, K.; Gray, J. Fan cultures and fan communities. In *The Handbook of Media Audiences*; Blackwell Publishing: Oxford, UK, 2011; pp. 425–443.
61. Gray, J. *Show Sold Separately: Promos, Spoilers, and Other Media Paratexts*; NYU Press: Manhattan, NY, USA, 2010.
62. Halverson, E.R.; Halverson, R. Fantasy baseball: The case for competitive fandom. *Games Cult.* **2008**, *3*, 286–308. [CrossRef]
63. Hills, M. Sherlock's Epistemological Economy and the Value of 'Fan' Knowledge. How Producer-Fans Play the (Great) Game of Fandom. In *Essays on the BBC Series. Sherlock and Transmedia Fandom*; Stein, L.E., Busse, K., Eds.; McGraw Hill: New York, NY, USA, 2012; pp. 27–40.
64. Hills, M. The expertise of digital fandom as a 'community of practice' Exploring the narrative universe of Doctor Who. *Convergence* **2015**, *21*, 360–374. [CrossRef]
65. Huettermann, M.; Uhrich, S.; Koenigstorfer, J. Components and outcomes of fan engagement in team sports: The perspective of managers and fans. *J. Glob. Sport Manag.* **2019**, *7*, 1–32. [CrossRef]
66. Cheung, C.-K.; Yue, X.D. Identity achievement and idol worship among teenagers in Hong Kong. *Int. J. Adolesc. Youth* **2003**, *11*, 1–26. [CrossRef]
67. Karlin, J.G. Through a looking glass darkly: Television advertising, idols, and the making of fan audiences. In *Idols and Celebrity in Japanese Media Culture*; Springer: London, UK, 2012; pp. 72–93.
68. Katayama, R. Idols, celebrities, and fans at the time of post-catastrophe. *Celebr. Stud.* **2021**, *12*, 267–281. [CrossRef]
69. Yoon, K. Korean Wave | Cultural Translation of K-Pop Among Asian Canadian Fans. *Int. J. Commun.* **2017**, *11*, 17. Available online: https://ijoc.org/index.php/ijoc/article/view/6303/2053 (accessed on 10 January 2023).
70. Carr, N.; Broom, D.M. *Tourism and Animal Welfare*; CABI: Wallingford, UK, 2018.
71. Mulier, C.; Müller, H. *Draft Horses in Viticulture: Conditions for the Co-Creation of Value*; The Deutsce Nationalbibliothek: Frankfurt, Germany, 2022.
72. Serpell, J.A. How happy is your pet? The problem of subjectivity in the assessment of companion animal welfare. *Anim. Welf.* **2019**, *28*, 57–66. [CrossRef]
73. Verga, M.; Michelazzi, M. Companion animal welfare and possible implications on the human-pet relationship. *Ital. J. Anim. Sci.* **2009**, *8*, 231–240. [CrossRef]
74. Fernandes, J.G.; Olsson, I.A.S.; de Castro, A.C.V. Do aversive-based training methods actually compromise dog welfare? A literature review. *Appl. Anim. Behav. Sci.* **2017**, *196*, 1–12. [CrossRef]

75. Lefebvre, D.; Diederich, C.; Delcourt, M.; Giffroy, J.-M. The quality of the relation between handler and military dogs influences the efficiency and welfare of dogs. *Appl. Anim. Behav. Sci.* **2007**, *104*, 49–60. [CrossRef]
76. Burn, C.C. What is it like to be a rat? Rat sensory perception and its implications for experimental design and rat welfare. *Appl. Anim. Behav. Sci.* **2008**, *112*, 1–32. [CrossRef]
77. Neville, V.; Mounty, J.; Benato, L.; Hunter, K.; Mendl, M.; Paul, E.S. Thinking outside the lab: Can studies of pet rats inform pet and laboratory rat welfare? *Appl. Anim. Behav. Sci.* **2022**, *246*, 105507. [CrossRef]
78. Sanders, C.R. The impact of guide dogs on the identity of people with visual impairments. *Anthrozoös* **2000**, *13*, 131–139. [CrossRef]
79. Leite, Â.; Ramires, A.; Costa, R.; Castro, F.; Sousa, H.F.P.E.; Vidal, D.G.; Dinis, M.A.P. Comparing psychopathological symptoms in Portuguese football fans and non-fans. *Behav. Sci.* **2020**, *10*, 85. [CrossRef]
80. Reysen, S.; Plante, C.N.; Roberts, S.E.; Gerbasi, K.C. A social identity perspective of personality differences between fan and non-fan identities. *World* **2015**, *2*, 91–103. [CrossRef]
81. Sun, Y.; Lu, X.; Williams, M.; Thompson, W.F. Implicit violent imagery processing among fans and non-fans of music with violent themes. *R. Soc. Open Sci.* **2019**, *6*, 181580. [CrossRef]
82. Teng, W.; Su, Y.; Liao, T.-T.; Wei, C.-L. An exploration of celebrity business ventures and their appeal to fans and non-fans. *J. Retail. Consum. Serv.* **2020**, *54*, 102004. [CrossRef]
83. Banz, M.; Grosswiele, L.; Huber, J.; Probst, F. More fans at any cost? Analyzing the economic effects of the ratio of fans to non-fans in a customer portfolio considering electronic word of mouth. *J. Decis. Syst.* **2016**, *25*, 193–213. [CrossRef]
84. Wilson, M.; Kelling, A.; Poline, L.; Bloomsmith, M.; Maple, T. Post-occupancy evaluation of zoo Atlanta's Giant Panda Conservation Center: Staff and visitor reactions. *Zoo Biol. Publ. Affil. Am. Zoo Aquar. Assoc.* **2003**, *22*, 365–382. [CrossRef]
85. Kennedy, J.S. *The New Anthropomorphism*; Cambridge University Press: Cambridge, MA, USA, 1992.

animals

MDPI

Article

Mandan, Hidatsa, and Arikara Nation Perspectives on Rez Dogs on the Fort Berthold Reservation in North Dakota, U.S.A.

Alexandra Cardona [1], Sloane M. Hawes [1], Jeannine Cull [2], Katherine Connolly [1], Kaleigh M. O'Reilly [1], Liana R. Moss [1], Sarah M. Bexell [1], Michael Yellow Bird [2] and Kevin N. Morris [1,*]

[1] Institute for Human-Animal Connection, Graduate School of Social Work, University of Denver, Denver, CO 80208, USA
[2] Department of Social Work, University of Manitoba, Winnipeg, MB R3T 2N2, Canada
* Correspondence: kevin.morris@du.edu

Simple Summary: Free-roaming dogs, also referred to as reservation dogs or rez dogs, hold important and unique roles in Indigenous communities. The purpose of this study is to document the cultural significance of rez dogs, challenges related to rez dogs, and community-specific solutions to rez dog issues affecting community health and safety from the perspective of 14 members of the Mandan, Hidatsa, and Arikara (MHA) Nation, also referred to as the Three Affiliated Tribes (TAT), who live on the Fort Berthold reservation in North Dakota, U.S.A. The primary intervention areas described by the participants included: culturally relevant information sharing, improved animal control policies and practices, and improved access to veterinary care and other animal services.

Abstract: The research on the relationships between free-roaming dogs, also referred to as reservation dogs or rez dogs, and Indigenous communities is extremely limited. This study aimed to document the cultural significance of rez dogs, challenges related to rez dogs, and community-specific solutions for rez dog issues affecting community health and safety from members of the Mandan, Hidatsa, and Arikara (MHA) Nation, also referred to as the Three Affiliated Tribes (TAT), who live on the Fort Berthold reservation in North Dakota, U.S.A. One hour semi-structured interviews with 14 community members of the MHA Nation were conducted in 2016. The interviews were analyzed via systematic and inductive coding using Gadamer's hermeneutical phenomenology. The primary intervention areas described by the participants included: culturally relevant information sharing, improved animal control policies and practices, and improved access to veterinary care and other animal services.

Keywords: indigenous knowledge; colonization; rez dogs; reservation dogs; free-roaming dogs; cultural responsiveness; companion animals; access to care; animal control

1. Introduction

Domestic dogs (*Canis familiaris*) and humans have cohabitated for thousands of years, and dogs have been readily integrated as companions, guardians, and working partners in many societies around the world [1–4]. According to recent estimates, there are between 700 million and 1 billion dogs globally, and about 70–75% of dogs are considered free roaming [5–7]. The World Society for the Protection of Animals defines free-roaming dogs as "dogs that are in public areas and not currently under direct control" [8]. This definition is typically applied to owned dogs who are allowed to roam freely, dogs who were previously owned but have since become lost or abandoned, and dogs who may have never been owned [9]. Some of these free-roaming dogs live within and alongside Indigenous communities in North America and hold a unique and important status in this context. In an Indigenous context, free-roaming dogs can also be referred to as "rez dogs", short for reservation dogs, because of the land on which they roam [10]. While the imposition of settler colonialism and the impact of modernization have evolved and changed traditional practices related to rez dogs in Indigenous communities, rez dogs continue to be an integral part of many Indigenous Peoples' social environment, carrying deep cultural meaning, spiritual significance, and roles that extend further than companionship [2,11,12].

There is a critical need to address concerns regarding free-roaming dogs' impact on community health and safety (e.g., dog bites, aggressive behavior, food competition or predation on native and endemic wildlife, livestock predation, and zoonotic disease) [2,12–17]. Current management strategies have been limited by their inability to address the complex social, cultural, and structural factors that have informed historical and present practices related to free-roaming dogs. For example, recent literature has supported increasing access to veterinary care as a strategy to improve the health and well being of free-roaming dogs [18]. However, efforts to increase access to veterinary care for Indigenous communities have been significantly restricted due to financial constraints, the remote or rural location of Indigenous communities, and the prioritization of other issues, such as "inadequate housing, water supply, and sanitation" [19]. In the face of this resource gap, non-profit animal sheltering and rescue organizations have been the primary providers of animal services in Indigenous communities around the globe [3,20]. Unfortunately, the literature indicates that this approach is ineffective, resource intensive, and unsustainable as a long-term strategy for managing rez dogs within Indigenous communities [21].

To improve the health and safety of their communities and the growing rez dog population, many Indigenous communities in North America have developed population management strategies specific to their communities [22]. Without other effective or sustainable alternatives, some communities have resorted to lethal strategies for management, including culling their free-roaming dog population. However, this practice has been negatively received in most communities [9]. Furthermore, while culling may quickly reduce the population of dogs, it cannot stabilize the population without also being paired with a fertility control program that can reduce the rate of new births [23]. Initial research indicates that population management strategies that incorporate culturally responsive communication and culturally specific education may be more effective and regenerative than culling or fertility control alone [11]. However, community-specific knowledge and perspectives on rez dog population management within Indigenous communities remain poorly documented in the scientific literature.

This study aimed to document the traditional and contemporary beliefs, attitudes, and strategies for management and care of the rez dog population within the Mandan, Hidatsa, and Arikara (MHA) Nation, known together as the Three Affiliated Tribes (TAT), who are currently residing on the Fort Berthold reservation in North Dakota, U.S.A. [24]. Historically, the Fort Berthold reservation has been the home for a large and growing population of rez dogs with few laws or policies that address dog health and dog-related risks for community members [25]. This study originated from MHA Nation members identifying the need for a coordinated effort to address the issue of free-roaming dogs in their community. By documenting these perspectives, this study can be utilized to develop culturally appropriate and community-specific initiatives for the MHA Nation.

2. Materials and Methods

2.1. Design

Gadamer's hermeneutic phenomenological approach was used to gain a better understanding of how the cultural significance of rez dogs within the MHA Nation could be incorporated into future free-roaming dog management strategies. Phenomenology is typically used when there is limited research on a topic and researchers want to gain a collective understanding of a phenomenon based on the lived experiences of a group of individuals [26,27]. Gadamer's hermeneutic phenomenology uses a reciprocal interpretation process to further integrate the knowledge and interpretations of the researchers with the lived experiences of the participants to generate a nuanced description of the phenomenon. Due to the diversity of perspectives included in the analysis process, the findings of studies conducted using this method have cultural and historical accuracy as well as increased scientific credibility [26,27].

Gadamer's hermeneutical phenomenology is based on three philosophical principles: recognizing preconceived notions, pursuing a "fusion of horizons", and developing a "hermeneutic circle" [26,27]. Prior to engaging in the research process, the research team shared and reflected on their own lived experiences around rez dogs ("preconceived notions"). The research team members represent various disciplinary backgrounds (animal sheltering; social work and Indigenous studies; academic research; community organizing; and community outreach) which formed the basis of the preconceived notions that were brought into the study. The "fusion of horizons" principle of Gadamer's hermeneutical phenomenology was operationalized in this study through the development of the interview questions and through the data analysis process (see Sections 2.3 and 2.4). Finally, the "hermeneutic circle" principle was achieved through the peer debriefing sessions of the research team and through the presentation of direct quotes from the participants in this manuscript (see Section 2.5).

The researchers also evaluated the research procedures prior to data collection, striving to minimize bias and potential harm to participants who engaged in the study. Interview questions were evaluated for their capacity to recognize and validate the lived experiences of tribal members by acknowledging privilege and power dynamics in researcher–participant relationships. During consent administration, researchers informed participants of their right to decline to answer any questions and their right to cease participation in the study at any time.

2.2. Participant Recruitment

This study was conducted in partnership with Indigenous community members and tribal leaders from the MHA Nation. The research team collaborated closely with Michael Yellow Bird, MSW, PhD, and a member of the MHA Nation, on most phases of this study, including participant recruitment and the preparation of this manuscript. Dr. Yellow Bird is a Professor and the Dean of the Faculty of Social Work at the University of Manitoba, a scholar of Indigenous studies, and a respected and integrated member of his Indigenous community. Dr. Yellow Bird directed the participant recruitment and provided the research team with a list of MHA Nation community members consisting of his personal and professional connections to contact for the research study.

Recruitment for the study was carried out under a University of Denver Institutional Review Board (IRB) protocol (DU IRB Protocol #767926). Eligibility for the study required participants to be over the age of 18 and members of the MHA Nation. Fourteen participants were selected to participate in the study. The tribal members recruited for this study were community elders, animal control officers, and general community members residing on the Fort Berthold reservation.

The demographics of the participants have been included in Table 1. Of the 14 participants, 6 (43%) self-identified their gender as women and 8 (57%) as men. In addition, all 14 participants (100%) self-identified as Native American, American Indian, or First Nations. Finally, 10 (72%) were 30–59 years old and 4 (28%) were 60 years or older.

Table 1. Demographic characteristics of the interview participants (n = 14).

Participant Demographics		Frequency	Percentage
Gender	Woman	6	43%
	Man	8	57%
	Other (Transgender, Nonbinary, Two-Spirits, etc.)	0	0%
Race/Ethnicity	Native American, American Indian, First Nations	14	100%
	Asian	0	0%
	Black	0	0%
	Hispanic/Latino	0	0%
	White	0	0%
	Two or More Races	0	0%
Age	18–29 years	0	0%
	30–59 years	10	72%
	60+ years	4	28%

2.3. Data Collection

Participants completed a one hour semi-structured interview in August 2016 with four members of the research team. The participants did not receive compensation for their participation. To facilitate the hermeneutic process, the questions for the interview were developed so that participants could reflect deeply and openly on their experiences with rez dogs [26–28]. First, the researchers asked participants about their experiences with rez dogs in the present. Second, participants were asked about the cultural significance of rez dogs. Third, the participants were asked to share their suggestions for how to improve the community relationship with and elevate the cultural significance of rez dogs. The interviews were conducted in the setting that was preferred by the participants (e.g., their personal homes, their place of work, or a public location). Interviews were audio recorded and de-identified by the research team. The de-identified audio files were then transcribed by a second-party vendor (GoTranscript, Ltd., Harrow Middlesex, United Kingdom) and stored in a password-protected database. To ensure the accuracy of the transcriptions, a member of the research team reviewed the transcripts while listening to the recording and made corrections if there were any inconsistencies between the transcripts and the recording. These transcripts were then archived prior to analysis due to staffing changes on the research team.

2.4. Data Analysis

In 2022, the archived transcripts were systematically and inductively coded for themes by five members of the research team. To begin, the researchers read each transcript to determine each individual's overall experience ("immersion") [26]. The second phase involved developing interpretive summaries of the themes that emerged during the interviews. A codebook was then developed based on the identified themes. The codebook was compiled based on the collective consensus among the researchers. A total of 11 primary codes (themes) and 32 subcodes (sub-themes) were included in the codebook. In the fourth phase, interviews were divided among the researchers, and the codebook was applied to the interviews using qualitative coding software (ATLAS.ti, Version 3.19.0; ATLAS.ti, GmbH, Dartmouth, NS, Canada). To reduce researcher bias, each interview was verified for coding accuracy by a second researcher after the initial application of the codebook. The coded data were then reviewed, and themes were organized and grouped according to their common meaning ("aggregation" and "illumination") [26]. Throughout the analysis process, researchers developed an iterative understanding of the phenomenon by evaluating how they have come to experience as a result of listening to the participants' lived experiences ("fusion of horizons") [26,27,29].

Finally, to fulfill the final philosophical principle of the method, researchers continuously reflected on how their meanings connected to and differed from those of the participants ("the hermeneutic circle") [26]. The research team then generated a summary of the participants' shared experiences related to the phenomenon. The resulting list of themes and participant quotes most representative of the interview and the participant's experience have been reported here. Participant interviews are cited at the end of each quote with "P" (participant) followed by the number associated with that participant's interview. For example, Participant 4's interview would be cited as "P4".

2.5. Validity and Trustworthiness

To improve the validity of the findings, the research team engaged in several peer debriefing sessions to critically assess their own interpretations of the data and to identify how assumptions and biases were informing their interpretations of the data [30]. As many direct quotes as possible have been included in the results to facilitate the reader's validation of the findings [26].

3. Results

3.1. Cultural Significance of the Dogs

Throughout the interviews, the MHA Nation members who participated in the study shared the variety of ways in which rez dogs have current and historic significance in their culture. Participants described rez dogs as "sacred" and "part of our culture". This cultural significance manifests in traditional oral stories passed down from generation to generation and in present-day interactions with dogs.

Throughout the interviews, participants discussed rez dogs' role in cultural ceremonies and stories. One tribal member emphasized "We have origins that include the dog. There were songs for it" (P8). Another tribal member shared "I believe we used the dog in ceremony [...] They revived some of our Sun Dances with the Mandan/Hidatsa/Arikara people" (P5). For the MHA Nation, dogs are a living symbol of courage, loyalty, perseverance, protection, and wisdom.

Moreover, participants detailed the reciprocal relationship between dogs and humans. They explained that dogs played several important roles, including serving as protection, guardians, pulling travois to transport materials, and as partners in hunting efforts. One participant highlighted these roles:

> *"In the books on Maxhidiac or Waheenee, there are some stories about the dogs and they talk about the importance of the dogs to us here. Number one for protection, because many times ... the old stories tell of people coming in to raid us. Other tribes would raid us. The dogs helped sound the alarm. The dogs will help in the fight when it occurred. But they were also useful to travois" (P8).*

One tribal member summarized how the cultural significance of dogs translates to an emphasis on the humane treatment of the dogs: "Well, the stories that we heard was, it was how we treated [dogs]. They say if they mistreat [a dog] and killed [a dog] that they will live a terrible life and death" (P6). Another tribal member noted, "Indians don't like to kill dogs [...] That's why they are all over [...] I tell you, what my Uncle's telling me, they're powerful. Dogs protect us from evil" (P9).

Participants also described the present-day relationship between people and dogs on the reservation as deeply interconnected. Participants noted, "You can gauge the health of the community by the health of their animals and vice versa," (P1) and "When people suffer animals suffer too" (P1). Many participants described continuing the reciprocal relationship between dogs and humans in the present day by acknowledging their spiritual significance and by demonstrating care and respect for the dogs.

3.1.1. Role of Dogs

Several participants stated that dogs continue to have a substantial role in the culture of the MHA Nation and they are considered a source of "protection" and/or "part of the family" (P4). Multiple participants referred to the dogs' ability to anticipate danger: "They know when anything bad is going to happen. We should all be taking care of these animals because these animals are going to protect us" (P11). Another participant shared the belief that dogs "know when people are going to die. They're messengers" (P11).

Another tribal member described the integral role of dogs in child care:

> *I raised my brother's kids. The dogs were my babysitters. Through the years, I think, I've had maybe 4 or 5 dogs, and when I was in the house cooking or doing whatever, if I went outside and I look for the kids and I didn't see them, I'd whistle for my dog and whatever house she came around, I know where the kids were because that's where they were. She was my babysitter and she did take really good care of them (P6).*

Participants also noted the role of dogs in ceremonies and traditional stories. One participant tells the story of a village surrounded by soldiers that was led to safety by a female dog and her litter by going under a river (P11). Another traditional story detailed nations of cats and dogs pitted against each other in a generational conflict in which the cats stole the tails of the dogs. This traditional story was used to explain the origin of dog greeting rituals by saying "It looks like they're smelling each other's you-know-whats, but what they are really doing is looking for their tail that they lost" (P1).

3.1.2. Colonization and Cultural Disruption

Many tribal members detailed the effects that colonization and cultural disruption had on the community's relationship with dogs, describing colonization as a "cultural clash" that causes a "disconnect" and that it meant that "we got so distanced from who we were" (P10). One participant summarized, "It's understood that a lot of our culture is not the way it used to be" (P11).

Several participants spoke about the implications of colonization on tribal members' relationships with the dogs: "You're talking about serious culture disruption. Animals, dogs, people, culture. All that sort of stuff, relationships break down" (P14). Additionally, participants stated "The dogs got set aside," (P14) and "People used to be more connected to take care of the dogs" (P8).

3.2. Challenges Related to Free-Roaming Dogs

The interview participants had various perspectives on the extent to which rez dogs represented a challenge for their community. The majority of the participants agreed that dogs were roaming in their community. However, there was disagreement on how many there are and if they pose a negative issue for the community. While some tribal members felt strongly that the dogs were an issue for their community, others felt that the dogs were not an issue at all. Some participants reported that there are fewer roaming dogs than there used to be, while for others, it felt like the number of dogs was the same or increasing. Most tribal members noted that the number of dogs and severity of issues varied by location on the reservation.

Community members listed several factors that contribute to the challenges related to free-roaming dogs on the reservation. Many tribal members cited growth in the human population and an overwhelmed animal control agency as possible reasons for the number of free-roaming dogs on the reservation. One tribal member felt that poverty was the "main link", explaining that the areas with a high prevalence of roaming dogs were also "the areas that tend to have more economic difficulties" (P13). Meanwhile, other community members identified "irresponsible pet ownership" and a small number of "repeat offenders" who choose not to spay/neuter their dogs and/or let their dogs run loose as the cause for the number of free-roaming dogs (P7). Moreover, some tribal members felt that community members might be abandoning puppies once they get older, saying, "they think they are cute when they are tiny, but after they get older, then they just don't care about them" (P4).

Many participants shared a concern that the free-roaming dog population was growing from sources outside the reservation, such as oilfield workers, casino patrons, or other non-enrolled members who abandon their dogs on tribal land. One participant shared their ideas about where the free-roaming dog population originates:

> *[The free-roaming dogs] live in the abandoned houses and trailers, old vehicles, just pretty much anywhere that they can seek shelter. That is when they start gathering in packs and start getting that pack mentality because they're trying to stay warm, they're trying to survive that cold winter (P3).*

Furthermore, there is a lack of consensus on how or if dogs should be contained when they have a designated family or owner. One tribal member explained:

> *[I] heard it stated that some families don't think that animals should be contained. They should have the ability to roam and the freedom to go about their business. They don't think that they should contain them. And whether that's just personal views or social views, I don't know, it's hard to say (P13).*

Lastly, one participant discussed how the condition of the kenneling facility on the reservation serves as a barrier to housing dogs:

> *One of our major limiting factors is our kennel—the condition of our kennel. The director before me had this kennel built out of Quonset, and it's by no means adequate to be a kennel, but we made do with what they have (P7).*

Animal Control

Several participants highlighted their desire for improved animal control policies that could address the community's health and safety concerns. In particular, participants shared their worry about the free-roaming dogs' potential to harm humans as a key issue to address. Some tribal members likened the rez dogs to "wolf packs who attack people as well as other dogs and animals, but back off if you stand your ground" (P2). Another participant said that the rez dogs were "constantly running on the road" (P11), while others described the rez dogs as "hungry" and said "they devour everything" (P12). Many tribal members characterized the rez dogs as "aggressive" (P8), sharing that some dogs bark at people or chase them which prevents community members from being able to walk outside without fear. One participant shared:

> *I used to be an avid outside walker. We've since purchased an indoor treadmill because depending on the breed and the size, the dogs are dangerous. If they don't know you, they may not be that nice. I've had family members bitten, chased where they had to jump on top of something. But as far as recreational activities being able to walk and having that freedom of your community, I see the dogs as being difficult (P5).*

Participants discussed how the breeds of the rez dogs in the community have shifted over time and attributed these changes to the rising safety concerns. One participant shared: "I don't think the rez dogs were as bad growing up" (P12). Another tribal member elaborated that some breeds, such as pit bull type dogs and Rottweilers "are not necessarily what I consider the animal you should have in my town" (P10). Another member echoed distrust of certain breeds because dogs, such as German Shepherds were "attacking people and hurting them" and that they are "not Indian dogs" (P9). As a result of these safety concerns connected to specific breeds of dog, animal control officers who participated in the study noted that in some areas of the reservation, breeds, such as Rottweilers, pit bull-type dogs, and Mastiffs, were banned by laws or ordinances.

Conversely, other tribal members categorized the behavior of free-roaming dogs more positively. Animal control officers who participated in the study estimated that they see only 10–20 cases of dog bites per year and that they had come across rez dogs that "were actually well trained" (P7).

In addition to safety concerns about dog bites and attacks, tribal members noted a few additional concerns related to the rez dogs. Diseases, such as canine brucellosis, mange, parvovirus, and flea/tick-borne diseases, were all listed concerns regarding free-roaming dogs. Notably, the participants did not describe rabies as an issue in their community.

Participants described several limitations of existing animal control policies to address the free-roaming dog challenges on the reservation. Several participants noted that while there is a leash law in place that should limit the number of free-roaming dogs, that law is not well known by community members. One participant described the need to improve public knowledge of the leash laws:

> *[Animal control officers] try to educate them [when they come to pick up their dog] because education is the most important thing. A lot of people claim that they had no idea that there was a Leash Law, so whenever they come in to get their dog, [animal control officers] explain to them the reason why we picked up your dog; it was at-large and then, here's the fees for kenneling and everything. You have a certain amount of days to come and retrieve your dog. [Animal control officers] try to educate them as best as we can (P7).*

Other participants were opposed to laws that would restrict the free movement of dogs. One participant discussed the historic role of dogs for the MHA Nation, the concept of ownership traditionally, and how these ideas "clash" with the policies that have been put in place:

> *Well you know that's kind of how people lived back then. They didn't have fences up. They didn't understand that—hey there's a restriction here—horses roam, dogs roam. But this law is saying now you live in a different world that says you can't do that any longer. Right? And so it's like a cultural clash (P14).*

Interview participants shared that their most desired outcome for free-roaming dogs is that those dogs are returned to their owner or to the community. One participant described informal approaches for getting the dogs back to their owners that did not require getting animal control involved:

> *[When we pick up a dog] the local tribal newspapers started putting them in the paper and just word of mouth, we get it on the tribal email system, and people say, 'Oh that's someone else's dog'. So they let them know (P7).*

Participants detailed other approaches for managing free-roaming dogs. One participant noted

> *"Not everybody thinks that they are sacred people roaming the earth. There are people who say 'the heck with that idea.' They want them dead" (P12).*

A central theme of many interviews was the challenges that animal control experiences in the community. One interviewee said:

> *I think it's frustrating for the animal control workers. They come in very excited to help. They want to do their best. They want to help animals. They just get burned out and run down by this constant source of picking up these dogs (P13).*

A few animal control officers discussed the difficulties of the job: "You not only have enrolled tribal numbers, but you also have non-members too, so it's hard to enforce that over the jurisdiction of non-members" (P7). They emphasized one of the problems in the community is people who repeatedly leave their dogs out to roam freely: "We spend so much time going after repeat offenders, and we don't get time to reach out to the community very often" (P7).

Another challenge the officers shared was the stigma faced by animal control in the community:

> *"The huge misunderstanding or misconception within our community on the reservation that animal control is perceived as dog catchers, or dog killers even. They think that once [animal control] gets your dog, they shoot the dog" (P7).*

3.3. Community-Specific Solutions

3.3.1. Culturally Relevant Information Sharing

When reflecting on how to best address the current challenges related to free-roaming dogs, many participants identified a need for culturally relevant information and education on how to care for the rez dogs. Many participants suggested that this education focuses on young people. One tribal member summarized the most important topics for these information-sharing efforts: "Number one they need to know how important these animals are and that we're here to take care of them. They're given to us for a reason" (P11).

Given the identification of the behavior of some of the dogs being one of the primary concerns, one tribal member suggested offering animal handling and training classes for the community:

> *I'd like to see classes on training animals and showing young kids. They're going to see somebody show them how to respect and how to get certain things from an animal. They will just eat it up and love it and show off with it. And be proud of their animals (P11).*

Another participant suggested these information-sharing interventions occur in schools and believed those lessons could take a story format: "We could have like dogs tales. Where we tell old stories or something, those old stories since I have been here" (P14).

Another participant felt that information-sharing efforts should not just focus on young people and, instead, focus on the family unit as a whole: "It has to be family for me. Teaching begins from the family. Education [in schools] has played a big role, but for me, anything I learned in my culture starts at home" (P5).

Tribal members shared that information sharing on the cultural significance of rez dogs would not only help the community with caring for the free-roaming dogs but also work to combat cultural disruption and erasure caused by colonization:

> *I think that a lot of cultures are going extinct and we hear about that a lot. It would be fun to get a beautiful artist to do the drawings, and some of the elders to tell the stories and we write them down and publish them. And do it in different languages too. That is one thing I see missing, in children books in the two languages. It's something you do with your mom and dad, and it's kind of interactive. You have discussions about it (P14).*

3.3.2. Community Care and Responsibility

The participants expressed varying perspectives about caring for dogs in their homes and community. When describing the general approach to taking care of dogs on the reservation, study participants expressed a common theme: "You gotta take care of them, the dogs have to be maintained. They have to be loved (P9)".

Multiple study participants described community care and ownership of dogs on the reservation. Community care was provided most often by providing food and water for free-roaming dogs. Participants shared that "there are a lot of people willing to take in animals", and "We don't have any [dogs], but they come to our house all the time. We feed them. We have bags of dog food. Feed and water them" (P11).

According to the participants, some individuals have even taken on the responsibility of providing veterinary care and facilitating the adoption of dogs. One tribal member described how their daughter has "out of her own pocket has made sure they had their shots, were fixed, and then she doesn't give them to just anybody either, or she makes sure they're going to good people. There are so many people out there like that in our community" (P11).

3.3.3. Access to Veterinary Care and Other Animal Services

Access to care was an essential theme for many of the participants. Nearly all participants emphasized the importance of improved access to veterinary care and animal services in improving the health and well being of rez dogs on the Fort Berthold reservation (P13). Multiple interviewees discussed the relationship between the free-roaming dogs on their reservation and the gaps they experience in access to veterinary care and other pet support services, including spay/neuter and basic veterinary care. When asked why many of the dogs on the reservation are not spayed or neutered, one respondent answered "access and money" (P11). A general theme expressed by participants was that "it takes a lot to take care of a dog, and you have to make that commitment" (P2).

Mobile veterinary clinics and/or private practice clinics located in adjacent cities are the primary forms of veterinary care for tribal community members' animals. One interview participant highlighted the overall need to increase access to low-cost spay/neuter, vaccinations, and other veterinary care:

> *[A barrier] is just the resources that might be needed in the community for vaccination and for a spay/neuter and all those things that are needed to keep the dogs healthy. It's, you know, one thing to educate people, but then you have to back it up with the resources so that we can provide the right level of care (P2).*

Participants shared that mobile veterinary clinics come to the community intermittently and offer free services. However, "it takes a lot for the owner to take [dogs] down to these little vet clinics" (P3), and "some people can't get to them" (P8). Another individual indicated that the veterinary services are inconsistent and unpredictable: "This year, I think, is the first year they didn't come up this way [...] I wanted to get my dogs a few shots, but they didn't come around this year" (P5).

Similarly, the participants expressed a lack of capacity for veterinary services to serve the community. One participant described one of the nearby veterinarian's capacity as a limiting factor:

> *[He] comes down and spends a day in Parshall or New Town and he spays and neuters out of his mobile clinic. He is only one surgeon though versus with that mobile clinic that could get thirty or forty done in a day. He can do five or six (P13).*

Even with the services provided by the existing clinics, community members expressed frustration with other barriers to accessing the services, such as scheduling processes and appointment restrictions. One community member recalled, "It's on a first-come, first-serve basis. So if you're not there right away at 5 o'clock in the morning signing up, you probably aren't going to get seen" (P3). Another tribal member summarized their understanding of the communities' experience attempting to access the traveling clinics:

> *A lot of people had to call ahead of time to say they were bringing [in an animal]. And there were a lot of people that got turned away. Because they were hearing "just bring them in" and they were going to spay, neuter, whatever [...] now they are making everybody do appointments (P11).*

Another participant described a program in the area that provides free vaccinations for animals adopted through the local animal control. The participant explained that being able to provide low-cost vaccinations was important. However, the current voucher-based system was ineffective and underutilized:

> *We do a voucher system with animal control here for vaccinations. So, for an animal adopted out of animal control, they get a voucher saying that animal control will pay for the vaccinations. They take the voucher, bring the voucher down to [the veterinarian] and the vaccination is free. [The veterinarian] sends the voucher back to animal control. They reimburse [the veterinarian] for the vaccination. But [the veterinarian has] only seen two vouchers in six months (P13).*

Participants frequently shared that tribal community members were open to receiving veterinary care services if they could access them. One interviewee shared the high demand for mobile veterinary services:

> *Judging by our waiting list and even now we are out of funding for those mobile vet services, but people stalk our office non-stop. A majority of our calls at the Game and Fish Office are animal control calls, because they want to know when the next clinic is. So, people want to do it (P14).*

One participant also highlighted challenges related to the lack of access to behavior training support in their community:

> *It takes a lot of effort to get up to Minot for dog training classes. It's a day because you have to drive up there, do your stuff, and drive back. It takes a lot of effort to do that (P13).*

Tribal community members noted several concerns around neglect of the free-roaming dogs. Some participants described issues around the lack of adequate food, shelter, and water for dogs or containing dogs using a tether or in a yard. Some described instances of dogs passing away "because nobody was feeding it. Nobody took that responsibility" (P8) or that certain dogs they personally cared for did "belong to somebody else, but evidently they are not taking care of them" (P4).

4. Discussion

This study identified cultural perspectives, traditions, and current or proposed strategies for the management of rez dogs living on the Fort Berthold reservation in North Dakota, U.S.A. By centering the perspectives and proposed interventions of Indigenous community members, this study represents an important step toward developing culturally responsive and community-specific strategies for addressing rez dog-related issues on the Fort Berthold reservation. The primary intervention areas described by the participants included: culturally relevant information sharing, improved animal control policies and practices, and improved access to veterinary care and other animal services.

4.1. The Need for Culturally Responsive Information Sharing

Similar to the findings of the present study, previous studies on the cultural significance of dogs in Indigenous communities have found that the rez dogs of the MHA Nation are viewed as guardians and companions, they have spiritual and cultural significance, and rez dogs are accepted and celebrated in society [2,3,11]. For many members of the MHA Nation who participated in this study, rez dogs are more than companion animals; they are spiritual allies and protectors of the land and people.

Participants shared how colonization has impacted the flow of cultural knowledge from tribal elders to the younger generations and, therefore, impacted the younger generations' relationships with dogs. In this study, participants shared how interventions that support decolonization are critical to restoring both the traditional role of dogs and the dogs' sovereignty within animal welfare practices and policies. Decolonization is defined as "the process of reversing colonization and the damage done by reclaiming Indigenous Knowledge and implementing that knowledge in the lands and communities of Indigenous Peoples physically, psychologically, and emotionally" [31]. Decolonization in the animal welfare context should center Indigenous Knowledge (the knowledge of Indigenous and local communities about their own interaction with ecosystems) in the development of animal welfare policies and programs in Indigenous communities [32]. For example, strategies recommended by participants included building free-standing shelters and feeding stations for free-roaming dogs which would allow the community to preserve and optimize their tradition of community care and management of the free-roaming dog population. Research on decolonization has also identified additional strategies that could support the health and well being of both Indigenous communities and rez dogs, including broader interventions like recognized sovereignty, self-determination, and reparations [3]. Future research could also investigate how lateral violence has influenced free-roaming dog management practices [33].

By demonstrating cultural responsiveness and centering the community-specific needs in their work with Indigenous communities, animal welfare service providers can establish more effective and sustainable programs and policies [34]. Developing programs that utilize culturally responsive communication and provide accurate culturally specific information when engaging with Indigenous communities requires an understanding of the dynamic nature of Indigenous Knowledge and how it should constitute the foundation of community-specific health and regenerative practices [35,36]. This knowledge is grounded in culturally distinctive concepts and diverse notions of health, well being, and resilience that bridge relationships among people, other animals, and the environment [37]. Indigenous Peoples have a rich history of diverse languages, strategies, and values in areas of identity, the land, and responsibilities of looking and taking care of each other that are tied to their traditional territory [38]. Therefore, animal service providers should not extrapolate community-specific programs to other communities. By engaging in their own relationship-building with their focus communities to understand the community-specific strengths, priorities, and needs, service providers can support the implementation of community-led and culturally relevant programs.

For the MHA Nation, many participants prioritized the need for culturally relevant information sharing for the young people in their community. The participants were excited about the potential solution of building a center for humane education that could also function as an animal shelter and resource center. This "MHA Nation Center for Humane Education" would also be a place where the community, particularly young people, could engage in dog training classes. Several participants shared how having a physical building dedicated toward cultural revitalization and preserving the cultural significance of rez dogs would greatly enhance the community's efforts to care for the dogs.

4.2. The Need for Community-Based Animal Control

The role of animal control/field services is to "provide a wide array of services to their communities, including saving pets in danger, protecting human health and safety, enforcing laws and ordinances, providing support and education to community members, disaster response, helping lost pets get home, and helping wildlife, livestock, and exotic animals, in addition to cats, dogs, and other pets" [39]. Free-roaming dogs fall within the purview of animal control/field services due to the ongoing discussion in the veterinary epidemiology and public health fields about the potential risk of free-roaming dogs to community health [14–17]. This discussion includes their negative impacts on native and endemic wildlife, they are common vectors for disease transmission (rabies, parvovirus, parasites, and canine distemper virus), and can be perpetrators of aggression towards humans [12,13].

However, research on the relationship between free-roaming dogs and various public health and safety concerns has produced mixed results. For example, studies conducted in urban/non-reservation environments have documented people being more likely to be attacked when approaching dogs that are chained up, fenced, or owned [40–42]. In Bali, free-roaming dogs were noticeably calmer, less likely to attack or chase other animals, and less excitable or active than owned/kept dogs [43]. Further, there are clear benefits to the dogs themselves in allowing them to roam freely [44–48]. In a study on animal sheltering in India, unrestricted movement (the capability to move freely about a neighborhood regardless of human control) was identified as the most important factor in upholding good welfare for free-roaming dogs [45]. Similarly, ethnographic interviews with the Indigenous Tofa communities of southern Siberia highlighted that dogs need the freedom to roam in order to live their fullest lives [46]. Future research should focus on developing a more complete understanding of how free-roaming dogs can be effectively managed so that equitable health for all components of a communities' ecosystem is respected [3,7,45,47,49,50].

Due to the multi-faceted and community-specific nature of rez dog-related challenges, community engagement represents a promising approach to improving the effectiveness and sustainability of animal control/field services policies and practices on the Fort Berthold reservation [49,50]. A limited number of studies have discussed how community engagement strategies have been used in the animal welfare field [51–54]. Community engagement provides community members the opportunity to collaboratively identify issues and offer critical perspectives on the policies and practices that directly affect them. The approach can also identify individual and collective strengths, challenges, assets, and lived experiences that can be leveraged to improve the effectiveness and sustainability of policies and programs [55,56]. Findings from this study could be used to form the foundation of future Fort Berthold animal control officers' community engagement efforts. For example, most participants in this study identified a need for improved information-sharing efforts around the current animal control policies of the Fort Berthold reservation, particularly the leash laws and shelter requirements for dogs who live outdoors. Community engagement approaches could facilitate animal control officers in identifying culturally relevant and community-specific approaches for information sharing that would be most effective at improving the health and well being of both the people and the rez dogs, while also gathering input on strengths or limitations of the current animal control policies.

4.3. The Need for Access to Veterinary Care and Other Animal Services

The majority of participants explained that they either participated in or witnessed the community care of dogs on the Fort Berthold reservation. This community care typically took the form of providing food and water to the dogs. Many MHA Nation members were known to connect rez dogs with veterinary care and permanent families and homes. These community care practices of feeding and sheltering are similar to those documented through research conducted in India, where "low barrier" forms of community care (i.e., feeding) are commonly practiced, with a few community members engaging in acts for long-term welfare (e.g., vaccinations, spay/neuter, shelter, etc.) [44]. These community care efforts represent a critical source of mutual aid in the community, which could reduce the burden on the already under resourced and overburdened animal control system on the Fort Berthold reservation. However, the MHA Nation members noted several barriers to engaging in these ongoing community care efforts, including insufficient access to veterinary care and other pet support services.

Access to veterinary care and other pet support services (e.g., grooming, behavior training, and pet supplies) was highlighted by most of the study participants as a challenge related to caring for the rez dog population on the Fort Berthold reservation. Research on barriers to accessing veterinary care is extensive. Research conducted in non-Indigenous communities throughout the U.S. has identified factors that exacerbate barriers to accessing pet support services, including affordability of services, geographic proximity to pet care services, transportation barriers, clinic hours of operation, bilingual services providers, positive and trusting service provider–client relationships and communication, client values concerning the importance of veterinary and other pet support services, and general access to pet care information [57–61]. Of these barriers identified across the U.S., geographic location of services, clinic hours of operation, and scheduling processes were identified as the most significant barriers to accessing services for MHA Nation members.

Only a few studies have documented access to veterinary care as a barrier to community-wide health in Indigenous communities [3,10,62]. Many of the participants in this study highlighted a need for animal welfare service providers to be responsive to community-specific needs around how they structure their programming. For example, the participants described that while the mobile clinics were a valued source of veterinary care, they were often limited in their effectiveness. The participants discussed challenges related to the mobile veterinary clinics, as they were inconsistently scheduled and did not meet the demand for services on the reservation. To this point, there is an important distinction in the human health care literature between "having access" to services, meaning the potential to access a particular service, and an "gaining access," referring to actual utilization of the service [63]. There are a variety of strategies currently being used in the animal welfare field to attempt to support remote communities' moving toward actually gaining access to veterinary care, including providing telehealth options for non-surgical procedures, offering transportation for pets and their owners to and from appointments, providing pet support care and supplies alongside existing human service providers or events (e.g., food and supply pantries), and delivering pet supplies (e.g., food, treats, litterboxes, collars, and leashes) directly to clients' homes at no cost [64]. Animal service providers working to serve the MHA Nation community members should consider what the participants have shared regarding the barriers they are experiencing to accessing the limited pet support services in their community and make efforts to facilitate the MHA Nation community in gaining access to their services.

Finally, it is important to highlight how most animal control policies and programs in the U.S. are heavily influenced by the highly subjective definition of "responsible pet (companion animal) ownership" [54]. Unfortunately, some of the animal welfare literature has equated the strength of an individual or a community's bond to animals to colonial and classist standards, such as their willingness to pay for services or allow the animal to sleep in their bedroom [65–68]. This narrative has reinforced implicit and explicit bias against socially and economically marginalized populations in the U.S., particularly Black, Indigenous, and People of Color (BIPOC) and communities experiencing poverty. These biases justify the assertion that companion animal ownership/caregiving is—or should be—only available for individuals who can afford all facets of companion animal ownership under any circumstances and who agree with and operate within the settler colonial definition of responsible pet ownership [69,70]. Fortunately, more rigorous studies have demonstrated that when structural barriers to accessing pet care services were addressed through a community-level intervention the pet owners' race and ethnicity was not predictive of seeking pet support services [71]. Future research should aim to develop a more inclusive and equitable definition of concepts such as "the human–animal bond" and "responsible pet ownership" that can then be used to assess the effectiveness of various animal welfare interventions in promoting the health and well being of people and companion animals alike.

4.4. Limitations

This study has several limitations to consider. The study had a small sample size of 14 MHA Nation community members, with all participants over the age of 30 years, which should be considered when evaluating its representativeness for the entire MHA Nation. Future research could explore variations in perspectives with greater representation of the young people in the community. The use of phenomenological methods in this study was intended to gather an in-depth foundational understanding of free-roaming dog issues on the Fort Berthold reservation, and future research could utilize quantitative approaches to begin to generate evidence that would support concepts of generalizability. Additionally, Indigeneity is not a monolithic cultural identity, and all tribal communities have distinct cultural values and traditions. Only members from the MHA Nation were included in this study, which impacts the generalizability of the study's findings to other Indigenous communities.

It is also important to note that the data were collected seven years before the analysis took place and that the data analysis in this study relied on archived data. While valuable, archived data may be less relevant as these perspectives may have changed since 2016, and the gap in time may make member checking less relevant [72]. The study's findings would be strengthened with a follow-up study on current perspectives and needs regarding interventions. This study also highlights an opportunity and need for community-driven and participatory action research that is conducted through an ongoing partnership with the focus population.

Finally, the positionality of the research team should be noted when considering these findings. While one of the senior members of the research team (Dr. Michael Yellow Bird), who led the design and recruitment for this study, is a member of the MHA Nation, there was a lack of positionality within the data analysis team, as the majority are not Indigenous, and none identify as members of the MHA Nation. Three of the authors who conducted data analysis for this study identify as white women, one identifies as a Latina woman, and one identifies as a Denesuline Indigenous woman. There is an opportunity for future studies to explore similar issues and questions but with increased Indigenous representation in the positionality of the authors.

5. Conclusions

Future efforts to develop or evaluate community-specific free-roaming dog management strategies should integrate the community's historical and contemporary understandings of rez dogs. The effectiveness and sustainability of the interventions will be improved by prioritizing cultural responsiveness and promoting capacity for community caregiving and increasing access to pet support services.

Author Contributions: Conceptualization, K.N.M., M.Y.B. and S.M.B.; funding acquisition, K.M.O., M.Y.B. and S.M.B.; methodology, S.M.H.; investigation, K.N.M., M.Y.B, and S.M.B.; formal analysis, A.C., J.C., K.C., K.M.O. and S.M.H.; writing—original draft preparation, A.C., K.C., K.M.O. and S.M.H.; writing—review and editing, J.C., K.N.M., L.R.M., M.Y.B. and S.M.B.; supervision, S.M.H.; project administration, A.C. and S.M.H. All authors have read and agreed to the published version of the manuscript.

Funding: This study was funded by the University of Denver's Interdisciplinary Research Institute for the Study of (In)Equality (IRISE) [IRISE Research Grant]. MHA Nation tribal leadership funded the research team's travel expenses. Alexandra Cardona's Research Fellowship and Liana R. Moss's Research Associate position are funded through a grant provided by an anonymous donor to the University of Denver's Graduate School of Social Work. Sloane M. Hawes and Kevin N. Morris' positions are partially funded by the latter's American Humane Endowed Chair research fund.

Institutional Review Board Statement: The study was conducted in accordance with the Declaration of Helsinki and approved by the Institutional Review Board of the University of Denver (Protocol #: 767926. Date of approval: 28 June 2016).

Informed Consent Statement: Informed consent was obtained from all subjects involved in the study.

Data Availability Statement: Not Applicable.

Acknowledgments: We thank the fourteen MHA Nation community members who participated in the study and shared their stories and for the MHA Nation Council's support of this study. Thank you to Philip Tedeschi for his support in building the partnerships involved with obtaining funding for this study and for his input on the design of this study. We would also like to express our appreciation to Melissa Janiszewski who participated in gathering these interviews. Finally, thank you to Valli-Laurent Fraser-Celin for her thoughtful input on this manuscript.

Conflicts of Interest: The authors declare no conflict of interest. The funders had no role in the design of the study; in the collection, analyses, or interpretation of data; in the writing of the manuscript; or in the decision to publish the results.

References

1. Anza-Burgess, K.; Lepofsky, D.; Yang, D. "A part of the people": Human-dog relationships among the northern coast Salish of SW British Columbia. *J. Ethnobiol.* **2020**, *40*, 434–450. [CrossRef]
2. Jervis, L.L.; Warren, D.; Salois, E.M.; Ketchum, S.; Tallbull, G.; Spicer, P. Protectors, Aggressors, and Kinfolk: Dogs in a tribal community. *Anthrozoos* **2018**, *31*, 297–308. [CrossRef]
3. Ma, G.C.; Ford, J.; Lucas, L.; Norris, J.M.; Spencer, J.; Withers, A.; Ward, M.P. "They reckon they're man's best friend and I believe that." Understanding relationships with dogs in Australian Aboriginal communities to inform effective dog population management. *Animals* **2020**, *10*, 810. [CrossRef] [PubMed]
4. Ní Leathlobhair, M.; Perri, A.R.; Irving-Pease, E.K.; Witt, K.E.; Linderholm, A.; Haile, J.; Lebrasseur, O.; Ameen, C.; Blick, J.; Boyko, A.R.; et al. The evolutionary history of dogs in the Americas. *Science* **2018**, *361*, 81–85. [CrossRef] [PubMed]
5. Gompper, M.E. The dog-human-wildlife interface: Assessing the scope of the problem. In *Free-Ranging Dogs and Wildlife Conservation*; Gompper, M.E., Ed.; Oxford University Press: New York, NY, USA, 2014; pp. 9–54, ISBN 978-01-9181-018-3.
6. Hughes, J.; Macdonald, D.W. A review of the interactions between free-roaming domestic dogs and wildlife. *Biol. Conserv.* **2013**, *157*, 341–351. [CrossRef]
7. Smith, L.M.; Quinnell, R.; Munteanu, A.; Hartmann, S.; Dalla Villa, P.; Collins, L. Attitudes towards free-roaming dogs and dog ownership practices in Bulgaria, Italy, and Ukraine. *PLoS ONE* **2022**, *17*, e0252368. [CrossRef] [PubMed]
8. World Society for the Protection of Animals. Surveying Roaming Dog Populations: Guidelines on Methodology. 2008. Available online: https://www.worldanimalprotection.org.cn/sites/default/files/media/cn_files/cn_attachment/surveying_roaming_dog_populations_guidelines_on_methodology.pdf (accessed on 5 March 2023).
9. Boissonneault, T.; Epp, T. Reflection on the provision of veterinary services to underserved regions: A case example using northern Manitoba, Canada. *Can. Vet. J.* **2018**, *59*, 491–499.

10. Griffith, C.L. Reservation Dogs: Occupancy, Community Beliefs, and Lakota Ways of Knowing. Ph.D. Dissertation, Purdue University, Lafayette, IN, USA, 2022. [CrossRef]
11. Constable, S.; Dixon, R.; Dixon, R. For the love of dog: The human-dog bond in rural and remote Australian Indigenous communities. *Anthrozoos* **2010**, *23*, 337–349. [CrossRef]
12. Constable, S.E.; Dixon, R.M.; Dixon, R.J.; Toribio, J.A. Approaches to dog health education programs in Australian rural and remote Indigenous communities: Four case studies. *Health Promot. Int.* **2013**, *28*, 322–332. [CrossRef] [PubMed]
13. Djordjevic, M.M.; Nenadovic, K.M.; Vucinic, M.M. Initial study on free-roaming dogs in Serbian hunting grounds. *J. Vet. Beh.* **2022**, *57*, 67–73. [CrossRef]
14. Ksirshager, A.R.; Applebaum, J.W.; Randrian, Z.; Rajanoarivelo, T.; Rafaliarison, R.R.; Farris, Z.J.; Valenta, K. Human-dog relationships across communities surrounding Ramomafana and Andaside-Mantadia National Parks, Madagascar. *J. Ethnobiol.* **2020**, *40*, 483–498. [CrossRef]
15. Young, J.K.; Olson, K.A.; Reading, R.P.; Amgalanbaatar, S.; Berger, J. Is wildlife going to the dogs? Impacts of feral and free-roaming dogs on wildlife populations. *BioScience* **2011**, *61*, 125–132. [CrossRef]
16. Muinde, P.; Bettridge, J.M.; Sousa, F.M.; Durr, S.; Dohoo, I.R.; Berezowski, J.; Mutwiri, T.; Odinga, C.O.; Fevre, E.M.; Falzon, L.C. Who let the dogs out? Exploring the spatial ecology of free-roaming dogs in western Kenya. *Ecol. Evol.* **2021**, *11*, 4218–4231. [CrossRef]
17. Dorji, T.; Tenzin, T.; Rinzin, K.; Phimpraphai, W.; de Garine-Wichatitshy, M. Community perceptions of free-roaming dogs and management practices in villages at the periphery of a protected area in Bhutan. *J. Nat. Sci.* **2020**, *19*, 130–150. [CrossRef]
18. Vargo, D.; DePasquale, J.M.; Vargo, A.M. Incidence of dog bite injuries in American Samoa and their impact on society. *Hawai'i J. Med. Public Health* **2012**, *71*, 6–12. Available online: https://www.ncbi.nlm.nih.gov/pmc/articles/PMC3298428/ (accessed on 8 January 2023).
19. Dhillon, J.; Favel, D.; Delorme, D.; Ratt, A.; Epp, T. Finding pathways for bite prevention and decreasing dog populations: The process of animal control for Indigenous communities in Canada. *J. Indig. Wellbeing* **2016**, *1*, 82–92. Available online: https://journalindigenouswellbeing.co.nz/media/2022/01/47.39.Finding-pathways-for-bite-prevention-and-decreasing-dog-populations-The-process-of-animal-control-for-indigenous-communities-in-Canada.pdf (accessed on 5 March 2023).
20. Baker, T.; Rock, M.; Brook, R.; van der Meer, F.; Kutz, S. Indigenous communities perspective on dogs in Northern Canada after 10 years of veterinary services indicates animal and human welfare. *Prev. Vet. Med.* **2020**, *181*, 105061. [CrossRef]
21. Smith, L.M.; Hartmann, S.; Munteanu, A.M.; Dalla Villa, P.; Quinnell, R.J.; Collins, L.M. The effectiveness of dog population management: A systematic review. *Animals* **2019**, *9*, 1020. [CrossRef] [PubMed]
22. Ma, G.C.; Withers, A.M.; Spencer, J.; Norris, J.M.; Ward, M.P. Evaluation of a dog management intervention: Measuring indicators of impact. *Animals* **2020**, *10*, 1061. [CrossRef] [PubMed]
23. Native American Humane Society. Caring for Your Dogs. 2016. Available online: https://www.nativeamericahumane.org/uploads/6/4/2/5/64257269/nahs_dog_care_brochure_print_final_10-6-16.compressed_2.pdf (accessed on 5 March 2023).
24. MHA Nation. Our Tribe. Available online: https://www.mhanation.com/ (accessed on 8 January 2023).
25. Yoak, J.A.; Reece, F.J.; Gehrt, D.S.; Hamilton, M.I. Optimizing free-roaming dog control programs using agent-based models. *Ecol. Model.* **2016**, *341*, 53–61. [CrossRef]
26. Creswell, J.W.; Poth, C.N. *Qualitative Inquiry & Research Design: Choosing Among Five Approaches*; SAGE Publications, Inc.: New York, NY, USA, 2018; ISBN 978-1-5063-3020-4.
27. Alsaigh, R.; Coyne, I. Doing a hermeneutic phenomenology research underpinned by Gadamer's philosophy: A framework to facilitate data analysis. *Int. J. Qual. Methods* **2021**, *20*, 160940692110478. [CrossRef]
28. Van Manen, M. *Phenomenology of Practice: Meaning-Giving Methods in Phenomenological Research and Writing*; Routledge: New York, NY, USA, 2016; ISBN 978-1-61132-765-6.
29. Dowling, M. From Husserl to van Manen. A review of different phenomenolical approaches. *Int. J. Nurs. Stud.* **2007**, *44*, 131–142. [CrossRef]
30. Oerther, S. Analysis methods in hermeneutic phenomenological research: Interpretive profiles. *Front. Nurs.* **2020**, *7*, 293–298. [CrossRef]
31. Yellow Bird, M. Decolonizing social work education: Indigenizing the academy. *Can. J. Nativ. Stud.* **2013**, *32*, 139–157. [CrossRef]
32. Gomez-Baggethun, E.; Corbera, E.; Reyes-Garcia, V. Traditional ecological knowledge and global environmental change: Research findings and policy implications. *Ecol. Soc.* **2013**, *18*, 72. [CrossRef]
33. de Finney, S.; Shezell-Rae, S.; Adams, C.; Keenan, A.; McLeod, K.; Lewis, A.; Lewis, G.; Louis, M.; Haiyupis, P. Rekinning our kinscapes: Renegade Indigenous stewarding against gender genocide. *Girlhood Stud.* **2019**, *12*, 80–98. [CrossRef]
34. LaFrance, J.; Nichols, R. Reframing evaluation: Defining an Indigenous evaluation framework. *Can. J. Program Eval.* **2010**, *23*, 13–31. Available online: https://onlineacademiccommunity.uvic.ca/bcneihr/wp-content/uploads/sites/4766/2020/11/LaFrance-2010-reframing-indigenous-evaluation-copy.pdf (accessed on 8 January 2023). [CrossRef]
35. Cornstassel, J.; Hardbarger, T. Educate to perpetuate: Land based pedagogies and community resurgence. *Int. Rev. Educ.* **2019**, *65*, 87–116. [CrossRef]
36. Simpson, L.B. *As We Have Always Done: Indigenous Freedom through Radical Resistance*; University of Minnesota Press: Minneapolis, MI, USA, 2017; ISBN 978-14-5295-600-8.

37. Redvers, N.; Yellow Bird, M.; Quinn, D.; Yunkaporta, T.; Arabena, K. Molecular decolonization: An Indigenous microcosm perspective of planetary health. *Int. J. Environ. Res. Public Health* **2020**, *17*, 4586. [CrossRef] [PubMed]
38. Hart, M. For Indigenous People, by Indigenous People, with Indigenous People: Towards an Indigenist research paradigm. In *Wicihitowin: Aboriginal Social Work in Canada*; Sinclair, R., Hart, M.A., Bruyere, G., Eds.; Fernwood Publishing: Halifax and Winnipeg, NS, Canada, 2009; pp. 153–170, ISBN 978-155-266-317-2.
39. National Animal Care & Control Association. About. 2023. Available online: https://www.nacanet.org/about (accessed on 8 January 2023).
40. Tuckel, P.S.; Milczarsk, W. The changing epidemiology of dog bite injuries in the United States, 2005–2018. *Inj. Epidemiol.* **2020**, *7*, 57. [CrossRef]
41. McNeely, C.A.; Linquist, S.A. Dangerous dog laws: Failing to give man's best friend a fair shake at justice. *J. Anim. Law* **2007**, *3*, 99.
42. Overall, K.L.; Love, M. Dog bites to humans-demography, epidemiology, injury and risk. *JAVMA* **2001**, *218*, 1923–1934. Available online: https://www.k9behavioralgenetics.net/resources/Articles/Dog%20Bites%20to%20Humans.pdf (accessed on 8 January 2023). [CrossRef]
43. Corrieri, L.; Adda, M.; Kubinyi, E. Companion and free-ranging Bali dogs: Environmental links with personality traits in an endemic dog population of South East Asia. *PLoS ONE* **2018**, *13*, e0197354. [CrossRef]
44. Srinivasa, D.; Mondal, R.; Von Rentzell, K.A.; Protopopova, A. Interviews with Indian Animal shelter staff: Similarities and differences in challenges and resiliency factors compared to Western counterparts. *Animals* **2022**, *12*, 2562. [CrossRef]
45. Kittisiam, T. Social Network Analysis of Free-Roaming Dogs in Kasetsart University, Kamphaeng Saen Campus. Master's Thesis, Graduate School of Veterinary Public Health, Kasetsart University, Bangkok, Thailand, 2021. Available online: https://ethesis.lib.ku.ac.th/dspace/bitstream/123456789/1085/1/6224900019.pdf (accessed on 8 January 2023).
46. Oeler, A. Humans and dogs of mountainous inner Asia: Sensory collaboration and personhood. *Hum. Ecol.* **2021**, *49*, 765–778. [CrossRef]
47. Warembourg, C.; Wera, E.; Odoch, T.; Bulu, P.M.; Berger-Gonzalez, M.; Alvarez, D.; Abakar, M.F.; Sousa, F.M.; Silva, L.C.; Alobo, G.; et al. Comparative study of free-roaming domestic dog management and roaming behavior across four countries: Chad, Guatemala, Indonesia, and Uganda. *Front. Vet. Sci.* **2021**, *8*, 617900. [CrossRef] [PubMed]
48. Lord, K.; Feinstein, M.; Coppinger, R. Barking and Mobbing. *Behav. Process.* **2009**, *81*, 358–368. [CrossRef]
49. Mostafa, A.H.; Matoock, M.Y.; Khalil, H.M.A. A modified and simplified behavioral system associated with management procedures for assessing and managing the aggression level of free roaming dogs. *Adv. Anim. Vet. Sci.* **2022**, *10*, 369. [CrossRef]
50. Brookes, V.; Ward, M.P.; Degeling, C. The ripple-effects of free-roaming community dogs. In Proceedings of the Conference of International Society for Economics and Social Sciences of Animal Health, Bogor, Indonesia, 17–18 October 2019; p. 28. Available online: https://researchoutput.csu.edu.au/en/publications/the-ripple-effects-of-free-roaming-community-dogs (accessed on 11 March 2023).
51. Moss, L.R.; Hawes, S.M.; Connolly, K.; Bergstrom, M.; O'Reilly, K.; Morris, K.N. Animal control and field services officers' perspectives on community engagement: A qualitative phenomenology study. *Animals* **2023**, *13*, 68. [CrossRef]
52. Reese, L.A.; Vertalka, J.J.; Richard, C. Animal cruelty and neighborhood conditions. *Animals* **2020**, *10*, 2095. [CrossRef]
53. Covner, A.L.; Ekstrom, H.; Kalaf, N.; Sherman, B. Creating evidence driven practices to enhance human care services for unhoused and low-income pet owners. *J. Nurs. Healthc.* **2022**, *7*, 1–11. [CrossRef]
54. Hawes, S.M.; Hupe, T.; Morris, K.N. Punishment to support: The need to align animal control enforcement with the human social justice movement (Commentary). *Animals* **2020**, *10*, 1902. [CrossRef] [PubMed]
55. Saltmarsh, J. A collaborative turn: Trends and directions in community engagement. In *Learning Through Community Engagement*; Sachs, J., Clark, L., Eds.; Springer: Singapore, 2017; ISBN 978-981-10-09999-0.
56. Dempsey, S.E. Critiquing community engagement. *Manag. Commun. Q.* **2009**, *24*, 359–390. [CrossRef]
57. LaVallee, E.; Mueller, M.K.; McCobb, E. A systematic review of the literature addressing veterinary care for underserved communities. *J. Appl. Anim. Welf. Sci.* **2017**, *20*, 381–394. [CrossRef]
58. Gandenberger, J.; Hawes, S.M.; Wheatall, E.; Pappas, A.; Morris, K.N. Developing the Animal Welfare Cultural Competence Inventory (AWCCI) to assess cultural competence in animal welfare. *J. Appl. Ani. Welf. Sci.* **2021**, *24*, 1–12. [CrossRef]
59. Park, R.M.; Gruen, M.E.; Royal, K. Association between dog owner demographics and decision to seek veterinary care. *Vet. Sci.* **2021**, *8*, 7. [CrossRef] [PubMed]
60. King, E.; Mueller, M.; Wolfus, G.; McCobb, E. Assessing service-learning in community-based veterinary medicine as a pedagogical approach to promoting student confidence in addressing access to veterinary care. *Front. Vet. Sci.* **2021**, *8*, 644556. [CrossRef]
61. Hawes, S.M.; Rhodes, T.; Hupe, T.M.; Flynn, E.; Morris, K.N. Measuring perceptions of One Health: Development and validation of the One Health Community Assessment. *One Health* in preparation. **2023**.
62. Schurer, J.M.; Phipps, K.; Okemow, C.; Beatch, H.; Jenkins, E. Stabilizing dog populations and improving animal and public health through a participatory approach in Indigenous communities. *Zoonoses Public Health* **2015**, *62*, 445–455. [CrossRef]
63. Gulliford, M.; Figueroa-Munoz, J.; Morgan, M. What does 'access to health care' mean? *J. Health Serv. Res. Policy* **2002**, *7*, 186–188. [CrossRef]
64. Rauktis, M.E.; Rose, L.; Chen, Q.; Martone, R.; Martello, A. "Their pets are loved members of their family": Animal ownership, food insecurity, and the value of having pet food available in food banks. *Anthrozoös* **2017**, *30*, 581–593. [CrossRef]

65. Brockman, B.K.; Taylor, V.A.; Brockman, C.M. The price of unconditional love: Consumer decision making for high-dollar veterinary care. *J. Bus. Res.* **2008**, *61*, 397–405. [CrossRef]
66. Brown, S.E. Ethnic variations in pet attachment among students at an American school of veterinary medicine. *Soc. Anim.* **2002**, *10*, 249–266. [CrossRef]
67. Krause-Parello, C.A.; Wesley, Y.; Campbell, M. Examining pet attitude in relationship to loneliness and parenthood motivation in pet-owning adults. *Health* **2014**, *6*, 598–606. [CrossRef]
68. Human-Animal Bond Research Institute. HABSCORE: Quantifying the Strength of the Human-Animal Bond. 2023. Available online: https://habri.org/hab-score/ (accessed on 8 January 2023).
69. Lue, T.W.; Pantenburg, D.P.; Crawford, P.M. Impact of the owner-pet and client-veterinarian bond on the care that pets receive. *J. Am. Vet. Med. Assoc.* **2008**, *232*, 531–540. [CrossRef]
70. Harding, M.J. Access to veterinary care for low-income Canadians. *Can. Vet. J.* **2018**, *59*, 121–1122. Available online: https://www.ncbi.nlm.nih.gov/pmc/articles/PMC6135307/ (accessed on 11 March 2023).
71. Decker Sparks, J.L.; Camacho, B.; Tedeschi, P.; Morris, K.N. Race and ethnicity are not primary determinants in utilizing veterinary services in underserved communities in the United States. *J. Appl. Ani. Welf. Sci.* **2018**, *21*, 120–129. [CrossRef] [PubMed]
72. Long-Sutehall, T.; Sque, M.; Addington-Hall, J. Secondary analysis of qualitative data: A valuable method for exploring sensitive issues with an elusive population? *J. Res. Nurs.* **2011**, *16*, 335–344. [CrossRef]

animals

MDPI

Article

Recognizing Zooeyia to Promote Companion Animal Welfare in Urban Bangladesh

Abu-Hena Mostofa Kamal [1,2,*,†], Colleen Anne Dell [1,†] and Timothy Kang [1]

1 Department of Sociology, University of Saskatchewan, Saskatoon, SK S7N 5A2, Canada; colleen.dell@usask.ca (C.A.D.); tim.kang@usask.ca (T.K.)
2 Department of Humanities, Khulna University of Engineering and Technology (KUET), Khulna 9203, Bangladesh
* Correspondence: m.kamal@usask.ca
† These authors contributed equally to this work.

Simple Summary: The practice of companion animal keeping is growing in urban Bangladesh. People are beginning to identify how the addition of a companion animal in their lives can benefit their health. There are also longstanding concerns about animal welfare in Bangladesh. Interviews with companion animal owners, companion animal sellers and veterinarians in two major urban Bangladesh cities offer insights into the benefits and challenges of companion animal ownership. We suggest four ways to apply this understanding to promote companion animal welfare in Bangladesh.

Abstract: The One Health concept of zooeyia refers to the benefits of companion animals in human health and is gaining global research attention. This exploratory study aimed to understand contemporary experiences and perceptions of the social benefits and challenges of living with a companion animal in urban Bangladesh. Thirty-five qualitative interviews were conducted with companion animal owners (20), animal sellers (10), and livestock service department officers (5) from two major cities in Bangladesh, Dhaka and Khulna. Thematic analysis found that historically, animals had a utilitarian purpose, such as livestock for food and dogs for security. The role and perceptions of companion animals began to change for some around the turn of the century. Today, companion animal caretakers report social, psychological and physical health benefits from integrating companion animals into their lives. They also report that companion animal ownership can contribute to social problems due to the prevailing stigma against companion animals. This is rooted in the continued utilitarian role attached to companion animals by the majority of the Bangladesh population as well as religious-based non-acceptance. As a result, the *Animal Welfare Act* (2019) is not well implemented, posing a key concern for companion animal welfare. To tackle this, we propose various ways in which the emerging concept of zooeyia can help promote the welfare of companion animals by challenging the stigma associated with them in Bangladesh.

Keywords: companion animal; animal welfare; Bangladesh; zooeyia; stigma

1. Introduction

The benefits of companion animal ownership on human health is gaining global attention [1,2]. We use the term ownership in this paper because it indicates the possession of companion animals by humans, but at the same time it is important to recognize this is problematic for several reasons and may be a contributor to companion animal welfare concerns. In 2011, the concept of zooeyia was introduced to the One Health Framework; One Health recognizes the interconnection between human, animal, and environment health [3]. Zooeyia refers to the "positive benefits to human health from interacting with animals, focusing on the companion animal" [4]. Awareness of zooeyia was amplified during the COVID-19 pandemic, where forced isolation of many people increased the time they spent with a companion animal in their homes [5,6].

Research indicates that over half of North American households, where the majority of research is concentrated, have at least one companion animal and most owners identify them as a family member [4,7,8]. This is a growing trend within developed countries [9–11]. Hodgson and Darling (2011) identify four key ways companion animals can benefit human health: as builders of social capital, agents of harm reduction, facilitators of behaviour change and as a part of treatment plans [12]. For example, an emerging body of study suggests companion animals can help increase physical activity among aging individuals [13,14] and improve the immune function of children in a household [15]. Research has also identified the specific role of companion animals in reducing human loneliness [1,16,17].

With the rapid increase in companion animal ownership across the globe, there is an increased need to focus on companion animal welfare, especially given that cruelty toward companion animals is a serious global concern [18]. Research shows that thousands of companion animals are abused each day, although most abuse is undetected because it occurs in private spaces [19]. Dogs, cats, and horses tend to be the most common documented victims of violence [20]. That said, the types of abuse against companion animals and the reasoning behind them varies across demographics and countries. In the United States, for example, negligence and abandonment have been identified as common types of violence against companion animals, with the primary reason being lack of awareness and knowledge about companion animal maintenance [20]. For instance, a 2018 statistic indicates that approximately 32% of the deaths of companion animals in the United States were due to negligence [19]. Negative stereotyping and myths about companion animals (e.g., black cats are bad luck, bully breed dogs are aggressive) and the resulting stigma can contribute to a lack of education and resources to safeguard for proper animal care and treatment [21].

Asian countries have a storied history of companion animal abuse, with much of it unreported [22]. Studies show that poverty and a lack of animal cruelty laws in the majority of Asian countries are the key reasons behind animal abuse [22]. Specific to Bangladesh, there is no exact statistic on the companion animal population in the country, but it appears to have increased in the past few decades [23]. The Bangladesh *Business Post* newspaper indicates that live pet animal and bird importation increased in 2020–2021, as has spending on companion animals and the number of pet shops across the country [24].

There is also no accurate statistic associated with companion animal abuse or welfare in Bangladesh [25]. This is the case even though the 1920 Bangladesh *Cruelty to Animals Act* was revised and updated to the *Animal Welfare Act* in 2019. The new *Act* aims to "ensure the proper treatment and responsible rearing of animals and of preventing cruel treatment" [26]. Newspaper articles indicate a lack of awareness about animal welfare among companion animal owners [25,27,28]. Animal welfare groups such as Obhoyaronno-Bangladesh Animal Welfare Foundation and Guardians of Paws and Claws have documented companion animal abuse [29,30]. Animal shelters offer additional insights on the concerning state of animal welfare in Bangladesh, including a general lack of empathy among community members for companion animals [31]. Harmful stereotypes and the resulting stigma can contribute to a lack of legal protections and/or implementation for companion animals [32]. Educating the wider population about animal welfare is a major challenge as well as a dire need.

To our knowledge, this exploratory study is the first of its kind to attempt to gain insight on contemporary companion animal welfare in Bangladesh by investigating contemporary experiences and perceptions of the social benefits and challenges of owning companion animals in two urban centres, namely Dhaka and Khulna. Based on this insight, we offer several suggestions for how companion animal welfare may be promoted through the emerging recognition of zooeyia in Bangladesh, that is, the positive impact of companion animals in human health.

2. Materials and Methods

2.1. Study Areas

We purposively selected Dhaka and Khulna—both metropolitan areas of Bangladesh—for this study given that the international literature suggests there is a high prevalence of loneliness and fragile social interactions among people living in densely populated cities, which may result in a greater desire for animal companionship [33–35]. Dhaka is the capital of Bangladesh and the most densely populated city in the country. It is also the 6th most populous city in the world, with approximately 23,234 people living per square kilometre [36,37]. Khulna is the 3rd largest and fastest-growing city in Bangladesh [37].

2.2. Data Collection

We applied a combination of snowball and purposive sampling techniques to recruit a total of 35 participants for the study. Our main recruitment strategy was to select participants based on referral from other participants [38]. To begin, the first author on this paper (AHMK) identified potential participants for the study from his own network. These participants then referred others who might be interested. We then applied a purposive sampling technique to include participants with diverse socio-demographic characteristics such as age, gender, religion, education, marital status and economic status. We also shared our recruitment poster on the first author's Facebook site. The inclusion criteria for companion animal owners were: having an animal in their care for at least one year, living in Dhaka or Khulna during the data collection period, being a Bangladeshi citizen, and being at least 18 years of age. Animal sellers included either owners or employees of a pet animal store and professional companion animal breeding farm owners. All animal seller participants had at least one year of experience selling companion animals. Livestock service department officers had to have been in their employment positions for at least one year. Informal observations at animal stores, farms and households were also made by the interviewer, including non-verbal cues and/or gestures of the participants. These were recorded by hand in the interview notes, both during and immediately after the interview. These observations helped to document attachment with and attitudes of the participants toward companion animals, specifically among the companion animal owners. It was also an effective strategy to document the conditions of companion animal seller and breeder spaces, including stocking density, cleanliness and types of care towards companion animals.

Data were collected between 20 December 2021 and 2 February 2022 through face-to-face in-depth interviews with companion animal owners (20) and key informants, including animal sellers (storekeepers and breeders) (10) and livestock service department officers (veterinary surgeons, 5). The interviews were conducted by the primary author on this paper (AHMK), a resident of Bangladesh with no prior companion animal ownership himself.

A semi-structured interview guide with open-ended questions was developed for each of the three target participant populations of the study. Given the lack of published information on companion animal ownership and welfare in Bangladesh, we formulated the interview guide based on related literature and studies from across the globe [39–41]. The guide's main focus is the contemporary role of companion animals in urban Bangladesh and the perceived benefits and challenges of human–companion animal interaction (see Supplementary Materials). The three population-specific interview guides were pre-tested with companion animal owners (3), animal sellers (2) and a livestock service department officer.

2.3. Data Management and Analysis

All the interviews were conducted in Bangladeshi and audio-recorded by the first author (AHMK). The average lengths of the interviews were 54.25 min (ranging from 20.51 min to 124 min) with companion animal owners, 29.40 min (ranging from 15 min to 61 min) with sellers and 25.54 min (ranging from 18.48 min to 31.44 min) with livestock service department officers. The interviews were transcribed verbatim in Bengali and then the interview transcriptions were translated into English. We used NVivo (version 12) as the primary organizational software. Data analysis was guided by Braun and Clarke's (2006) six-phase framework for thematic analysis, including familiarization, coding, generating themes, reviewing themes, defining and naming themes and writing up. The primary author (A.-H.M.K.) undertook the data analysis. An open coding approach was applied to analyse the data, with no prior code list identified. Rather, the codes were developed based on the research focus and as we worked through the coding process. The second and third authors (C.A.D. and T.K.) were consulted throughout the coding process for feedback and insights. Additional codes were included based on field notes and unstructured observation. Each interview was analysed separately and the conclusions were drawn collectively based on the perceived benefits and drawbacks of companion animal ownership on human health.

2.4. Ethical Considerations

The research protocol [Beh#2973] was reviewed and approved by the Behavioural Research Ethics Board at the University of Saskatchewan. The consent form was sent to each participant by email at least one day before the interview. The first author also explained the objectives and ethical considerations of the study to each of the participants in-person immediately prior to conducting the interview, and informed written consent was collected. It is important to note that the interviews were conducted during the COVID-19 pandemic. The interviewer supplied facemasks and hand sanitizers to all participants. The use of a facemask, hand sanitation at frequent intervals and the maintenance of social distance (2 m apart) was mandatory in each interview scenario.

3. Results and Discussion

3.1. Participant Characteristics

Of the 20 companion animal owners, 11 were male, 10 were between 18 and 30 years of age, 9 were married, majority (15) had attained a Bachelor of Arts or higher degree, 11 were employed and 15 were Muslim. Of the 20 owners, 12 reported owning their companion animals between 1 and 3 years and 13 reported renting their houses. The average monthly income of the participants was USD 1369.81 (see Table 1). As a note, the average per capita monthly income in Bangladesh is USD 226.52 [36,37].

Among the animal sellers (10), six represented animal breeding farm owners (i.e., wholesale) and four were employees of an animal store (i.e., retail). All of the sellers were male and seven were Muslim. Two animal sellers had completed a Bachelor of Arts or higher level of education. Out of the six animal breeding farms, three bred dogs and the remaining bred cats. The storekeepers reported selling various types of animals, including dogs, cats, rabbits and aquarium fish. All the livestock service department officers were male veterinary surgeons. The average duration of time they had worked in the sector was 21 years, with a minimum of 8 and a maximum of 33 years (see Table 2).

3.2. Changing Role and Perception of Companion Animals in Bangladesh

3.2.1. The Past and Its Influence on the Present

Our participants contextualized their experiences with and perceptions of the contemporary role of companion animals in Bangladesh by sharing their thoughts about the historical role of animals more generally, as well as companion animals. They collectively shared that prior to around 2000, the majority of Bangladesh households owned livestock, including cattle, buffalos, goats, chicken, ducks and hens as working animals in agricultural fields (e.g., to tend the land) as well as to meet the nutritional demands of family members (e.g., milk, egg and meat). Some participants shared that women, in particular, owned ducks, chickens and goats as a source of household income. This finding echoes the existing literature on the historic role of animals in Bangladesh, which outlines their utilitarian purpose [25,42–45].

Table 1. Socio-economic characteristics of companion-animal-owner study participants.

Characteristics	No. of Participants (%)
Age (years)	
18–30	10 (50)
31–50	8 (40)
50 and above	2 (10)
Gender	
Male	11 (55)
Female	9 (45)
Marital status	
Never married	9 (45)
Married	9 (45)
Divorced	1 (5)
Widower	1 (5)
Education	
Grades 10–12	5 (25)
Bachelors and above	15 (75)
Religion	
Muslim	15 (75)
Hindu	4 (20)
Christian	1 (5)
Occupation	
Employed	11 (55)
Unemployed	3 (15)
Homemaker	2 (10)
Student	4 (20)
Duration of owning companion animals (years)	
1–3	12 (60)
4–6	3 (15)
6 and above	5 (25)
Household ownership status	
Own house	7 (35)
Rented house	13 (65)
Average monthly family income (USD)	1369.81 (range: 171.41–10,713.49)

All of the companion animal owners in our study reported that in the past, almost every rural household in Bangladesh had one or more stray dogs or cats associated with it. They shared that people fed the dogs and cats so they would stay around to protect the households from intruders and rodents. Like livestock, these animals had a utilitarian purpose. All companion animal owners shared that this utilitarian role continues today [46]. A 58-year-old veterinarian said, "[many] people, especially the non-companion animal owners, argue that animals cannot be companion to humans. For them, animals, like dogs can be utilized to guard households. Likewise, cats can help to safeguard grains and other valuable stuffs at home by killing rats and other insects." He said that the Bangladesh media had recognized a fairly recent transition in the role of stray community dogs in particular, with them being identified by some as companion animals [47].

Table 2. Socio-demographic information of Animal Sellers and LSD officers.

Characteristics	Animal Sellers (N)	LSD Officers (N)
Age (mean in years)	31.70	47
Sex		
Male	10	5
Female	0	0
Religion		
Islam	7	4
Hinduism	2	1
Christianism	1	0
Education		
Below Grade 10	1	0
Grades 10–12	7	0
Bachelors and above	2	5
Average duration at the current occupation (in years)	8.4	21

Our participants shared that many people owned caged wild birds as companion animals in the past in both rural and urban Bangladesh. The owning of birds was due to space limitations in dwellings as well as the influence of religious beliefs. All the companion animal owners, sellers and livestock service department officers mentioned that people did not own dogs because they were haram (forbidden) in Islam. A 36-year-old female companion animal owner stated:

> " ... historically individual houses are very rare in our urban settings. It was one of the reasons why people living in apartments did not own companion animals. Also, there are some socio-cultural barriers, such as religious beliefs, that prohibited people from owning non-human animals in living rooms."

Religious forbidding of the ownership of companion animals, especially dogs, continues to the present day in Bangladesh [10,25,48]. Similar to other Muslim-majority countries, the people of Bangladesh, and especially older adults, do not allow dog ownership or dogs inside a dwelling, except for security purposes [49–52]. There is growing debate in the literature among pro-dog and anti-dog Islamic scholars. In contrast to the anti-dog scholars, drawing on the Hadith, the pro-dog scholars argue that Islamic sentiments are being incorrectly applied against dogs [53–55]. The anti-dog scholars capitalize on the religious beliefs and sentiments of the majority of people and claim that owning a dog as a companion animal has been prohibited in Islam.

3.2.2. Current Day

Participants in our study shared that since the turn of the century, the companion-animal role has evolved in Bangladesh, with some recognition of their potential positive benefits to human social, psychological and physical health. This aligns with research findings in North America and other parts of the world [56,57]. Still, almost all of our study participants mentioned that the vast majority of people in Bangladesh are of the position that it is wasteful to invest money in non-human animals while humans suffer in the country.

Reflecting on their experiences with and perceptions of companion animals, the majority of companion animal owners in our study shared that interaction with their companion animals benefits their psychological health. They specifically shared that it reduces their sense of loneliness and controls stress, as well as depression, in their lives. A 38-year-old cat owner shared:

> "I am a service holder. Most often I have extreme workload at office. Further, I stay alone at my residence given that my wife lives in another town. So, interaction with my cats help me to de-stress myself as well as gives a feelings like someone is around me."

Some participants also reported that owning a companion animal benefits their physical health by requiring them to be active because of the associated caretaking duties. A 62-year-old companion animal owner shared: "I am retired person. I have also diabetes. My dog helps me to go for walk everyday in the morning and evening. I also wash them, clean her potty, and take her to veterinarian hospital for regular check-up and vaccination. All these helps me to be physically active." A few participants shared that they quit smoking in consideration of the harmful second-hand impacts on their companion animals.

All of the participants in the study reported that owning companion animals is beneficial to their social networks and interactions, both online and in person. For example, they shared that there are several "animal lovers' pages" on social media platforms, specifically Facebook and WhatsApp. Members of these online groups organize animal shows and meet and greet events. Some share information about cases of violence against animals, including the killing of stray dogs and cats by city corporations and individuals, and the violation of the animal welfare law in Bangladesh. A 36-year-old companion animal owner commented:

> "We the companion animal owners organized several processions against mass culling of stray dogs in our town. Companion animal owners and activists from other cities also participated in those movements. Finally, we did a press conference where we recommended updating the hundred years old animal welfare act in Bangladesh. Our movements were successful and we believe the *Animal Welfare Act 2019* is one of the successes of our movements."

Participants identified the evolving role of companion animals in Bangladesh as being particularly evident during the COVID-19 pandemic. Most veterinary surgeons and some companion animal owners observed that the number of people owning companion animals increased at the start of the pandemic to combat isolation. A 45-year-old female dog owner shared that she, along with her family members, purchased a dog during the pandemic because social interactions were restricted. She reported that her companion dog helped reduce her loneliness and mental depression. A 35-year-old veterinary surgeon shared:

> "I have seen YouTube videos where psychiatrists are saying companion animals can help to reduce loneliness. They also suggest that animals' companionship can help children to get rid of from substance use, addiction to social media, smartphone and computers. These are the reasons, I think, why parents encouraged their children to own companion animals during the pandemic."

While studies across the globe have identified a positive impact of human–companion animal interactions on the mental wellbeing of owners during the COVID-19 pandemic [58,59], others have found limited or no impact [60,61]. Although larger-scale studies have yet to be conducted in Bangladesh, our exploratory results suggest there may be important, albeit qualified, perceived positive effects of companion animal ownership in the Bangladeshi context.

Our study participants shared that Western culture, specifically through social media channels, has influenced the changing role of companion animals in Bangladesh. In fact, all of the veterinary surgeons in our study shared that traditional views toward companion animals are gradually changing due to the availability of information about human–companion animal interactions in Western countries. Research originating in Scotland notes that individuals' perceptions of animals can be influenced by social media, with other researchers identifying that the emotional reactions animal elicit on social media is a facilitator of this [62–64]. Some companion animal owners reported that they felt their goodness as people was recognized for owning companion animals. A 61-year-old cat owner said, " ... people admire seeing my photo with my cats on Facebook. One day a person said, you must be so kind since you are kind towards animals." One of the companion animal owners extended this to suggest that owning an animal helped to create a social media identity for them, for example, as an "animal lover".

3.3. *Challenges of Owning Companion Animals in Urban Bangladesh*

The most common companion animal ownership challenge mentioned by the majority of our study participants is the harmful impact of stigma. This is foremost related to zoonotic disease transmission and living with companion animals.

Our study participants shared that there is fear of zoonotic disease transmission (from animals to humans) in Bangladesh. Two dog owners commented that people consider animals and animal owners to be "dirty" and "unhygienic." All but one participant mentioned a concern about the risk of disease transmission from humans to animals, and in fact, one companion animal owner and one seller erroneously shared that disease cannot be transmitted from humans to animals. All the veterinary surgeons shared that the possibility of zoonotic disease transmission is a valid concern in Bangladesh given the lack of consistent and preventative hygienic practices among companion animal owners. One veterinary surgeon claimed that only about one-third of companion animal owners in Bangladesh are knowledgeable about zoonotic disease transmission and vaccinate their companion animals. A 58-year-old veterinary surgeon said, "making both animal owners and non-owners [aware] is important, in order to limit the transmission rate of zoonotic diseases as well as to control the fears about animals". There is a need for companion animals in Bangladesh to be vaccinated to counteract zoonotic-related fears and the consequent poor treatment and stereotypes of companion animals [65].

All the companion animal owners mentioned conflicts with landlords and neighbours because of their companion animal ownership. They reported that neighbours were concerned about noise and the cleanliness of their dwellings. A 23-year-old male dog owner mentioned that one of his neighbours poisoned and killed his dog because of his dog's barking at night. A 35-year-old female dog owner said that she ignores what others think about her and does not react because it could impact her relationship with her neighbours. Further, a 26-year-old female dog owner said, "in my town, it is almost impossible to find an apartment to stay in with companion animals. It seems like having a companion animal is like a sin." Indeed, companion animal owners' difficulty with renting because of the stigma attached to companion animal ownership is true in many parts of the world [66,67].

Many companion animal owners mentioned that the negative stigma attached to companion animal ownership limited their social interactions. They shared that some close relatives and friends do not visit their dwellings because an animal lives in it. Others shared that people will not eat food at their home because of the possibility of contamination with their companion animal's fur. A 35-year-old animal companion owner of a cat stated:

> " ... my brothers do not come to my house because of owning cats. I cannot organize any social gathering at my home given that there are many people who are afraid of [my] dogs. Thus, I feel uncomfortable inviting anyone at my home or offering them food because they might deny."

Over half of the companion animal sellers mentioned that they do not sell dogs given that people criticize "earning a livelihood by selling an animal like a dog". Two animal sellers shared that they do not sell dogs at the local market because buying and selling dogs is prohibited by the market committee. A 49-year-old companion animal seller shared that "animals are not treated with respect. Many people hate animals claiming that they are dirty. As a result those who are involved in animal related businesses are treated as the lower class citizens." Bangladesh is not the only country with negative views of dogs and other companion animal ownership [68,69], but the socio-historic context of Bangladesh nonetheless presents companion animal owners and sellers with significant structural challenges.

3.4. Companion Animal Welfare Concerns in Bangladesh

Our study participants were specifically asked about companion animal welfare concerns in Bangladesh. Informal observations were also made in the households, animal markets and farms where the interviews took place. Two primary and interrelated concerns were identified: the challenges of social–structural limitations in Bangladesh and the occurrence of animal abuse. Both can amplify the negative stigma attached to companion animals and contribute to the consequent inadequate implementation of the *Animal Welfare Act*.

Majority of the animal owners and sellers mentioned social–structural limitations as negatively impacting animal welfare in Bangladesh. They identified limited open spaces in urban areas to walk companion animals, limited number of animal shelters to house stray animals and inadequate veterinary treatment facilities alongside a lack of veterinarian expertise to treat companion animals. This is not surprising given the concerning poor treatment of farm animals in utilitarian roles [70], which would be similar to the current primary role of companion animals. In fact, most of the companion animal owners reported that veterinarians are often careless when treating sick and injured animals. They shared that veterinarians infrequently touched the animals in their care. One of the animal owners shared: " ... my dog died due to the wrong treatment of a veterinarian. Since then I do not go to veterinary hospital. Rather, I feed my dogs homeopathic medicine in case they become sick." Lack of expertise among veterinarians is a problem observed in many countries [71,72]. Consequent homeopathic treatment of animals is also evidenced in other countries [73,74]. Professional and adequate assessment and treatment is important for companion animal welfare and zoonotic disease management [70].

A key social–structural limitation influenced by stigma and identified by our study participants is that the *Animal Welfare Act* is not well implemented in Bangladesh. All the veterinary surgeons suggested that because of this people do not rely on the *Act*'s implementation. To explain, the livestock staff shared that companion animals are required to be registered at the sub-district level with the livestock service department, but the majority of owners do not. A 32-year-old veterinary surgeon said the reasons for not doing so are that there is no attached penalty and owners do not find any benefit in registering their companion animals. However, most of the veterinary surgeons emphasized the need to register companion animals because it could assist the livestock service department with keeping track of the number of companion animals, animal illnesses and could support investigating reasons for companion animal deaths. A 58-year-old veterinary surgeon shared, "[we] often hear about brutality against stray animals. Meanwhile, we barely hear anything about violence against companion animals because it remains unreported. Also, it occurs at the owners' personal space. Therefore, our department, or animal activists, cannot interfere."

In fact, the *Animal Welfare Act* indicates that proven violence against an animal can result in six months of imprisonment [26]. All the veterinary surgeons recommended strict implementation of the *Act*. Meanwhile, most of the companion animal owners reported the mishandling and abuse of companion animals by breeders and sellers. They shared that in order to make a profit, sellers too frequently bred companion animals, fed them poorly, and did not treat them when ill. A 35-year-old female dog owner shared, "I went to the largest animal wholesale market and found most of the dogs and cats cannot stand properly because they are frequently used for breeding purposes. I could not control my tears seeing the inhuman conditions of the animals in the market. I have never visited that market since then given that it is not possible to tolerate the inhuman conditions of the animals". The abuse of animals was also mentioned as a result of anti-dog religious Muslim sentiment. However, as pro-dog scholars are increasingly sharing, Islam does not permit cruelty toward any living animal [75,76].

Further, informal observations during the interviews documented that the majority of companion animal owners kept their dogs and cats properly in separate locked rooms or iron cages. During the interviews with the sellers and breeders, it was observed that all the sellers, and all but one breeder, kept different types of animals (i.e., birds, cats, dogs, rabbits) in side-by-side iron cages. The animal sellers reported that the livestock services department requires them to separate the cages of different animal species. However, due to space limitations in their stores, they keep different species in side-by-side cages. During the interviews, the sellers restricted the interviewer from taking any photographs of their stores, with one study participant indicating " ... it is not allowed [to take photographs and videos] by the market committee."

4. Conclusions and Suggested Next Steps

This study focused on understanding companion animal welfare in Bangladesh by exploring the benefits and challenges of owning companion animals in two urban centres. We documented the evolving role of companion animals in Bangladesh based on our study participants' experiences with and perceptions of companion animals. It appears that the number of people owning companion animals is increasing in Bangladesh, along with a recognition of zooeyia. Majority of our participants mentioned the social, psychological and physical health benefits of companion animal ownership. Our participants also identified key challenges to owning companion animals in urban Bangladesh due to the negative stigma associated with companion animal ownership. According to our participants, this is rooted in the continuing utilitarian role attached to companion animals by majority of the Bangladesh population and religion-based non-acceptance. The consequent lack of implementation of the *Animal Welfare Act* was identified as a key companion animal welfare concern.

According to the American Veterinary Medical Association, "[p]rotecting an animal's welfare means providing for its physical and mental needs"[77]. Recognition of zooeyia in Bangladesh offers an opportunity to combat stigma and the harmful consequences associated with companion animal ownership. Quite simply, a companion animal cannot benefit human health if they are not well cared for themselves. For example, it would be contradictory to consider the positive impact of petting a dog for human benefit (e.g., to increase stress-reducing hormones), while the dog themself is not well. In fact, companion petting an unhappy dog would most likely not result in a human health benefit. According to Proctor (2012), "[d]emonstrating objectively what animals are capable of is key to achieving a positive change in attitudes and actions towards animals, and a real, sustainable difference for animal welfare". The findings of this study helped identify four ways in which recognition of zooeyia, that is, the ability of companion animals to benefit human health, can help promote the welfare of companion animals in Bangladesh by challenging the stigma associated with them.

1. Zooeyia is a newer and growing concept in North America. For example, visiting companion dogs are being integrated into hospital environments to promote patient health. A recent controlled trial found a clinically significant decrease in patient pain following a 10-min visit with a dog in a hospital emergency department [78]. The trajectory toward recognition of the positive impact of companion animals in human health has likely been influenced in-part by the COVID-19 pandemic as well as Western social media (e.g., companion animals in human product advertisements). Companion animal-related stigma remains a concern though, especially in agriculture-based states and provinces. The lingering effects of the COVID-19 pandemic may offer a window of opportunity in Bangladesh to consider animal welfare as a result of attention being paid to their benefit in human lives during this particular time of need. This can be coupled with promotion of Bangladesh's newly released 2019 *Animal Welfare Act*. A potential starting point is stray animal societies amenable to sharing frontline stories about the impacts of zooeyia. A key collaborator to also consider is the Livestock Service Department in Bangladesh because it requires assistance in promoting its work that is dependent on the *Animal Welfare Act*.
2. A specific human health area that highlights the presence of zooeyia is interpersonal violence. We know from the literature in North American that the important relationships individuals form with their companion animals can prevent them from leaving unsafe living conditions. A recent Canadian study of domestic violence shelter

workers found that 77% of individuals did not leave abusive situations because they did not want to leave their companion animals behind [79]. It is well-established that there is "a link" between the abuse of humans and the abuse of animals [80]. Family and community violence is a concern in Bangladesh [81]. Recent statistics indicate that Bangladesh ranks 4th in the world in terms of violence against women by intimate partners [82]. Other research shows that over 70% of women face domestic violence in their lifetime [83]. Research also suggests that only 4% of victims report their abuse to police [84]. To our knowledge, there are no published studies in Bangladesh about violence against women and the link with companion animals. There is, however, literature suggesting that religious belief can influence behaviour and that companion animal ownership is lower among church attendees [85]. Paying attention to "the link" may be a novel way of increasing attention in the Bangladesh community about interpersonal and domestic violence and human health, and consequently influence discussion about animal welfare.

3. One Health is increasingly recognized today because of the COVID-19 pandemic, with near exclusive attention to zoonotic disease transmission, including in Bangladesh. For example, resent research has examined hygiene practices in Bangladesh refugee camps during the pandemic [86]. As shared, zooeyia is an important but under-recognized component of One Health. More specifically, the related One Welfare framework specifically considers animal welfare. According to Animal Health Canada, "One Welfare builds on the One Health concept and is a way to recognize the many social interconnections between human welfare, animal welfare and the integrity of the environment" [87]. Attention to One Welfare in Bangladesh alongside pandemic-related One Health work and scholarship may be an important means to consider animal welfare. One suggestion is to explore the influence of social media as there seems to be an existing platform to build off of.
4. There is much to be learned from across the globe about the potential to improve animal welfare by promoting practices that recognize zooeyia. For example, street dogs in Kampala, Uganda, are often killed to control diseases and their population [88] and the history of dogs is similar to that of Bangladesh—for hunting and security [89]. Meanwhile, in Northern Uganda, a project has trained street dogs to assist "survivors of war [to] cope with their trauma", with the beneficial impacts of the human–animal bond highlighted [89]. In Canada, as another example, there is a human rights designation of emotional support animals across many provinces and territories. These are animals that support the mental health of their owners via the human–animal bond and these animals are allowed in rentals that are not designated to allow companion animals [90]. The human–animal bond is defined by the American Veterinary Association as "a mutually beneficial and dynamic relationship between people and animals that is influenced by behaviours essential to the health and wellbeing of both. This includes, among other things, emotional, psychological, and physical interactions of people, animals, and the environment" [90]. It recognizes the reciprocal nature of the human–companion animal relationship and thereby supports and extends the concept of zooeyia.

Despite the novelty and importance of the insights gathered in this exploratory study and their potential to inform companion animal welfare in Bangladesh, we recommend considering them with caution. Foremost, the small sample size of this study may not be representative of the entire companion-animal owning population in Bangladesh, or even the two targeted urban centres. However, this study is one of the first efforts of its type to identify the contemporary beneficial role of companion animals in human health and its potential to promote companion animal welfare in Bangladesh.

Supplementary Materials: The following supporting information can be downloaded at: https://www.mdpi.com/article/10.3390/ani13091523/s1.

Author Contributions: Conceptualization: C.A.D. and A.-H.M.K.; methodology: C.A.D., T.K. and A.-H.M.K.; formal analysis: A.-H.M.K.; investigation: A.-H.M.K.; data curation, A.-H.M.K.; writing—original draft preparation: A.-H.M.K. and C.A.D.; writing—review and editing, T.K., C.A.D. and A.-H.M.K.; supervision: C.A.D. All authors have read and agreed to the published version of the manuscript.

Funding: This research was partially funded by the Office of One Health & Wellness, University of Saskatchewan, to support data collection and publication costs.

Institutional Review Board Statement: The study was conducted as part of AHMK's MA thesis at the Department of Sociology at the University of Saskatchewan. The research protocol [Beh#2973] was reviewed and approved by the Behavioural Research Ethics Board at the University of Saskatchewan.

Informed Consent Statement: Informed written consent was obtained from all participants involved in the study.

Data Availability Statement: The data presented in this study are available on request from the corresponding author and the senior author. The data are not publicly available in deference to participants' privacy in accordance with the approval of the Ethics Committee.

Acknowledgments: We wish to acknowledge all the participants for their participation and valuable information. We are grateful to Sadekur Rahman, Umme Umama, Sajal K. Roy and Debashis Mondal for their support during data collection.

Conflicts of Interest: The authors declare no conflict of interest.

References

1. Fox, R.; Gee, N.R. Great expectations: Changing social, spatial and emotional understandings of the companion animal–human relationship. *Soc. Cult. Geogr.* **2019**, *20*, 43–63. [CrossRef]
2. Marshall-Pescini, S.; Schaebs, F.S.; Gaugg, A.; Meinert, A. The Role of Oxytocin in the Dog–Owner Relationship. *Animals* **2019**, *9*, 792. [CrossRef] [PubMed]
3. WHO. One Health. Available online: https://www.who.int/news-room/questions-and-answers/item/one-health (accessed on 8 April 2023).
4. Hodgson, K.; Darling, M. Zooeyia: An essential component of "One Health". *Can. Vet. J.* **2011**, *52*, 158–161.
5. Williamson, L.E.; Dell, C.A.; Chalmers, D.; Cruz, M.; De Groot, P. Strengthening Zooeyia: Understanding the Human-Animal Bond between Veterans Living with Comorbid Substance Use and Posttraumatic Stress Disorder and their Service Dogs. *Hum.-Anim. Interact. Bull.* **2022**, *2022*, 20–47. [CrossRef]
6. Pinello, K.C.; Niza-Ribeiro, J.; Palmieri, C.; Ruiz, J.; Zaidan Dagli, M.L. Risks and benefits of the interaction with companion animals. *One Health* **2022**, 113–153. [CrossRef]
7. Carlisle, G.K.; Johnson, R.A.; Wang, Z.; Brosi, T.C.; Rife, E.M.; Hutchison, A. Exploring Human–Companion Animal Interaction in Families of Children with Autism. *J. Autism Dev. Disord.* **2020**, *50*, 2793–2805. [CrossRef] [PubMed]
8. Owens, N.; Grauerholz, L. Interspecies Parenting: How Pet Parents Construct Their Roles. *Humanit. Soc.* **2019**, *43*, 96–119. [CrossRef]
9. Wong, P.W.C.; Yu, R.W.M.; Ngai, J.T.K. Companion animal ownership and human well-being in a metropolis—The case of Hong Kong. *Int. J. Environ. Res. Public Health* **2019**, *16*, 1729. [CrossRef]
10. Kamal, A.-H.M. *Examining the Social Benefits of Human-Animal Interaction in Urban Bangladesh*; University of Saskatchewan: Saskatoon, SK, Canada, 2022.
11. Brooks, H.L.; Rushton, K.; Lovell, K.; Bee, P.; Walker, L.; Grant, L.; Rogers, A. The power of support from companion animals for people living with mental health problems: A systematic review and narrative synthesis of the evidence. *BMC Psychiatry* **2018**, *18*, 31. [CrossRef]
12. Hodgson, K.; Barton, L.; Darling, M.; Antao, V.; Kim, F.A.; Monavvari, A. Pets' impact on your patients' health: Leveraging benefits and mitigating risk. *J. Am. Board Fam. Med.* **2015**, *28*, 526–534. [CrossRef]
13. Lee, H.; Collins, D.; Creevy, K.E.; Promislow, D.E.L. Age and Physical Activity Levels in Companion Dogs: Results From the Dog Aging Project. *J. Gerontol. Ser. A Biol. Sci. Med. Sci.* **2022**, *77*, 1986–1993. [CrossRef] [PubMed]
14. Zhao, X.; Huang, H.; Du, C. Association of physical fitness with cognitive function in the community-dwelling older adults. *BMC Geriatr.* **2022**, *22*, 868. [CrossRef] [PubMed]
15. Hesselmar, B.; Hicke-Roberts, A.; Lundell, A.C.; Adlerberth, I.; Rudin, A.; Saalman, R.; Wennergren, G.; Wold, A.E. Pet-keeping in early life reduces the risk of allergy in a dose-dependent fashion. *PLoS ONE* **2018**, *13*, e0208472. [CrossRef] [PubMed]

16. Johnson, A.; Bruneau, L. Pets and relationships: How animals help us understand ourselves and our connections with others. In *Clinician's Guide to Treating Companion Animal Issues*; Academic Press: Cambridge, MA, USA, 2019; pp. 173–191. ISBN 9780128129623.
17. Brooks, H.; Rushton, K.; Lovell, K.; McNaughton, R.; Rogers, A. He's my mate you see': A critical discourse analysis of the therapeutic role of companion animals in the social networks of people with a diagnosis of severe mental illness. *Med. Humanit.* **2019**, *45*, 326–334. [CrossRef]
18. Cardoso, S.D.; Faraco, C.B.; de Sousa, L.; Da Graça Pereira, G. History and evolution of the European legislation on welfare and protection of companion animals. *J. Vet. Behav.* **2017**, *19*, 64–68. [CrossRef]
19. Hrubenja, A. 35 Disturbing Animal Abuse Statistics (2022 UPDATE). *Petpedia*, 14 December 2022.
20. Schlutz, A.M. Facts, Statistics, and Stories About Animal and Pet Abuse. *Pet Helpful*, 26 December 2022.
21. Hoy-Gerlach, J.; Ojha, M.; Arkow, P. Social Workers in Animal Shelters: A Strategy Toward Reducing Occupational Stress Among Animal Shelter Workers. *Front. Vet. Sci.* **2021**, *8*, 1–10. [CrossRef]
22. Laybourne, A. 50 Alarming Animal Cruelty Statistics–Enough to Crush One's Heart. Available online: https://worldanimalfoundation.org/advocate/animal-cruelty-statistics/ (accessed on 10 February 2023).
23. Haroon, J.U. Of Man and Animals as Companion. *Daily Financial Express*, 2022.
24. Ali, M.A. Pet Culture Grows, Business too. *Business Post*, 22 January 2022.
25. Sayem, S.N. The Dire State of Animal Rights in Bangladesh. *The Business Standard*, 6 January 2022.
26. Daily Star. An Overview of the Animal Welfare Act 2019. 2019. Available online: https://www.thedailystar.net/law-our-rights/news/overview-the-animal-welfare-act-2019-1813918 (accessed on 21 August 2022).
27. Sultana, R.N.; Uddin, A.S.; Asmaul, H.; Yesmin, R.N.; Sabina, Y.; ATM, B.; Sadikul, I.M.; Monira, N.; Jahengir, A.K. Prevalence of diseases in pet animals at Dhaka city of Bangladesh. *Ann. Vet. Anim. Sci.* **2016**, *3*, 1–5.
28. Obhoyaronno Obhoyaronno-Bangladesh Animal Welfare Foundation. Available online: https://www.facebook.com/obhoyaronnobangladesh/ (accessed on 28 March 2023).
29. GPC Guardians of Paws and Claws. Available online: https://www.facebook.com/groups/530950651223157/ (accessed on 28 March 2023).
30. Sinclair, M.; Lee, N.Y.P.; Hötzel, M.J.; de Luna, M.C.T.; Sharma, A.; Idris, M.; Derkley, T.; Li, C.; Islam, M.A.; Iyasere, O.S.; et al. International perceptions of animals and the importance of their welfare. *Front. Anim. Sci.* **2022**, *3*. [CrossRef]
31. Stucki, S. Towards a theory of legal animal rights: Simple and fundamental rights. *Oxf. J. Leg. Stud.* **2020**, *40*, 533–560. [CrossRef]
32. Hammoud, R.; Tognin, S.; Bakolis, I.; Ivanova, D.; Fitzpatrick, N.; Burgess, L.; Smythe, M.; Gibbons, J.; Davidson, N.; Mechelli, A. Lonely in a crowd: Investigating the association between overcrowding and loneliness using smartphone technologies. *Sci. Rep.* **2021**, *11*, 24134. [CrossRef]
33. Tan, J.S.Q.; Fung, W.; Tan, B.S.W.; Low, J.Y.; Syn, N.L.; Goh, Y.X.; Pang, J. Association between pet ownership and physical activity and mental health during the COVID-19 "circuit breaker" in Singapore. *One Health* **2021**, *13*, 100343. [CrossRef] [PubMed]
34. Wang, J.; Mann, F.; Lloyd-evans, B.; Ma, R.; Johnson, S. Associations between loneliness and perceived social support and outcomes of mental health problems: A systematic review. *BMC Psychiatry* **2018**, *18*, 156. [CrossRef]
35. BBS Dhaka Population 2022 (Demographics, Maps, Graphs). Available online: https://worldpopulationreview.com/world-cities/dhaka-population (accessed on 17 June 2022).
36. Opu, M.H. Dhaka Ranks World's Sixth Most Populous City. *Dhaka Tribune*. 2022. Available online: https://www.dhakatribune.com/bangladesh/2022/01/14/dhaka-ranks-worlds-sixth-most-populous-city (accessed on 17 June 2022).
37. Parker, C.; Scott, S.; Geddes, A. Snowball sampling. *SAGE Res. Methods Found* **2020**, 1–14. [CrossRef]
38. Pavelich, A.R. *Exploring Mattering and the Human-Animal Bond: The Impact of Service Dogs for Military Veterans at High Risk for Suicide*; University of Saskatchewan: Saskatoon, SK, Canada, 2021.
39. Staller, K.M. Confusing questions in qualitative inquiry: Research, interview, and analysis. *Qual. Soc. Work* **2022**, *21*, 227–234. [CrossRef]
40. Buys, T.; Casteleijn, D.; Heyns, T.; Untiedt, H. A Reflexive Lens on Preparing and Conducting Semi-structured Interviews with Academic Colleagues. *Qual. Health Res.* **2022**, *32*, 2030–2039. [CrossRef]
41. Deb, A.K.; Emdad Haque, C. "Every mother is a mini-doctor": Ethnomedicinal uses of fish, shellfish and some other aquatic animals in Bangladesh. *J. Ethnopharmacol.* **2011**, *134*, 259–267. [CrossRef] [PubMed]
42. Hamid, M.A.; Hossain, K.M. Role of private sector in the development of dairy industry in Bangladesh. *Livest. Res. Rural. Dev.* **2014**, *26*.
43. Swain, B.B.; Teufel, N. The impact of urbanisation on crop–livestock farming system: A comparative case study of India and Bangladesh. *J. Soc. Econ. Dev.* **2017**, *19*, 161–180. [CrossRef]
44. Ali, Z.; Hossain, I. Barriers to Development of Livestock Sub-Sector in Bangladesh. *Bangladesh Institute of Development Studies*. Available online: https://bids.org.bd/page/researches/?rid=36 (accessed on 28 March 2023).
45. Chowdhury, M.S.H.; Izumiyama, S.; Nazia, N.; Muhammed, N.; Koike, M. Dietetic use of wild animals and traditional cultural beliefs in the Mro community of Bangladesh: An insight into biodiversity conservation. *Biodiversity* **2014**, *15*, 23–38. [CrossRef]
46. Alam, H.; Basher, N. Pay for Pets! *The Daily Star*. 2022. Available online: https://www.thedailystar.net/news/bangladesh/news/pay-pets-3202771 (accessed on 28 March 2023).

47. Tamanna, R. Dog Relocation Restarts Quietly in Dhaka South, Defying Laws. *Business Standard*. 2021. Available online: https://www.tbsnews.net/bangladesh/dog-relocation-restarts-quietly-dhaka-south-defying-laws-332035 (accessed on 28 March 2023).
48. Islam, M.K.; Alim, M.A.; Tsuji, N.; Mondal, M.M.H. An investigation into the distribution, host-preference and population density of Ixodid ticks affecting domestic animals in Bangladesh. *Trop. Anim. Health Prod.* **2006**, *38*, 485–490. [CrossRef]
49. Horsfall, J. *Signifying Difference: Muslim Experiences with Dogs in Canada Johanne*; Carleton University: Ottawa, ON, Canada, 2014.
50. Bergeaud-Blackler, F.; Fischer, J.; Lever, J. *Halal Matters: Islam, Politics and Markets in Global Perspective*; Taylor & Francis: New York, NY, USA, 2016; ISBN 9781317597384.
51. Demello, M. *Animals in Literature and Film*; Columbia University: New York, NY, USA, 2012; Volume 7, ISBN 9780231152945.
52. See, A.W.-L. The Dog Debate in Malaysia. *Hsuan Chuang J. Buddhism Stud.* **2015**, *23*, 119–138.
53. Costu, M.D. Pets in the interfaith family homes: The case of turkish speaking migrants in the northwest of england. *J. Int. Soc. Res.* **2022**, *15*, 1–14.
54. Kozlowski, G.C. Shariat and Ambiguity in South Asian Islam by. *Int. J. Middle East Stud.* **2023**, *22*, 120–123. [CrossRef]
55. Gartland, L.A.; Firth, J.A.; Laskowski, K.L.; Jeanson, R.; Ioannou, C.C. Sociability as a personality trait in animals: Methods, causes and consequences. *Biol. Rev.* **2022**, *97*, 802–816. [CrossRef]
56. Barcelos, A.M.; Kargas, N.; Packham, C.; Mills, D.S. Understanding the impact of dog ownership on autistic adults: Implications for mental health and suicide prevention. *Sci. Rep.* **2021**, *11*, 23655. [CrossRef] [PubMed]
57. Hunjan, U.G.; Reddy, J. Why Companion Animals Are Beneficial During COVID-19 Pandemic. *J. Patient Exp.* **2020**, *7*, 430–432. [CrossRef]
58. Shoesmith, E.; Shahab, L.; Kale, D.; Mills, D.S.; Reeve, C.; Toner, P.; de Assis, L.S.; Ratschen, E. The influence of human–animal interactions on mental and physical health during the first COVID-19 lockdown phase in the U.K.: A qualitative exploration. *Int. J. Environ. Res. Public Health* **2021**, *18*, 976. [CrossRef] [PubMed]
59. Lukoševičiūtė, J.; Šmigelskas, K. Mental Health during COVID-19 Pandemic: Qualitative Perceptions among Lithuanian Adolescents. *Int. J. Environ. Res. Public Health* **2022**, *19*, 7086. [CrossRef]
60. Phillipou, A.; Tan, E.J.; Toh, W.L.; Van Rheenen, T.E.; Meyer, D.; Neill, E.; Sumner, P.J.; Rossell, S.L. Pet ownership and mental health during COVID-19 lockdown. *Aust. Vet. J.* **2021**, *99*, 423–426. [CrossRef]
61. Riddle, E.; Mackay, J.R.D. Social media contexts moderate perceptions of animals. *Animals* **2020**, *10*, 845. [CrossRef]
62. Lenzi, C.; Speiran, S.; Grasso, C. "Let me take a selfie": Implications of social media for public perceptions of wild animals. *Soc. Anim.* **2020**, *18*, 64–83. [CrossRef]
63. Martin Brown Why Pet-Related Social Media Content Is So Effective 2021. Available online: https://jarvee.com/why-pet-related-social-media-content-is-so-effective/ (accessed on 28 March 2023).
64. Id, Y.B.R.; Hoque, M.; Blanton, J.D.; Kennedy, E.D.; Id, S.R.; Tahmina, S.; Id, S.B.; Head, J.R.; Wallace, R.M. Rabies healthcare-seeking behaviors of urban and peri-urban residents: Results from a rabies knowledge, attitudes, and practices survey, Bangladesh. 2018. *PLoS Negl. Trop. Dis.* **2022**, *16*, e0010634. [CrossRef]
65. Power, E.R. Renting with pets: A pathway to housing insecurity? *Hous. Stud.* **2017**, *32*, 336–360. [CrossRef]
66. Graham, T.M.; Milaney, K.J.; Adams, C.L.; Rock, M.J. "Pets negotiable": How do the perspectives of landlords and property managers compare with those of younger tenants with dogs? *Animals* **2018**, *8*, 32. [CrossRef] [PubMed]
67. Payne, E.; Bennett, P.C.; McGreevy, P.D. Current perspectives on attachment and bonding in the dog–human dyad. *Psychol. Res. Behav. Manag.* **2015**, *8*, 71–79. [CrossRef]
68. Albertson, B.L. Dog-Whistle Politics: Multivocal Communication and Religious Appeals. *Political Behav.* **2015**, *37*, 3–26. [CrossRef]
69. Paul, A.K.; Shathi, I.J.; Biswas, D. People's Perception, Attitude in Relation to Animal Welfare and Constraint of Livestock Rearing in Bangladesh. *Int. J. Res. Rev.* **2020**, *7*, 506–514.
70. Christiansen, S.B. Assessment of animal welfare in a veterinary context—A call for ethologists. *Appl. Anim. Behav. Sci.* **2007**, *106*, 203–220. [CrossRef]
71. Loomans, J.B.A.; Van Weeren, P.R.; Vaarkamp, H.; Stolk, P.T.; Barneveld, A. Quality of equine veterinary care: Where can it go wrong? A conceptual framework for the quality of equine healthcare, based on court cases against equine practitioners in The Netherlands. *Equine Vet. Educ.* **2008**, *20*, 159–165. [CrossRef]
72. Morales-vallecilla, C.; Ram, N.; Villar, D.; Camila, M.; Bustamante, S.; Ferguson, D. Survey of Pain Knowledge and Analgesia in Dogs and Cats by Colombian Veterinarians. *Vet. Sci.* **2019**, *6*, 6. [CrossRef]
73. Chapman, S.F. Homeopathic and integrative treatment for feline hyperthyroidism e four cases (2006–2010). *Homeopathy* **2011**, *100*, 270–274. [CrossRef] [PubMed]
74. Fuseini, A.; Knowles, T.G.; Hadley, P.J.; Wotton, S.B. Food and companion animal welfare: The Islamic perspective. *CAB Rev. Perspect. Agric. Vet. Sci. Nutr. Nat. Resour.* **2017**, *12*, 1–6. [CrossRef]
75. Sulaiman, K.O. A Debate on the Islamic Rullings in the Natures and Functions of Dogs in the Light of Hadith. *Nucl. Phys.* **1959**, *13*, 104–116.
76. AVMA Animal Welfare: What Is It? Available online: https://www.avma.org/resources/animal-health-welfare/animal-welfare-what-it#:~:text=An%20animal%20is%20in%20a,pain%2C%20fear%2C%20and%20distress (accessed on 22 March 2023).

77. Carey, B.; Dell, C.A.; Stempien, J.; Tupper, S.; Rohr, B.; Carr, E.; Cruz, M.; Acoose, S.; Butt, P.; Broberg, L.; et al. Outcomes of a controlled trial with visiting therapy dog teams on pain in adults in an emergency department. *PLoS ONE* **2022**, *17*, e0262599. [CrossRef]
78. SaskSPCA The Relationship between Animal Abuse and Interpersonal Violence. Available online: https://www.saskspca.ca/resources/the-link/about-the-link/ (accessed on 22 March 2023).
79. American Humane Understanding the Link between animal abuse and family violence. *Int. J. Environ. Res. Public Health* **2003**, *17*, 3116. [CrossRef]
80. Sifat, R.I. Impact of the COVID-19 pandemic on domestic violence in Bangladesh. *Asian J. Psychiatry* **2020**, *53*, 102393. [CrossRef]
81. Rahman, K.F. Focus on Domestic Violence in Bangladesh: A Study from Criminological Perspectives. *J. Int. Womens Stud.* **2019**, *20*, 98–115.
82. Sultana, M. Why Is Bangladesh Still Seeing a Rise in Violence against Women. *Bdnews24.com*. 2022. Available online: https://bdnews24.com/society/93uzt307kg (accessed on 26 March 2023).
83. Study Finds Gender-Based Violence in Bangladesh Goes Unreported to Police. *Bdnews24.com*. 2016. Available online: https://bdnews24.com/bangladesh/study-finds-gender-based-violence-in-bangladesh-goes-unreported-to-police (accessed on 28 March 2023).
84. Perry, S.L.; Burge, R.P. How Religion Predicts Pet Ownership in the United States. *J. Sci. Study Relig.* **2020**, *59*, 190–201. [CrossRef]
85. Kamal, A.M.; Huda, N.; Dell, C.A.; Hossain, S.Z.; Ahmed, S.S. Translational Strategies to Control and Prevent the Spread of COVID-19 in the Rohingya Refugee Camps in Bangladesh. *Glob. Biosecur.* **2020**, *2*.
86. AHC One Welfare-Animal Health Canada. Available online: https://animalhealthcanada.ca/work-areas/one-welfare (accessed on 13 March 2023).
87. Nakisanze, S. Kampala Residents Divided Over Poisoning of Stray Dogs To Fight Rabies. *Anim. Welf.* **2016**, *1*, 1–5.
88. Lirri, E. In northern Uganda therapy dogs are helping the survivors of war cope with their trauma. *Equal Times*, 1 January 2020.
89. SHRC. Policy on Support Animals. 2022. Available online: https://saskatchewanhumanrights.ca/education-resources/policies-guidelines/policy-on-support-animals/ (accessed on 29 March 2023).
90. AVMA Human-Animal Bond. Available online: https://www.avma.org/one-health/human-animal-bond (accessed on 13 March 2023).

animals

MDPI

Article

Perceptions of Cross-Cultural Challenges and Successful Approaches in Facilitating the Improvement of Equine Welfare

Suzanne Rogers [1,2,*], Natasha Y. P. Lee [2,3], Jo White [2] and Catherine Bell [1]

1 Equine Behaviour and Training Association, Surrey GU8 6AX, UK
2 Human Behaviour Change for Life CIC, Norfolk NR9 4DE, UK
3 Asia Animal Happiness Consultancy Puchong, Selangor 447180, Malaysia
* Correspondence: suz@hbcforlife.org

Simple Summary: Many projects that aim to improve the welfare of equids worldwide involve people from different countries and cultures working together. Given that professionals involved with multi-stakeholder projects often work cross-culturally, this study examined their experiences regarding the challenges involved and their reflections on how to work in a culturally sensitive way. Semi-structured interviews were conducted with 14 participants working in a total of 29 countries and analysed using thematic analysis. Key themes emerged from the responses to questions covering the areas of perceptions of animal welfare, challenges working cross-culturally and embracing cultural sensitivity. Previous works have highlighted the importance of shared linguistic knowledge, interpersonal skills and cultural knowledge, and these elements also emerged in this research. As well as providing insights into the challenges of working cross-culturally, the findings of this study have enabled the development of suggestions for how this work could be taken forward in a practical way to be of use to professionals in this sector.

Abstract: Projects that aim to improve the welfare of equids worldwide usually involve people from different countries and cultures working together. Given that professionals involved with multi-stakeholder projects often work cross-culturally, this study examined their experiences regarding the challenges involved in, and their reflections on, how to work in a culturally sensitive way. Semi-structured interviews were conducted with 14 participants working in a total of 29 countries and analysed using thematic analysis. Key response themes emerged from the responses to questions covering the areas of perceptions of animal welfare, challenges working cross-culturally and embracing cultural sensitivity. The overriding theme regarding perceptions of animal welfare was that of barriers to animal welfare, under which emerged the subthemes of limited financial and material resources, limited understanding of the tenets of animal welfare, and attachment to traditional medicines and practices. Exploring the key challenges resulted in two themes: challenges regarding the local context and etiquette, and those regarding working with different stakeholders. Considering cultural sensitivity, again, two themes emerged: the importance of trust and respect, and of working with local partners. Previous works have highlighted the importance of shared linguistic knowledge, interpersonal skills and cultural knowledge, and these elements also emerged in this research. As well as providing insights into the challenges of working cross-culturally, the findings of this study have enabled the development of suggestions for how this work could be taken forward in a practical way to be of use to professionals in this sector.

Keywords: equine welfare; working equines; cross-culture; perceptions; animal welfare

1. Introduction

There are four major organisations based in the UK [1], and similar organisations in other countries, that are working internationally with the aim of improving the welfare of working equines. Many of the most significant funding organisations regarding this work are acting cross-culturally, as they manage and fund projects in different countries to those in which the key staff are based.

There are an estimated 100 million equids in low- to middle-income countries, most of which are working animals providing income and support for their owners and communities [2]. Working equines are often owned by people with low socioeconomic status and, as such, are particularly important to marginalised and vulnerable groups (reviewed in [3]). Owners and caregivers are often unable to afford to manage their animals in a way that safeguards their welfare (reviewed in [3]). A study from nine low- to middle-income countries found that 90% of equids were suffering from hoof and limb issues, and 85% were underweight, with lameness, poor body condition, wounds, disease, parasites and dental problems also commonly reported ([4–6] and reviewed in [3]).

Projects aiming to improve the welfare of working equines often involve many stakeholders such as local veterinarians, local project staff, project managers (often not local), the authorities, veterinary authorities, tourists, equine owners and people who rent equines. Such projects, therefore, often involve people managing or working with others who belong to a different culture, i.e., working and communicating cross-culturally. Such cross-cultural working can pose challenges regarding clarity of communication and effectiveness. This paper aims to explore the experiences and recommendations of several people who work in this way.

The term "culture" is used to describe the collective social behaviour, norms and institutions found in human societies and includes collective knowledge, beliefs, arts, laws, capabilities and habits of individuals [7]. The Cambridge English Dictionary states that culture is *"the way of life, especially the general customs and beliefs, of a particular group of people at a particular time."* [8]. Therefore, although culture is often thought of as being shared between people in the same geographic location, in practice, people might identify more strongly, in terms of culture, with people sharing their religion, socioeconomic status or scientific discipline [9] than with people from the same region or country.

Saville-Troike's work [10] on cross-cultural communication demonstrated that, for successful communication to take place between people of different cultures, a person must have the appropriate linguistic knowledge, interaction skills and cultural knowledge. Building on this work, specifically with regard to qualitative interview research, Stanton argued that, to avoid misunderstandings, the interviewer must try to walk in the other person's shoes [11]. This is relevant to the present study not only in terms of communication between the professionals working cross-culturally with equid owners but also regarding the cross-cultural communication between the researcher and interviewees.

Perspective taking is the process by which we can understand the viewpoint of another person. Learning to do so has been considered, to some extent, to be a normal part of child development [12]. However, the degree to which we deliberately implement perspective taking when working cross-culturally can impact the success of those interactions. For example, Todd et al. [13] found that perspective taking was successful in reducing racial bias and improved the interpersonal experience of the interaction. Similarly, perspective taking during mediation was able to reduce conflict and improve relations between groups [14]. In a study of intergroup threat, opposition to policies that were perceived as threatening and overly beneficial to others was reduced in the experimental group who experienced perspective taking [15]. However, there are also cultural differences in perspective taking and empathic concern, whereby countries measuring more highly on empathy (as distinct from, but related to, perspective taking) were also found to have higher levels of collectivism and prosocial behaviour [16].

Over the past 15 to 20 years, the approaches used by international non-governmental organisations (iNGOs) have changed from predominantly reactive approaches, focusing on providing access to veterinary resources, to a proactive approaches, focusing on community engagement and prevention [17]. This shift has perhaps driven a change in the way that iNGOs, local partners and working equid owners interact with each other and has arguably made the need for the understanding of cross-cultural working even more important.

However, as also recognised by Sinclair and colleagues [18], apart from lessons learned by trial and error and personal experiences informally shared regarding working cross-culturally in animal welfare projects, there is a gap in the research to support the challenging task of running impactful projects internationally. Aside from the work of Sandøe and colleagues [19], there is a lack in the scientific literature exploring the ethical aspects of cultural differences regarding working equid welfare. In addition, much of the research published regarding the welfare challenges for working equids and the impact of projects to address those challenges (for example, [20]) has not covered the interpersonal aspects between the people working on those projects; this paper seeks to address this gap.

This qualitative study consisted of a series of semi-structured interviews with experienced professionals working in several different roles with the common goal of improving the welfare of working equines. The interviews were used to provide insight into the challenges faced, and successful approaches used, by staff working cross-culturally in iNGOs.

There were four key objectives of this study: first, to gain an understanding of the similarities and differences regarding how animal welfare is considered by people from different cultures; second, to gain an understanding of the challenges associated with working with stakeholders across different cultures; third, to gain insight into some ways that professionals aim to work in a culturally sensitive way; and fourth, to synthesise information in a way that will be useful for people when planning and implementing programmes to improve equine welfare across different cultures.

Despite the variations of the contexts and locations of the interviewees included in this study, there were key similarities in their perceptions of the crucial points to consider when working cross-culturally. Participants' responses were considered by taking into account Saville-Troike's elements [9] for effective cross-cultural communication. Although there are academic networks focusing on equine welfare internationally, few studies have synthesised the experiences of project managers in the field. To the authors' knowledge, this is the first study in the sector to examine the cross-cultural interpersonal aspects of equine welfare professionals. It is hoped that the collective experiences documented can aid everyone working in the field when planning, implementing, assessing or even just visiting projects that aim to improve equine welfare.

2. Materials and Methods

Interviews were conducted throughout January and February 2023 with a total of 14 people involved with projects concerning working equines in 29 countries (Sri Lanka, Indonesia, Thailand, Cambodia, Nepal, Costa Rica, Nicaragua, Honduras, Panama, Guatemala, Colombia, Argentina, Chile, Mexico, Haiti, Cuba, Hawaii, Senegal, the Gambia, Ghana, Kenya, Burkina Faso, Ethiopia, South Africa, Lesotho, Zimbabwe, Jordan, Pakistan and India; note that some interviewees worked in several countries). Interviewees were selected from the authors' network and resulting recommendations. All interviewees are currently heavily involved with equine welfare projects and work across cultures in different ways; some are native to one region and work with equid owners in another region, some share nationality with the equid owners but are employed by funding organisations from a different region, some manage projects in several countries and so on.

The criteria for inclusion were that the interviewee had to be professionally involved with working equines and that there had to be a cross-cultural nature (where culture was initially considered to be a shared native country) of that work. Through this form of sampling, a volunteer sample of participants was recruited. Note that communications were conducted in line with the processes outlined during the ethical review approval procedure (see the Institutional Review section at the end of this paper). After 14 interviews, data saturation was reached, with no new themes emerging using thematic analysis.

Interviews were arranged using email and took place online using Zoom 5.11.1 (6602) video conferencing software. For the subsequent transcriptions, all interviews were recorded and ranged in length from 18 min 06 s (Participant 12) to 45 min 56 s (Participant 13). Participants received a briefing document detailing information about the project and regarding consent before the interviews were conducted. Participants gave their verbal consent to the process at the beginning of each interview.

Participants were asked five questions on the theme of their experience working cross-culturally in their professional roles. The interviews were semi-structured, so the question script was flexible, and clarification was sometimes requested. Semi-structured interviews were chosen as the qualitative research tool because they provide a flexible technique for small-scale research [21], provide useful data from small sample sizes and allow thematic analysis of the qualitative data [22].

The interviews were transcribed using an automated online service and edited to remove duplicate words, the phrases "I know", "You know", "So", "Uh" and "Um" and to change the spelling from American English to British English. The names of any organisations were also edited out to remove any conflicts of interest and to maintain some anonymity. The grammar was edited throughout; for example, adding punctuation to help the readability, especially 'full stops', because the automatic transcription programme often created very long sentences. These changes were made whilst listening to the audio recording to ensure that the meaning was not changed by the editing. Participants were assigned codes P1 to P14 according to the order in which they were invited for interview. Thematic analysis was conducted following Braun and Clarke's six-stage process [23]. As also described by Rogers and Bell [24], Stage 1 (familiarisation) involved reading the transcripts and listening to the recordings to remind the researcher of the interview contents. Stage 2 (coding) was performed on the transcripts from the 14 interviews to facilitate the generation of themes. This stage was performed manually using the highlighter function in MS Word (Microsoft® Word for Microsoft 365 MSO (Version 2209 Build 16.0.15629.20200) to colour similar concepts and phrases that were repeated throughout the interviews [25]. The documents were then clustered, putting similar contents together, to enable stage 3 (generation of themes). In stage 4, these draft themes were reviewed to enable themes to be combined or separated. In stage 5, the themes were then defined and named before stage 6 in which they were written up into a narrative as the research results. The process was conducted by the lead researcher with support from a second person. The transcripts were all read again to extract quotes that were especially relevant to the emerging narrative. Finally, the extracted quotes were arranged (in a Word document) so that the most representative quotes could be used in the write-up of the research paper.

3. Results

The interviews explored four key areas: the context of the participant's work, perceptions of animal welfare across cultures, the challenges of working cross-culturally and embracing cross-cultural sensitivity. This section outlines and discusses the results of the thematic analysis that was used to examine the responses to questions in each of these four areas.

3.1. Exploring the Context of Participants' Work

It was found that some interviewees work in the country that they are native to, some work in many countries in a region that they are native to, and some work in a country that they had no connection with before their current role. The nature of cross-cultural working was described, for example, some interviewees work in their native country for an iNGO based in another region of the world. Participants described projects that involve stakeholders with multiple nationalities, for example, regional or national project staff, people from the donor organisations, international staff and tourists.

The participants described the nature of their role. The participants were all involved with projects aiming to improve the welfare of working equids: horses, ponies and mules used for activities such as the transport of people, goods and water; agricultural work such as ploughing; work in brick kilns and in the tourist industry (e.g., transport of tourists, tourist leisure riding and one participant was involved with a business providing rides for tourists rather than an equine welfare project). In addition, one interviewee was also working on a project regarding donkey slaughter and another on a project with equines used by the army.

Using the term "culture" as an umbrella term to encompass nationality was found to be unsuitable, as it became clear that the term "culture" more accurately refers to people who share similar socioeconomic backgrounds, education, religion, language or areas of habitation. This meant that participants who work with equid owners with shared nationalities to themselves often found differences in their "cultures", e.g., *"My case is quite particular because I'm a returnee from Cambodia. I grew up in France, so from my accent they consider me as a foreigner. When I say, "I'm Cambodian", they say, "No, you're not Cambodian, you're Korean!" so I have already this barrier in the sense of language."* (P6). Another participant outlined similar language challenges. Although also Sri Lankan, her primary language was English and not the predominant language, Singhalese. She had to translate all the veterinary terms from English into Singhalese in her head and reported that *"They think that I'm like struggling with myself, but it's more that I'm struggling with the language."* (P8). Finally, participants from the city described struggling to resonate with rural owners, even though they both came from the same country: *"We are all from the same country, so the culture is the same. But the lifestyle, the way of living, is very different—some areas are very rural, so the lifestyle, the way of living, is different from ours in the city."* (P5).

3.2. Perceptions of Animal Welfare

All the participants used the terms "animal welfare" and "equine welfare" throughout the interviews. The interviewees seemed keen to share the contexts of the equine owners they work with, and P10's quote was representative of the discussions regarding perceptions of animal welfare: *"I think the bigger problem isn't that people can't understand, it's that if they don't have money, they can't do much better"*.

When exploring perceptions of animal welfare in the countries in which the participants work, the overarching theme that emerged was "barriers to welfare". Within this overarching theme, three subthemes were nested that relate to the causes of these barriers: "limited financial and material resources", "limited understanding of the tenets of animal welfare" and "attachment to traditional medicines and practices". The first subtheme (limited financial and material resources) is best represented by the following quote: *"The carriage horse owners here are living in extreme poor economic conditions. They live next to the stable, not even a stable, an enclosure that is far from the hygiene and sanitations, where they also live, side to side with the cow, goat, the horses as well. In one tiny house, 10 or 15 people. They cannot fulfil their own welfare and that's why it's very difficult to get across the message about giving or providing welfare for their animals."* (P9).

Regarding subtheme 2 (limited understanding of the tenets of animal welfare), it was found that there is a general shared understanding of equine health between project managers, equid owners and funders but that the welfare domains are not well understood. Examples of illustrative quotes for this theme include: *"You say welfare in Spanish, which is 'bienstar', and automatically people start thinking about health and they start talking about vaccines and dewormers. They don't consider that horses are social creatures, have emotions, or can be in pain."* (P11) and *"A lot of the owners certainly see animal welfare as a negative thing, thinking of it as an enforcement process or something that's going to stop them using animals for their livelihoods."* (P3). The interviewee involved with a riding tour company (P14) described a different context whereby the tour operator's horses were well-looked after and resourced, but welfare was still not understood as a concept.

Regarding subtheme 3 ("attachment to traditional medicine and practices"), the participants mentioned challenges with traditional medicine and practices, that create a barrier to safeguarding welfare. For example, *"It was very difficult at the beginning for us to change their traditional practices to the new ones."* (P7), and more specifically, *"Most have stopped using monkeys [monkey blood] to treat horses but they do use traditional medicine a lot—and also human medicines to treat ponies! They give 10 times human tablets, for example, which can cause huge problems."* (P5) and *"A key issue is traditions passed down from generations; for example, urinating on the hoof to cure 'spider bites' or using beer to cure colic."* (P11).

Under the main theme ("barriers to animal welfare"), some examples of controversial aspects of animal welfare and how they are interpreted between different cultures and countries were provided. P8 explained how people often think that, if they neuter their animal, it will affect the person responsible for the animal in the same way, and they will not be able to have children either. P4 explained that a key challenge for her organisation is that Thailand is predominantly Buddhist, posing some moral and ethical dilemmas regarding welfare. For example, Buddhists resist the concept of euthanasia, believing that animals should be allowed to die naturally: *"To me that's a conflict because the animal was never allowed to live naturally, so why should it have to die naturally?"* She further explained that animals in the wild are different but, when in our care, we have the responsibility to prevent suffering. She went on to say *"It's the whole idea of samsara—that all life forms are a circle—there's rebirth, and reincarnation. I sometimes explain to owner that you could come back and get reborn as a pony so do you want to be treated like that?"*. P12 described challenges regarding the use of the word "welfare": *"With some owners, we don't use the word 'welfare' because they are afraid of activist groups that are trying to ban working animals [...]—we talk about good management practices, behaviour, health, emotions of the horses."* Some participants reflected on how the different ways horses are thought of from various ethical viewpoints affects the way they respond to the subject of animal welfare. For example, P3 explained that *"Some Ghanaians view animals very differently than we do—for example, they believe that animals have been given to us to be used and subsequently animal welfare is just not a widely discussed concept, animals are a means to an end. They don't see animals as something to hurt deliberately or to treat with malicious intent, it's more that there's a lack of understanding about their sentience and their ability to feel emotions and pain."*.

Despite the extensive barriers to animal welfare, all participants expressed throughout the interviews that the equine owners they work with, directly or through partners, are keen to learn and are open to discussions about equine welfare (whether or not explicitly termed "animal welfare"). P4's quote illustrated this theme: *"When people actually know that something is affecting the animal badly, they are quick to ask how to do it differently? People do care"*.

Some participants reflected on the way different stakeholders navigate the different understandings of animal welfare. For example, P6 explained *"Our funding organisation is very respectful of the local communities and local NGO they're working with. They give us the notion of what is welfare, and we cannot compare the welfare standard in Europe and the welfare standard in our community, in our country. Its too much to tell people to reach the standard in the sky when you are on the earth. [The organization] do not implement a 'must do' approach; they have their guidelines, and we can do as much as we can to reach that standard but they do not push us to reach a certain point. And I'm very appreciable for them because at a certain point, if you put too much, it will break and then everything will collapse. And this is very sensitive."* P13 articulated that solutions require a systems-based approach to animal welfare, where the role of the working equine is recognised as providing an important service to the community and, therefore, that it is not only the owner's responsibility but also that of local service providers, authorities, NGOs and other stakeholders.

3.3. Challenges Regarding Working Cross-Culturally

Perhaps the most representative reflections on this issue were articulated by P3: *"I think the biggest challenge for me in working cross-culturally is trying to separate out my own emotions from what I'm witnessing because it is so different to how we treat animals or how we see animals in our own culture. Trying to avoid judging because trying to put yourself in put in their shoes is incredibly difficult when I recognise that I have this massive white privilege, that I take for granted and I have so many things that they don't have. [...] Just trying to stay motivated and positive is really difficult when you're faced with such an awful animal welfare but also human welfare that is restricting their access, their ability to be able to improve that animal welfare."*.

Two key themes emerged regarding challenges in working cross-culturally: "challenges regarding understanding local context and etiquette" and "challenges regarding working with different stakeholders".

Theme 1, "challenges regarding understanding local context and etiquette", was illustrated by P13, who described the importance of mapping out a project to ensure the local context is truly understood because the project will effectively be external support to an area that has its own social and economic structures. He also described how the way the project is framed is important: *"We explain that we are not here to tell you what you need to do but we are here to learn from you, to support what you are doing, to manage risks and potentially improve what is good"*. P13 also described the importance of understanding the cultural background regarding the history of working equines in the project area: *"In Latin America, the horse is seen, especially in the indigenous areas, as fairly new tool for the communities because before the introduction of motor vehicles equids were mainly used for the wealthy landowners. The indigenous communities usually had to transport things themselves. Since the beginning of the motor vehicles the use of horses for transport has been pushing into more into the indigenous rural communities more and more."* P9 described how, in parts of Indonesia, the status of horse cart drivers has changed over the past 30 years. He said that 30 to 40 years ago, having a horse and cart was a source of pride and that they were relied on for transport. However, now, people in the community have motorbikes, cars and public transport, so things have changed, and being a cart driver has lost status. He described how the pandemic exacerbated this, as the tourist industry stopped, and now, the situation for horse cart drivers is one of abject poverty.

P12 highlighted the need to identify the context regarding social cues even if working in your own country: *"The first visit is just to talk with them, not even to see the horses but to see what they expect for us from us and see how we can help. Sometimes we can't help, but we can make a link with the municipality or with other people that maybe can help. The other thing, we have learned is not to impose our position. We try to facilitate the work and see if they can come up with solutions because otherwise it's not sustainable."*.

The context regarding social structure is also relevant: *"Cultural sensitivity extends to being sensitive to [...] things like their cultural structure—like the structure of families and communities and the roles within them. Some of the more rural places in Ghana have very strict roles and responsibilities and hierarchies, not just within the family but within the village. They'll have village elders who they go to for advice. [..] Working those relationships out, being sensitive to them, and working with the structure, not against it, is really important."* (P3).

P11 emphasised that the importance of not going into a community with a preconceived idea of what needs to be changed but rather working with the community to find out the context in terms of their issues, concerns, priorities and so on and how this approach can nurture trust, because you are helping them with an issue about which they are concerned. Regarding funders, one participant (P9) appreciated that a funder asked them to fill in an extensive questionnaire to understand the context of the working equine owners.

The need to understand the local context was expressed by the majority of the interviewees and eloquently summarised by P13: *"It is important that equid welfare projects that are coming from donors in Europe or the USA have a good understanding of the social, cultural and economic context of the communities. [...]. But we need to be very realistic about our expectations about what we can achieve—in a five- or ten-year project."*. P13 also described how different stakeholders think along different timescales: *"Owners are thinking, day to day, about what they're going to do tomorrow or what they're going to do a couple of days later, local authorities are thinking in terms of two years or four years—the time that they are in office. [...]."*.

Several interviewees expressed that it takes time to change behaviours and make an impact. For example, *"Sometimes it is not appreciated how slow it is to shift deeply held beliefs and attitudes. [...] We are not talking a workshop or a, a pamphlet or an advert or a billboard, it takes long hours of discussions, contact, time, trust, and building a relationship."* (P2). P7 described how sometimes the donors who live some distance from the project want to implement things that just are not practical within the project's context, especially in the timeframe expected. He also described the challenges of working with funding organisations based in other countries: *"The organisations do understand when they have visited the communities. It is sometimes very difficult to convince them through a Zoom call but when you bring them into the field and provide the exposure with the communities it is very easy for them to understand."*.

Other reflections on the theme of understanding the context included topics such as meeting etiquette and communication challenges. For example, P10 explained that she has experienced cultural challenges regarding behaviours associated with working in an effective team (e.g., being on time and prepared for meetings) and feels that a global professional culture is emerging as more people work cross-culturally. P10 also described how sometimes she must manage the expectations of the international project funding organisation. For example, in cases where the project had not progressed as hoped, she acted as an intermediary to ensure that the language used was clear to the staff on the ground. This had not always been the case due to cultural differences regarding the considered polite ways of framing things; sometimes, a more direct approach had been needed for the message to become clear.

Regarding the second theme in this section, "challenges regarding working with different stakeholders", the two key stakeholders mentioned were tourists and veterinarians. Challenges regarding the interface between tourists and working equines were described by several participants. One participant (P14) described the challenges a riding holiday tour operator had experienced associated with the views and "needs" of some of their tourist clients. For example, European tourists have expectations regarding the equipment that should be used (e.g., bits), and so, the tour operator adopted the use of equipment they would not usually use so that clients' expectations were met. She described how, on one occasion, the operator changed the horse allocated to the rider to avoid conflict due to the rider's method of handling a "slow" horse, an example of addressing cultural differences sensitively for the welfare of the horse.

P5 explained that, in a riding stable for tourists, common practices have included tying horses up in the sun and hitting them as a means of control. Tourists had sometimes complained to the welfare organisation, who had passed on their concerns, but the owners did not always respond positively: *"some people think the pony is like a robot"* (P5). P10 described challenges due to differences in cultures between tourists and people providing horses to be used to climb a volcano (Guatemala) or be ridden on the beach (Panama). In both cases, tourists have regularly complained to the hotels about the conditions of the horses: *"You can really see the disconnect between how we feel about horses and how the owners feel about them"* (P10).

P4 mentioned that this pressure from tourists has sometimes resulted in negative outcomes for the animals: *"With tourists, there's a lot of righteousness and thoughts that are very judgmental. Some of them are justified and some of them are not. For example, the Thai pony is small, no more than 13 hands, but when fed well, in good body condition, and with good equipment can pull more than twice its weight. When tourists come and complain that the ponies are too small to be pulling the carts, it puts pressure on the owners to look for larger ponies, crossbreeds and breeds that are not suited to working in a hot climate, and they suffer as they cannot withstand the heat. So, tourists putting pressure on the operators without understanding the context, is a huge challenge."*.

P1 described challenges that her organisation faces with the behaviour of expatriates who shout abuse at donkey cart drivers for beating the donkeys: *"If I was just riding my horse along the road and somebody started shouting and screaming at me, what you immediately do is either take it out on your donkey or get angry and be rude to the person shouting at you. It doesn't help the situation."*.

Several participants outlined challenges associated with veterinarians working cross-culturally. For example, P1 described *"What I say to all the vets that come out is that if all you can do is administer pain relief and give a kind word when that animal is dying, you've given more than it would've had if you hadn't been there. A lot of them get very emotional and say "If I was in the UK I could do this, that, and the other". And you can't, you have to accept that, and you have to see what you can do in the situation with the resources you have."* Similarly,

P5 described that, sometimes, international veterinarans who have visited his project in Cambodia have contradicted each other regarding the way they do things. For example, one veterinarian said that there was no need to give antibiotics after castration, and the animal then developed an infection. P4 outlined challenges with Western professionals who have visited the projects she is involved with who have seemed to want to save the world and how such judgement is not conducive to working in a culturally sensitive way. *"I don't want to kill the passion people feel that they can contribute to the world, but that's the key word, contribute. It's not saving the world. And we need to find a way to elevate a community without making them wrong all the time. We're not always right."* P4 also outlined the challenges this has posed for the in-country veterinary profession: *"People parachuting in creates a double standard for the veterinary profession that works within that country and that community, which makes matters worse because the profession needs to be able to sustain itself and grow and develop."*.

3.4. Embracing Cultural Sensitivity

Two key themes emerged when the interviewees reflected on embracing cultural sensitivity: "the importance of developing relationships based in trust and respect between project staff and equine owners" and "the importance of working with local partners".

Regarding theme 1, P4 reflected that, *"Imposing western ideas on how things should be done and trying to bulldoze that concept or that idea to get to an action destroys the relationship or sometimes the work of an organization"*. Several participants outlined specific ways in which people show respect. For example, P6 explained *"In our culture [...] you cannot say, "You have to do that". We talk very gently, smoothly, and with respect. Many of our visitors keep smiling. This is very important, basic and fundamental. [...] Regarding the intercultural differences and how we got accepted, the only way is that you need to be very, very respectful."* P9 highlighted the importance of respect when working with horse owners who can sometimes be very shy and have self-confidence and self-esteem issues due to the low status of their work. A final example quote for this theme illustrates the importance of a respectful approach: *"I was talking to a group of elders about it and I said, how 10 years ago if you'd suggested euthanasia, everybody would say no. But now it's, it's happening. And they said, "One, you've always respected us in our beliefs and so on; and two, because you always try, we know that if we take our animal to horse and donkey, they will try everything, and if they can't save it, no one can save it." I'd never thought about it like that, it was amazing. So inadvertently that wasn't the plan. It has kind of changed a culture, just by having respect for their beliefs."* (P1).

Theme 2 (the importance of working with local partners and community representatives) was highlighted by all of the interviewees. P5 explained the approach used in the project he manages: *"When we work in a new community, we ensure we have a contact person or the leader of the pony owner group, to introduce us. We have found that where we have such contacts it is easy to build trust, but if we don't get introduced to community, it is very difficult to step in."* P2 also described working with carefully chosen partners: *"They speak the language, they understand the environment that they're working in and are used to challenges of working there. They understand how traditional cultures work, how to negotiate, and how to work constructively in these areas."*. P11 explained that it is vital to work with community leaders or with people who can "translate" your messages into not just the local language but ensure that it is framed in a culturally appropriate way as well: *"They listen to their peers more than they would listen to an instructor that's come in, even from the next town"*.

P2 described the challenges in building genuine relationships with people: *"In Southern Africa, particularly in the rural projects where the mode of delivery is mostly primary healthcare clinics, people attend the clinics, and you don't know if you will ever see them again or if you've seen them before."* To address this, the projects she is involved with are working through intermediaries, representatives of communities that they have more contact with and support closely, recognising that they are on the ground 24/7, as opposed to the project staff who live 400 km away.

P4 described how, in Thailand, her project facilitates different communities in interacting with each other and that by doing so she has observed camaraderie, exchanges of ideas, competition and positive outcomes: *"I think that in any culture, not just Thai culture, the power of community is very important because then you can drive action easier. When we had communities that were not doing so well, we invited representatives from communities that are doing well, to visit them and work with the ponies, do some assessment and talk about it. This worked—the first community got the message because they actually saw improvements could happen, it's possible, as it has happened in a village not far from where they are."*

P2 described how organisational culture is also relevant — the importance of recognising that organisations have specific cultures that, in partnership projects, need to be navigated for effective collaboration. She presented the example of a partner organisation with a very bureaucratic culture that stifled the activities on the ground at the community level. She also described how the teams in different countries of the international funding organisation she is working in all work in slightly different ways: *"I sometimes joke with the UK staff because sometimes British people will want to say something or assert themselves, but they want to be culturally sensitive—and what they are trying to say just goes straight over the heads of some of the project staff. I have to translate English into English and it's nobody's fault, it's just getting used to each other."*

4. Discussion

Each of the four key research objectives outlined in the introduction were met. The results of the thematic analysis, together with the summaries of the responses to each interview question, provide insight into the challenges and successful approaches regarding working cross-culturally to improve equine welfare. Throughout the interviews, responses from the different professionals were consistent, as were the case studies they described with shared key elements.

The responses supported Saville-Troike's findings regarding cross-cultural communication — that for successful communication to take place between people of different cultures, a person must have the appropriate linguistic knowledge, interaction skills and cultural knowledge [10].

Several participants expressed that the key to working cross-culturally is also the key to being socially literate in general. For example, P1 said *"It's not really about the culture, it's more about really common human principles that go across any situation, whether they're people in your own culture or in other cultures. There's so much that's similar."*. The prevalence of reflections on the importance of trust and respect also seem to be key to social interactions whether within or between cultures.

Inherent in more general social literacy, is the need for perspective taking and an ability to view animal welfare from the position of the animal owners. Whether deliberately implementing perspective taking or doing so intuitively, all participants commented on the need to see situations from a local and cultural perspective, from recognising the interconnectedness of poverty and animal welfare to having an understanding of the religious barriers to a Western notion of welfare. This is in contrast with overt judgement by well-meaning, concerned tourists putting pressure on equid owners and inadvertently causing bad feelings that do not help the animals. Such concerns from Western tourists also suggests a "us against them" mentality regarding animal welfare in developing countries, focusing on a narrow range of welfare indicators (such as load size or condition). Although their concerns might be justified, the baseline for welfare should not be arbitrary factors but an established model such as the Five Domains [26]. In that case, Western standards of welfare can also be criticised for high numbers of horses experiencing pain (e.g. [27]), being unable to engage in their full ethogram of normal behaviours [28] and experiencing unresolved stress [29]. Thus, it is likely that cultural bias plays a key role in the perception of what constitutes "welfare".

P12 linked the benefits of working cross-culturally with the associated ethical responsibilities, which is certainly pertinent for anyone working on projects such as those described in this paper: *"I think working cross-culture is really enriching for us. You get to learn a lot, but I think we always have to take into account how it can be enriching for both sides and not just an extraction of information and then leaving the communities alone."*. This was also reflected by P1: *"The international vets want to put everything to sleep, they don't like the idea of waiting and seeing. One of the things we learned, is that a lot of our preconceptions are wrong. Many of the animals that I would've sworn could not get better have got better because we couldn't put them to sleep—so we've learned a lot as well."*.

As outlined in the introduction, Stanton's work with regard to qualitative interview research described that, to avoid misunderstandings, the interviewer must try to walk in the other person's shoes [11]. This is relevant to the present study not only in terms of communication between the professionals working cross-culturally with equid owners (arguably often in a similar context to that of an interviewer or researcher) but also regarding the cross-cultural communication between the researcher and interviewees. The sampling method, in retrospect, served to ensure that communication between the interviewer and interviewee was based on trust and respect due to either previous professional interactions or a personal introduction from shared contacts that was done in a very open and approachable way. On a related note, the cultural background of the interviewer in this study, in relation to the culture of the interviewees, should also be considered [30]. Reflexive practice [31] was used by the interviewer to self-consiously critique and evaluate how their subjectivity and own background might be influencing the research process.

A limitation of the study was the small number of interviewees. As explained in the methodology, the authors wanted to limit participation to enable a manageable dataset for a qualitative analysis. Further work could be done exploring professionals from a greater number of countries and, therefore, cultures.

The study highlighted several areas for future research. It would be interesting to assess whether the perceptions of a greater number of interviewees across a greater range of countries follow the same themes and highlight the same topics as this study. It would also be interesting to gain a wider cultural representation, not just nationality, among the researchers themselves. In addition, different ways of defining culture considering the work by Weary and Robbins, which showed cultural variations in animal welfare conceptions even within communities, could be explored [32]. A deeper investigation into the details regarding how to work in a culturally sensitive way would generate more information that could inform projects.

Recommendations for the practical application of this research in the field could include the development and testing of guidance to aid organisations and individuals who are working cross-culturally in projects such as those described in this work or in other animal welfare areas. This could be operationalised as, for example, a checklist of things to research before starting the work and ensuring the people are sufficiently trained in the skills needed for effective cross-cultural work (e.g., interpersonal skills, how to take an anthropological approach to asking questions by having an inquisitive mind and testing assumptions). Given the theme regarding the importance of working with local partners, a cocreation approach to project delivery, where partners in the focus communities are truly involved in project design and delivery, could be embedded in organisational guidelines or policies. Widespread adoption of these suggestions would enable the sector to truly evolve and avoid limitations and negative outcomes perhaps experienced in the past.

This study is the first of its kind to examine the cross-cultural interpersonal aspects of equine welfare professionals. It enables an increased understanding of the experiences of others and provide valuable information for students and professionals.

5. Conclusions

The study provides an informed insight into the cross-cultural challenges and successful approaches used in international equine welfare projects. Qualitative research has resulted in the documentation of a range of perspectives and experiences that can be shared between professionals in this field to improve the impact of multi-stakeholder projects. The synthesis of experience and reflection, together with the information gained regarding lessons learnt about cross-cultural working in this field, can be operationalised to provide practical tools.

Author Contributions: Conceptualisation, S.R. and C.B.; methodology, S.R.; formal analysis, S.R.; data curation, S.R.; writing—original draft preparation, S.R. and C.B.; writing—review and editing, S.R., C.B., J.W. and N.Y.P.L. and project administration, S.R. and N.Y.P.L. All authors have read and agreed to the published version of the manuscript.

Funding: This research received no external funding.

Institutional Review Board Statement: The study was conducted according to the guidelines of the The British Psychological Society (BPS) and approved by the Ethics Committee of the Equine Behaviour and Training Association (EBTA) (code 004; approved December 2022).

Informed Consent Statement: Informed consent was obtained from all subjects involved in the study.

Data Availability Statement: Anonymised transcripts of the interviews are available from the authors.

Acknowledgments: Thank you to Holly Roberts, who assisted with some of the administration of this project, and to the EBTA team for their wisdom and support. We would like to thank all the participants for their openness, willingness to share their insights and for volunteering their time to take part in this study.

Conflicts of Interest: The authors declare no conflict of interest.

References

1. Interntional Coalition for Working Equids. Available online: https://www.icweworkingequids.org/ (accessed on 26 February 2023).
2. FAOSTAT. FAO Statistical Year Book. Food and Agricultural Organisation of the United Nations. 2011. Available online: http://www.fao.org/faostat/en/#home (accessed on 26 February 2023).
3. Haddy, E.; Burden, F.; Prado-Ortiz, O.; Zappi, H.; Raw, Z.; Proops, L. Comparison of working equid welfare across three regions of Mexico. *Equine Vet. J.* **2021**, *53*, 763–770. [CrossRef] [PubMed]
4. Burn, C.C.; Dennison, T.L.; Whay, H.R. Environmental and demographic risk factors for poor welfare in working horses, donkeys and mules in developing countries. *Vet. J.* **2010**, *186*, 385–392. [CrossRef] [PubMed]
5. Curran, M.M.; Feseha, G.; Smith, D.G. The impact of access to animal health services on donkey health and livelihoods in Ethiopia. *Trop. Anim. Health Prod.* **2005**, *37*, 47–65. [CrossRef] [PubMed]
6. OIE Terrestrial Code Chapter 7.12 on the Welfare of Working Equids. Available online: https://www.woah.org/fileadmin/Home/eng/Health_standards/tahc/2018/en_chapitre_aw_working_equids.htm (accessed on 20 February 2023).
7. Tylor, E. *Primitive Culture*; J.P. Putnam's Son: New York, NY, USA, 1871; Volume 1.
8. "Meaning of "Culture"". Cambridge English Dictionary. Archived from the Original on August 15, 2015. Available online: https://dictionary.cambridge.org/dictionary/english/culture (accessed on 13 February 2023).
9. Lund, V.; Coleman, G.; Gunnarsson, S.; Appleby, M.C.; Karkinen, K. Animal welfare science—Working at the interface between the natural and social sciences. *Appl. Anim. Behav. Sci.* **2006**, *97*, 37–49. [CrossRef]
10. Saville-Troike, M. *The Ethnography of Communication: An Introduction*, 2nd ed.; Basil Blackweli: New York, NY, USA, 1989.
11. Stanton, A. *Reconfiguring Teaching and Knowing in the College Classroom*; Goldberger, N.R., Tarule, J.M., Clinchy, B.M., Beienky, M.F., Eds.; Knowledge, Difference, and Power, Basic Books: New York, NY, USA, 1996; pp. 25–56.
12. Piaget, J.; Inhelder, B. *The Child's Conception of Space*; Norton & Company, Inc.: New York, NY, USA, 1967.
13. Todd, A.R.; Bodenhausen, G.V.; Richeson, J.A.; Galinsky, A.D. Perspective Taking Combats Automatic Expressions of Racial Bias. *J. Personal. Soc. Psychol.* **2011**, *100*, 1027–1042. [CrossRef] [PubMed]
14. Boca, S.; Garro, M.; Giammusso, I.; Abbate, C.S. The Effect of Perspective Taking on the Mediation Process. *Psychol. Res. Behav. Manag.* **2020**, *11*, 411–416. [CrossRef] [PubMed]
15. Yang, Q.; Zhao, Y. The Effect of Perspective Taking on Negative Cognitive Responses Induced by Intergroup Threat. *Soc. Behav. Personal. Int. J.* **2013**, *41*, 91. [CrossRef]
16. Chopik, W.J.; O'Brien, E.; Konrath, S.H. Differences in Empathic Concern and Perspective Taking Across 63 Countries. *J. Cross-Cult. Psychol.* **2017**, *48*, 23–38. [CrossRef]
17. Rogers, S. Moving from a treatment-focussed to prevention-focussed approach. In *The 6th International Colloquium on Working Equids, 2010: Learning from Others: Drawing Inspiration from the Fields of International Development, Social Science, Human Health and Education to Break New Ground for Working Animals, Proceedings of the International Colloquium, New Delhi, India, 29 November–2 December 2010*; The Brooke: Atlanta, GA, USA, 2010; pp. 120–122.
18. Sinclair, M.; Phillips, C.J.C. Key Tenets of Operational Success in International Animal Welfare Initiatives. *Animals* **2018**, *8*, 92. [CrossRef] [PubMed]

19. Sandøe, P. Animal welfare for working equids: A case of cultural imperialism? In *The 6th International Colloquium on Working Equids, 2010: Learning from Others: Drawing Inspiration from the Fields of International Development, Social Science, Human Health and Education to Break New Ground for Working Animals, Proceedings of the International Colloquium, New Delhi, India, 29 November–2 December 2010*; The Brooke: Atlanta, GA, USA, 2010; pp. 64–66.
20. Haddy, E.; Brown, J.; Burden, F.; Raw, Z.; Kaminski, J.; Proops, L. "What can we do to actually reach all these animals?" Evaluating approaches to improving working equid welfare. *PLoS ONE* **2022**, *17*, e0273972. [CrossRef] [PubMed]
21. Drever, E. *Using Semi-Structured Interviews in Small-Scale Research. A Teacher's Guide*; Scottish Council for Research in Education: Edinburgh, UK, 1995.
22. Alvarez, R.; Urla, J. Tell me a good story: Using narrative analysis to examine information requirements interviews during an ERP implementation. *DATABASE Adv. Inf. Syst.* **2002**, *33*, 38–52. [CrossRef]
23. Braun, V.; Clarke, V. Using thematic analysis in psychology. *Qual. Res. Psychol.* **2006**, *3*, 77–101. [CrossRef]
24. Rogers, S.; Bell, C. Perceptions of Fear and Anxiety in Horses as Reported in Interviews with Equine Behaviourists. *Animals* **2022**, *12*, 2904. [CrossRef] [PubMed]
25. Coolican, H. *Research Methods and Statistics in Psychology*, 6th ed.; Psychology Press: London, UK, 2014.
26. Mellor, D.J.; Beausoleil, N.J.; Littlewood, K.E.; McLean, A.N.; McGreevy, P.D.; Jones, B.; Wilkins, C. The 2020 Five Domains Model: Including Human–Animal Interactions in Assessments of Animal Welfare. *Animals* **2020**, *10*, 1870. [CrossRef] [PubMed]
27. Dyson, S.; Routh, J.; Bondi, A.; Pollard, D. Gait abnormalities and ridden horse behaviour in a convenience sample of the United Kingdom ridden sports horse and leisure horse population. *Equine Vet. Educ.* **2022**, *34*, 84–95. [CrossRef]
28. McDonnell, S. *The Equid Ethogram: A Practical Guide to Human Behaviour*; Eclipse Press: Lexington, KY, USA, 2003.
29. Bell, C.; Rogers, S.; Taylor, J.; Busby, D. Improving the recognition of equine affective states. *Animals* **2019**, *9*, 1124. [CrossRef] [PubMed]
30. Finlay, L. Negotiating the swamp: The opportunity and challenge of reflexivity in research practice. *Qual. Res.* **2002**, *2*, 209–230. [CrossRef]
31. Olmos-Vega, F.M.; Stalmeijer, R.E.; Varpio, L.; Kahlke, R. A practical guide to reflexivity in qualitative research: AMEE Guide No. 149. *Med. Teach.* **2023**, *45*, 241–251. [CrossRef] [PubMed]
32. Weary, D.; Robbins, J. Understanding the multiple conceptions of animal welfare. *Anim. Welf.* **2019**, *28*, 33–40. [CrossRef]

animals

MDPI

Article

Beyond the Unitary State: Multi-Level Governance, Politics, and Cross-Cultural Perspectives on Animal Welfare

Paul Chaney *, Ian Rees Jones and Nivedita Narayan

Wales Institute of Social and Economic Research (WISERD), SPARK, Cardiff University, Maindy Road, Cardiff CF24 4HQ, Wales, UK; jonesir4@cardiff.ac.uk (I.R.J.)
* Correspondence: Chaneyp@cardiff.ac.uk

Simple Summary: Existing cross-cultural research on animal welfare often overlooks the way that policy and law are not the exclusive domain of central government. This can result in an over-simplification or misrepresentation of the true situation. The political culture and institutional arrangements for governing the modern state are more complex than a "one-size-fits-all" approach. It is argued that cross-cultural research needs to give greater attention to differences *within* as well as between unitary states. Specifically, it needs to examine developments in constituent nations and territories. Here we illustrate this by drawing on new research in the United Kingdom, and examine how 'devolution'—or decentralized government for Wales and Scotland—is providing contrasting opportunities for NGOs, campaigners, and the public to lobby to improve animal welfare policy based on local practices, beliefs, and demands (collectively known as the "political culture"). Our findings show how this is important because it results in contrasting animal welfare policies and laws in the constituent nations of the UK.

Abstract: It is argued that extant cross-cultural research on animal welfare often overlooks or gives insufficient attention to new governance theory, civil society, politics, and the realities of devolved or (quasi-)federal, multi-level governance in the modern state. This paper synthesizes relevant social theory and draws on new empirical findings of civil society accounts of campaigning on animal welfare policies and law in the United Kingdom. It is presented as a corrective to arguably reductive, earlier unitary state-based analyses. Our core, evidence-based argument is that cognizance of civil society activism and the contrasting institutional governance structures and political cultures of constituent nations in unitary states—such as the UK—are providing opportunities for the territorialization of legally grounded animal welfare regimes, and culturally distinctive practices.

Keywords: animal welfare; law; multi-level governance; civil society; devolution; political cultures; policy divergence

1. Introduction

Existing cross-cultural research on animal welfare often focuses on unitary states and overlooks the way that animal welfare policy and law are not the exclusive domain of central government. The political culture and institutional arrangements for governing the modern state are often more complex than a "one-size-fits-all" approach. Accordingly, it is argued that cross-cultural research needs to give greater attention to differences *within* as well as between unitary states. Specifically, it needs to examine developments in constituent nations and territories. Thus, the literature gap this study aims to fill is the lack of attention to governance within a state due to an assumption that states are culturally homogenous. Failure to do this can result in an over-simplification or misrepresentation of the true situation. Here we illustrate this by drawing on new research in the United Kingdom, and examine how 'devolution'—or decentralized government—for Wales and Scotland is providing contrasting opportunities for "civil society" (viz. charities and NGOs, campaigners, and the public) to lobby for improved animal welfare based on local practices, beliefs, and demands. The UK is a state, but animal welfare is largely a devolved issue to the four constituent nations; thus, Wales, Scotland, England, and Northern Ireland are effectively unitary states for animal welfare, not the UK. Our findings show how the post-devolution governance shift is important because it results in contrasting animal welfare policies and laws in the constituent nations of the UK. This is based on our interviews with animal welfare activists and NGOs and secondary data analysis of their campaigning 'grey literature' [1] and parliamentary proceedings.

The following discussion addresses two principal research questions based on developments in Wales and Scotland: (1) What are the views and experiences of civil society campaigners lobbying for animal welfare gains? And (2) What evidence is there of the "territorialization" of animal welfare policy driven by civil society activism? As we shall see, the findings show how *inter*-state analysis is important to cross-cultural research on animal rights because it furthers understanding of the different political opportunity structures for advancing animal welfare in the constituent nations of the UK and advances our knowledge of how the lobbying and activism of local policy communities results in contrasting animal welfare policies and laws in the devolved nations.

To address the foregoing questions, the remainder of this paper is structured as follows: first, we briefly outline the research context and explain why the UK is a suitable case study. Next, relevant strands of social theory are synthesized to make the conceptual case for cross-cultural analysis to examine differences *within* as well as between states. In the following section, the research methodology is summarized. Subsequently, the research findings are discussed. The concluding section reflects on the significance of the results.

2. Research Context

Extant cross-cultural research on animal welfare often overlooks the way that animal welfare policy and law are not the exclusive domain of central government [2–4]. This can result in an over-simplification if it fails to examine the issue from a multi-level governance perspective. The United Kingdom provides an example of this. The UK is a political creation based on the union of England with Wales (in 1536), Scotland (in 1707), and Ireland (in 1801, subsequently, Northern Ireland in 1921). It is a comparatively recent example of the 'devolutionary trend [that] has swept the world... [involving widespread] transference of power, authority, and resources to subnational levels of government' (Rodriguez-Pose and Gill 2003, p. 334) [5]. Yet it was not until 1998/99 that elected legislatures were established for Wales, Scotland, and Northern Ireland. Prior to that juncture, animal welfare policy and law were largely determined by Westminster for Wales and England. Scottish legislation has always had a different legal base and often passed different laws to England and Wales on animal welfare. In the Twentieth Century, minor variations to Westminster legislation on animal welfare were introduced by the territorial ministries of the UK government: the Welsh and Northern Ireland Offices [5]. After 1999, prior to Brexit, Wales and Scotland passed their own devolved legislation on animal welfare implementing EU Regulations and Directives. The UK leaving the EU has meant that Wales, Scotland, and England are able to pass laws on issues such as farm animal welfare that had previously been the exclusive preserve of the EU. It is also the case that, post-Brexit, Northern Ireland remains a member of the EU via the Single Market and Customs Union and it implements EU legislation on animal experiments, not Westminster. As earlier analysis has revealed, in the face of lobbying by Welsh and Scottish policy communities, the post-1999 devolved era has seen the rise of political competition as parties have advanced distinctive policies in their manifestos for Welsh and Scottish parliamentary elections (Chaney 2022) [6]. While limited aspects of animal welfare powers remain reserved to Westminster (e.g., the use of animals in scientific experimentation in Great Britain), the majority are devolved and as we shall see, the past quarter century has seen increasing territorialization of animal welfare policy and law at the devolved level.

3. The Theoretical Case for a Multi-Level Governance Approach

The aim of this part of the present discussion is to synthesize relevant strands of social theory to make the conceptual case for future cross-cultural analyses of animal welfare to examine differences *within* as well as between states. Our starting point is the concept of 'new governance'. Rod Rhodes' original characterization (1997: 53) of it posits that it should be conceived of as a system characterized by interdependence between organizations. Governance (or the structures and processes of government) is broader than government alone. Crucially, it involves different tiers of government and includes attention to the actions of non-state actors. This matters, for the changing form of the state in the modern era means that the boundaries between public, private, and voluntary sectors are ever shifting and often opaque. For non-state actors, there is a significant degree of autonomy from the state. Activists, social movements, and campaigners—such as those concerned with advancing animal welfare—are not accountable to the state; they are self-organizing. Although the state does not occupy a privileged, sovereign position, it can indirectly and imperfectly steer networks [7]. Crucially, Rhodes highlighted new forms of governing and state structures caused by devolution or government decentralization. This he referred to as 'the hollowing out of the state' (Rhodes 2007: 1258) [8]. The result he concluded was a 'differentiated polity' (Rhodes and Weller 2005) [9] (i.e., one not solely predicated on administration by central government but where constituent nations may have their own governance arrangements). All these factors are important to cross-cultural analyses of animal welfare (and other policy issues) because they underline the need to look beyond analysis of central government policy and explore regional structures of governance and the role of civil society. As Rhodes (2007: 1258) proceeds to explain, this perspective 'opens new avenues of exploration on key issues confronting policymaking and policy implementation... including: the sectoral character of policymaking; the mix of governing structures; the "philosopher's stone" of central coordination; devolution to the constituent territories of the UK... [in short, we can] use it to develop a new way of seeing state authority in its relationship to civil society'. It is a point more recently articulated by Bob Jessop (2015: 20), who argues that 'state powers... are activated through changing sets of politicians and state officials located in specific parts of the state apparatus, in specific conjunctures... the state's structural powers and capacities cannot be fully grasped by focusing on the [central] state alone' [10]. Accordingly, 'multi-level governance' (or alternatively, 'multi-spatial governance') aligns with the concept of new governance and emphasizes the tiered—or multiple layers—governance that exists in nominally unitary states [11]. As Rhodes' (2017: 23) also underlines, 'governance poses questions about the shifting boundaries between state and civil society' [12].

Put simply, civil society is the space outside of the family, business, and the state (Cohen and Arato 1992) [13]. However, it should be noted that 'these elements are intertwined such that their boundaries are effectively seamless' (Eto 2012: 78) [14]. Thus, civil society comprises a diverse range of associational activities extending beyond the family, to encompass non-governmental organizations (NGOs—or alternatively, civil society organizations—or CSOs), pressure groups, charities, community groups, social movements, and campaigning organizations [15]. A broad literature attests to the contested nature of 'civil society' (Edwards 2004) [16]. For example, De Tocqueville (1835) distinguishes between 'political' society and civil society, asserting that civilian and political associations act as a counterbalance to liberal individualism and the state [17]. Later Twentieth Century thinkers such as Gramsci (1948) [18] and Habermas (1962) [19] also underlined the fluidity of relations between the state and civil society. Notably, they claimed that the relationship is not unidirectional, but that civil society both resists *and reinforces* hegemonic ideas—whether about economic and social life—or, as in the present case, human-non-human relations. Subsequently, writers such as Putnam (1993) [20] stressed how the production and accumulation of social capital is integral to the effective functioning of society and democracy. Social capital here refers to the 'features of social organization, such as trust, norms, and networks, that can improve the efficiency of society by facilitating coordinated actions' (Putnam 1993: 167) [20]. As Hindess (2004: 40) explains, governance and governmentality 'should be understood more widely than the supreme authority in states, [but instead] as action aimed at influencing the way individuals regulate their own behaviour... the aim of modern government of the state is to conduct the affairs of the population in the interests of the whole. This is not restricted to the government but is performed also by agencies in civil society' [21]. As noted, the key point in all of this is that state decentralization or 'devolution' (in the present case, to Wales and Scotland) redefines governance and (re-)creates polities (or territorial political systems) for constituent nations and regions in unitary states. These are headed by their own governments and parliaments and provide political engagement opportunities for civil society activism to promote animal welfare claims and campaign for legal and policy change at the devolved level based on local needs and demands. International examples include (but are not limited to) regional indigenous legislatures in the Russian Federation [22], as well as in European Union members states, such as the Autonomous Communities (Regions) of Spain, Regions of Italy, German Federal States such as Bavaria, and the Autonomous Regions of the Azores [23], as well as the Canadian provinces and territories (notably Quebec and First Nations) [24].

Allied to the foregoing, classic social movements theory underlines the importance of what are termed "political opportunity structures" [25]. These are the structures and mechanisms that allow civil society organizations to press their policy claims on government [26]. They are allied to the academic literature on neo-institutionalism [27]. In short, this shows us how the design and operation of political institutions shapes the policy process. As will be seen, our findings suggest that devolution has made governance more "structurally allowing" for the territorial animal welfare lobbies in Wales and Scotland.

Lastly, governance and civil society are integral to the literature on "political culture". The term was initially proposed by Gabriel Almond [28] and subsequently developed with Sidney Verba [29]. As Stephen Chilton explains, it refers to the reciprocal and iterative relationship between institutions and governance structures and culture in a society by noting 'the term promised to solve in a scientific, cross-culturally valid way the micro-macro problem: the classic problem of specifying how people affect their political system, and vice-versa' [30]. Underlining the present argument that current research needs to adopt a multi-level governance approach to cross-cultural studies of animal welfare, a burgeoning literature traces how devolution in the UK: (1) was born out of the historical and cultural differences between the Welsh, Scottish, English, and Northern Irish peoples [31], (2) how Wales and Scotland possess their own policy communities on animal welfare [32], and (3) how in its wake, devolved politics are shaping distinctive laws, policies, and cultural practices in each nation [6].

4. Methodology

This study adopts a mixed methods approach. It is based on the findings of 30 semi-structured interviews with civil society organizations (or 'NGOs'—non-governmental organizations) that campaign for animal welfare (25 were undertaken in the post-Brexit period 2022–2023 and 5 were completed in 2013, the latter offering a longitudinal perspective). These were selected using purposive sampling to reflect the diversity of the sector and span different animal welfare fields including: wildlife, agriculture, and companion animals. Interviews were conducted in Wales, Scotland, and England [33]. Northern Ireland was not included because of the extended periods over recent years when devolved governance was suspended owing to political impasse (and direct rule from Westminster was imposed). The research interviews were transcribed and analyzed thematically. This dataset was complemented by secondary data analysis of the 'grey' literature of NGOs [34] (including newsletters, social media and websites), as well as public domain records of government policy, law, and parliamentary proceedings.

5. Results

5.1. What Are the Views and Experiences of Civil Society Animal Welfare Campaigners? How Does 'Devolved' Lobbying Compare to Campaigning at Westminster?

A key finding from our series of in-depth interviews with campaigners representing civil society organizations (CSOs) is their frustration with Westminster and generally positive views of the opportunities to engage and lobby parliamentarians in Wales and Scotland. Speaking about Westminster, this interviewee reflected the consensus: "The government definitely just can't prioritize anything above [and beyond] Brexit anymore. And so, I think that's probably the biggest problem we've got is that Brexit absorbs the oxygen from anything that isn't Brexit". Allied to this, civil society campaigners expressed their fury and frustration at the Conservative government's decision on 8 June 2023 to withdraw the Animal Welfare (Kept Animals) Bill [35]. This was going to deliver pledges set out in the Conservative Party's 2019 manifesto for greater protection of animals kept, imported into, and exported from the UK, including ending the unlicensed captivity of primates and export of live animals for slaughter and fattening. The rapid turnover of UK prime ministers and the political volatility of Westminster are also seen by CSOs as a key problem. As this interviewee explained: "Scotland's bit more [politically] stable [than Westminster]. And that's why we've had more success in getting [policy gains] ... [for example] they did put in new regulation for [dog] rescues... because they are working to a pretty regular schedule of elections, and it's a little bit more stable. It's a devolved administration... [that] feels better and more efficient".

In a similar vein, this interviewee spoke of the difference between Cardiff and Edinburgh on the one hand, and Westminster on the other: "I think because some animal welfare is devolved. It can be easier to deal with—[i.e.,] you know the devolved governments. I think when it comes to dealing with the UK Government it's always harder. They are quite good at stonewalling or shutting things down". This was a view shared by several interviewees, including this manager: "I think the devolved governments and devolved parliaments are much easier to engage with, much more open and are far more democratic. So, we sit on cross-party groups [of parliamentarians and external organizations] in both Wales and Scotland, and we've had success engaging with the Scottish Government relatively easily, so that, I mean, *they* actually approached us to give feedback and a heads up on launching a public consultation, which was quite a surprise! ... I think in comparison to the UK Government, it feels much more trusting of [civil society] organizations".

As noted, in conceptual terms, social movements theory and neo-institutionalism underline the importance of "political opportunity structures", or mechanisms and processes that facilitate or frustrate civil society organizations in pressing their policy claims on government. As the foregoing quotes attest, interviewees in this study spoke of how devolution has made governance more structurally allowing for the territorial animal welfare lobbies in Wales and Scotland. Notably, a number alluded to the work of the Scottish Animal Welfare Commission. Established under section 36 of the Animal Health and Welfare (Scotland) Act 2006, the Commission has been established to provide independent advice to Scottish Ministers on the welfare of sentient animals, primarily on wildlife and companion animal welfare [36]. While in Wales, interviewees highlighted Animal Welfare Network for Wales (AWNW). Also established in 2006, this is an independent network of organizations that engages with key stakeholders, including the Welsh Government. As this interviewee explained: "Other advantages we've got is we've got Animal Welfare Network Wales, [through this...] we've got a network that is accessible to smaller charities... and that means that there is a framework there that they can get involved in if they want" [37]. Such collaboration continues to shape policy, as in the case of the Welsh Government's Code of Practice for Animal Welfare Establishments [38].

Interviewees also spoke of the importance of the Third Sector Partnership Council. This is a unique feature of devolved governance in Wales that is designed to shape policy and lawmaking. It is based on a statutory requirement for collaboration in policy work between the Welsh government and the voluntary sector [39]. As this interviewee notes:

"then, of course, there's that direct link between the third sector and the obligation in law for the [Welsh Government] minister to meet with the third sector and animal welfare will be represented there through Animal Welfare Network Wales and Companion Animal Welfare Group Wales".

5.2. What Evidence Is There of the "Territorialization" of Animal Welfare Policy Driven by Civil Society Activism?

The past quarter century of devolved government for Wales and Scotland has resulted in increasing policy divergence between the nations of the UK. As Table 1 illustrates, there is clear evidence of the territorialization of animal welfare owing to the contrasting protections set out in the Welsh and Scottish Governments' policies and laws. Interviewees alluded to numerous examples of policy divergence, and we now turn to consider selected examples. In definitional terms, when we talk of "territorialization" of animal welfare, we mean civil society organizations (CSOs) successfully lobbying for distinctive laws and policies that convey contrasting legal protections in either Wales, Scotland, or England. Often, they place duties on named bodies, groups, and individuals, and/or make named practices unlawful and subject to criminal or civil proceedings and penalties, including fines and imprisonment. Whilst these new laws improve animal welfare in one jurisdiction, the practices they proscribe may remain lawful in other UK jurisdictions.

Table 1. Examples of the territorialization of animal welfare law in Wales, Scotland, England, and Northern Ireland.

Topic	Wales	Scotland	England	Northern Ireland
Hunting with Dogs	Under the Hunting Act (2004) [40], it is illegal to hunt wild mammals with dogs in Wales (and England). There are exemptions that allow hunting for certain types of humane control. This is called exempt hunting [41].	Protection of Wild Mammals (Scotland) Act 2002 [42], and Hunting with Dogs (Scotland) Act 2023 [43]. Inter alia, a person who deliberately hunts a wild mammal with a dog commits an offence, as does an owner or occupier of land knowingly permitting another person to do so.	Under the Hunting Act (2004) [40], it is illegal to hunt wild mammals with dogs in England (and Wales). There are exemptions that allow hunting for certain types of humane control. This is called exempt hunting [41].	Northern Ireland remains the only part of the UK where hunting with dogs remains legal [44].
Badger culling	In Wales, the Welsh Government has ended the use of badger culling as a technique to control bovine Tuberculosis (bTB) [45].	Badgers and their setts are protected under the Protection of Badgers Act 1992 (as amended by the Wildlife and Natural Environment (Scotland) Act 2011) [46].	In England, government policy and law permit the killing of badgers in an attempt to control bovine Tuberculosis (bTB). See the Protection of Badgers Act 1992, as amended [47].	Plans by Northern Ireland's Department of Agriculture, Environment and Rural Affairs (DAERA) to introduce an English-style badger cull in efforts to control bovine TB in cattle were thrown out by the High Court in a judgement on 25 October 2023 [48].

Table 1. *Cont.*

Topic	Wales	Scotland	England	Northern Ireland
Licensing of animal activities	Animal Welfare (Licensing of Activities Involving Animals) (Wales) Regulations 2021 [49]. This law concentrates on one issue and makes it illegal for a commercial seller to sell a puppy or kitten they have not bred themselves at their own premises and they must ensure the mother is present. Henceforth, puppies and kittens can only be purchased from where they were bred or from a rescue or rehoming centre.	The Animal Welfare (Licensing of Activities Involving Animals) (Scotland) Regulations 2021 [50]. Introduces a more modern and flexible licensing regime for pet selling, dog breeding, and other matters, such as extending licensing to a wider range of activities involving animals than currently found in existing pre-legislation.	The Animal Welfare (Licensing of Activities Involving Animals) (England) Regulations 2018 [51], as amended [52]. This law covers a broader range of topics than its Welsh and Scottish counterparts. Anyone wanting to get a new puppy or kitten in England must now buy direct from a breeder or consider adopting from a rescue centre instead.	Welfare of Animals Act (Northern-Ireland) 2011 [53] currently falls short of a ban on third party sales of puppies and kittens [54].
	Animal Welfare (Licensing of Activities Involving Animals) (Wales) (Regulations) 2021. New licensing requirement in relation to animal exhibits and animal establishments [55].	Animals and Wildlife (Penalties, Protections and Powers) (Scotland) Act 2020. This improves the penalties and powers available to enforcement agencies and the courts. It also introduces improved procedures for making permanent arrangements for animals taken into possession by the authorities to protect their welfare [56].	Animal Welfare Act 2006 extends to England and Wales, and to some provisions Scotland and Northern Ireland [57].	Welfare of Animals Act (NI) 2011 sets out provisions—all who own, or are responsible for an animal, are required by law to care for it properly and take reasonable steps to ensure its welfare needs are met [58].
Shock collars	The Animal Welfare (Electronic Collars) (Wales) Regulations 2010. Section 2 "It is prohibited for a person to—(a) attach an electronic collar to a cat or a dog; (b) cause an electronic collar to be attached to a cat or a dog; or (c) be responsible for a cat or a dog to which an electronic collar is attached" [59].	The use of electric shock collars is currently legal in Scotland. In 2018, the Scottish Government published non-statutory guidance advising against the use of these devices and other aversive training methods [60].	The Animal Welfare (Electronic Collars) (England) Regulations 2023 will make the use of shock collars unlawful from 1 February 2024 [61].	There are currently no legal restrictions on the use or sale of shock collars in Northern Ireland.

The first example of territorialization stems from civil society lobbying in Wales Table 2. It concerns the banning of snares [62]. This was achieved under the provisions of the Agriculture (Wales) Act passed by the Welsh Parliament (or *Senedd*) in 2023. Snares are wire or cord nooses that are placed in undergrowth with the intention of killing animals such as rabbits (although they are indiscriminate and catch all manner of creatures, including cats and dogs). Typically, the animal dies a slow and agonizing death as it is choked to death when trying to escape. Prior to the new enactment, CSOs lobbied for reform of the pre-devolution Westminster legislation covering Wales (and England). For example, a network of CSOs prepared an open letter to Members of the Senedd and asked their supporters to lobby for change [63]. It told its members: "Imagine a Land of Our Future where we restore nature, tackle climate change and secure healthy, sustainable food for future generations. For this to happen, we need to make sure the Agriculture (Wales) Bill is strong enough to deliver these changes. We can only influence the next phase effectively with your help" [64]. Subsequently, another CSO told its members: 'As you may be aware, we are celebrating yet another momentous victory for animals: Wales just banned the use of snares. In a few short months, shooting estates will no longer be able to lay death traps for animals. We got what we asked for: a straightforward ban on snaring, with no amendments, no loopholes. The bill was passed last night, meaning badgers, foxes, cats and dogs will be protected from a long and agonizing death in these cruel traps. But in England, snaring remains legal. Please donate today to help us take on the rest of the UK and protect more animals from the cruelty of snares' [65].

Our second example is a case of post-devolution policy transfer, where civil society organizations are using reforms in one jurisdiction to lobby for reform elsewhere [66]. Mandatory closed-circuit television (CCTV) was introduced in slaughterhouses in England in 2018. Amongst other things, the resulting video record reduces the chances of abuse and cruelty by abattoir staff. At present, this is not a requirement under Welsh law, although the Welsh Government is currently consulting on a ban. According to one animal welfare CSO, "polling has indicated that a huge 82% of the public in Wales supports the introduction of CCTV. It is clear that we are not alone in the belief that this is an important step in improving farm animal welfare" [67]. It proceeded to implore the public to "take action and help thousands of animals—getting this vital change is a huge reassurance that animal welfare standards are being delivered... We can't do this alone. The Welsh Government needs to hear from you". In the face of such demands, the Welsh Government has announced it will make CCTV mandatory from Spring 2024 [68].

Broadly defined, animal welfare policy extends beyond mammals and includes birds, reptiles, fish, and invertebrates, as well as habitat protection. Our third example of territorialization has again been driven by CSO activism. In June 2023, the courts found in favour of a welfare charity after it brought a judicial review case against the Scottish Government's licensing of scallop dredging and seabed trawling [69]. Whilst the sentience of shellfish and crustacea is debated, the sentience of lobsters, octopus, and crabs has recently been confirmed by scientists [70]. Accordingly, this case has wider welfare significance because the dredging and trawling causes habitat destruction for these forms of marine life. Following the judicial review, these practices have been ruled unlawful in Scotland. In contrast, over the past decade, and in the face of civil society campaigning [71], the Welsh Government has introduced new regulations, including the introduction of vessel tracking devices in Welsh scallop fisheries [72]. However, current Welsh legislation falls short of a ban [73]. A public consultation on scallop fisheries is presently gathering evidence [74].

Badger culling provides a further stark illustration of legal and policy divergence on animal welfare. In England, government policy and law permit the killing of badgers in an attempt to control bovine Tuberculosis (bTB), an infectious respiratory disease which affects cattle. The UK government estimates over 200,000 badgers have been killed over recent years [75]. According to one NGO that is opposed to the policy, "The solution to bovine TB is simple: move to the better types of cattle testing that are available; switch to cattle vaccination; stop killing badgers; significantly reduce cattle movements... The [Westminster] government has deliberately not invested in cattle vaccine research" [76]. Wales looked set to adopt a similar policy until civil society activism forced a rethink. Initially, in 2009, the Welsh Government legislated to allow a nonselective badger cull [77]. In response, an animal welfare NGO sought a judicial review and the Court of Appeal ruled that the 2009 Order [allowing the cull in law] should be quashed [78]. In consequence, in 2011 the Welsh Government ruled out a badger cull and replaced it with a five-year vaccination program [79]. As a result, Welsh herds are tested and cattle movements are more strictly controlled. In further contrast to England (where animals other than cattle are not routinely tested for bTB), in Wales, badgers, deer, and other livestock species are tested and badgers are vaccinated against bTB. In Scotland, culling is unlawful. Badgers and their setts are protected under the Protection of Badgers Act 1992 (as amended by the Wildlife and Natural Environment (Scotland) Act 2011) [46]. Northern Ireland had no badger cull from 2012, but introduced one in 2022, then replaced it with a badger vaccination programme in 2023 following a campaign by civil society organizations. According to UK government data, the incidence of bTB in cattle herds in Wales and Scotland is lower than in England [80].

Table 2. Selected Animal Welfare Manifesto pledges for the 2021 Welsh and Scottish Parliament Elections.

Wales	Scotland
"Develop a national model for regulation of animal welfare, introducing registration for animal welfare establishments, commercial breeders for pets or for shooting, and animal exhibits. We will improve the qualifications for animal welfare inspectors to raise their professional status. We will require CCTV in all slaughterhouses, we will ban the use of snares, and restrict the use of cages for farmed animals. We will not allow the culling of badgers to control the spread of TB in cattle" (Welsh Labour Party 2021, p. 37) [81].	"We will support farmers to produce more of our own food needs sustainably and to farm and croft with nature, including through enhanced animal welfare... Animal Welfare: We will adopt the highest possible animal welfare standards, including shifting to entirely free range, woodland or barn chicken and egg production. We will modernise and update the Animal Welfare Act from 2006 and implement the new livestock worrying legislation. We will seek to reflect so far as we can, new EU animal welfare labelling to promote food produced to higher than EU welfare standards. We will create a new Scottish veterinary service to ensure that we have enough people with the right qualifications in veterinary services, animal health and food safety to meet all our needs across the public and private sector for land and marine based animal health issues. We will ban live exports of animals for fattening and slaughter and only allow live transport of livestock to and from islands and the mainland with stringent welfare standards in place. We will legislate to close loopholes in the law protecting foxes and other wild mammals and remain committed to implementing the licensing of driven grouse shooting. We will implement the recommendations of the Deer Working Group and modernise deer management legislation" (SNP 2021, p. 56) [82].
"We will introduce a baseline support payment to offer the [farming] industry greater economic stability. This support will be used to encourage the highest standards of public health and animal health and welfare" (Plaid Cymru 2021b. p. 62) [83]. Animal Welfare: A Plaid Cymru Government will work with stakeholders to build upon the high level of animal welfare standards already in place in Wales. We will: • Improve the enforcement and delivery of licensing requirements relating to dog breeding establishments in Wales, building on the recent review of regulations by the Wales Animal Health and Welfare Framework Group. • Improve horse welfare by taking action on equine tethering. • Review pet vending, focusing especially on the regulation of animals sold online. • Issue model tenancy proposals on pets in social housing and work to reduce barriers between homeless pet owners and homeless shelters. • Support the development of statutory codes of practices for the keeping of exotic pets in Wales" (Plaid Cymru 2021b. p. 64) [83].	
"[We will] Enhance Animal Welfare and Protect Wildlife: • Establish an animal offender register in Wales • Ban the keeping of primates as pets • Review pet breeding standards and registration requirements to ensure adequate protection for animals and bring forward the ban on the third party sale of cats and dogs • Promote honest labelling to enhance consumer choice, including distinguishing between stunned and non-stunned slaughter methods and introduce CCTV in abattoirs • Establish a £20 million Wales Wildlife Fund to support conservation efforts across Wales" (Welsh Conservative Party 2021, p. 30) [84].	"Scottish Labour will carry out a full review of Scotland's outdated animal welfare legislation, with a view to strengthening wildlife protection law and animal welfare... We will introduce a National Animal Cruelty Register to support enforcement agencies. We will reform the law on keeping domestic pets in different tenures after life events, including domestic violence. The pandemic has highlighted concerns over the illegal importation of puppies, and we need to raise public awareness and ban imports of very young puppies, and other illegal pets not on the positive list of species that are suitable to keep as pets. We support a more comprehensive approach to public education on animal welfare. We will introduce a comprehensive ban on fox hunting and snares and the use of electric shock collars. There also needs to be more effective monitoring of raptor conservation and stronger penalties. Labour supports a ban on live animal exports for fattening and slaughter. Parliament should pay full regard to animal welfare requirements when formulating and implementing policies" (Scottish Labour Party 2021, p. 172) [85].

Table 2. *Cont.*

Wales	Scotland
"Work with the UK Government to ensure that it only enters into trade agreements under which imported goods meet the high environmental, food quality and animal welfare standards expected of home-produced food... Despite progress in recent years, there is much to be done to ensure the highest standards and protections for animals in Wales. We will: • Ensure that animal welfare standards are as least as good if not better than those we enjoyed as members of the European Union • Pass a Wildlife Act for Wales, creating clarity and consistency on the policy and legislation that protects wildlife. • Work with the sector to improve the welfare of farm animals, including live exports, and wider issues such as public sector food procurement and labelling. • Regulate all animal sales to protect the welfare of any pet traded, bought or sold in Wales. • Issue guidance to local authorities ensuring they do not use their powers under the Anti-Social Behaviour, Crime and Policing Act 2014 in a way that compromises animal welfare. • Work with Police and Crime Commissioners to take action against growing instances of dog thefts" (Welsh Liberal Democratic Party 2021, unpaginated) [86].	"Delivering the highest standards of animal welfare—The Scottish Conservatives stand for the highest standards in animal welfare. We are proud of our campaign to deliver Finn's Law in the last Parliament, giving proper protection to service animals like policedogs. We will... bring forward a new Animal Welfare Bill... As part of our Animal Welfare Bill, we would ban the sale of dogs with cropped ears in Scotland. We would follow Wales in banning the use of electric shock collars. We would amend the Dangerous Dogs Act, so that dogs are not automatically put down due to their breed. We also would take forward measures to improve the welfare of farmed animals in transportation. In doing so, we will take account of Scotland's geography and established farming models. We do not believe that the scheme as proposed in England is right for Scotland" (Scottish Conservative and Unionist Party 2021, pp. 42–43) [87].

As earlier work has noted, political parties' manifesto pledges on animal welfare tell us about contemporary political thinking on the relationship between humans and non-humans, evolving notions of sentience, and how these are related to structures and processes of democratic governance. There are several contrasting conceptualizations of electoral politics, yet a core, shared feature is their underlining of the fact that elections are intimately related to civil society policy demands and opinions. However, whilst earlier work has examined the electoral politics of animal welfare in state-wide elections (Chaney 2014) [88], internationally, there has been a dearth of attention to how parties use the new political spaces associated with the electoral politics of 'regional' legislatures to address animal welfare issues. One exception is research on devolved elections in the UK 1998–2017. This concluded that: 'the move to multi-level electoral politics provides new political spaces to advance animal welfare and how meso-ballots are increasingly attuned to the symbiosis of humans and animals. These factors are driving the territorialization of policy and leading to distinctive animal welfare regimes in the different countries of the UK' (Chaney et al., 2022: 116) [6]. The present analysis of the 2021 Welsh and Scottish Parliament elections supports this and underlines how animal welfare policy is grounded in the formal representational structures and processes of contemporary democracies and the dynamic relationship between political actors, parties, civil society, and government. As Table 2 reveals, animal welfare has become a mainstream political issue in parties' electoral competition. In Wales and Scotland, all main parties contesting the 2021 elections set out detailed and extensive policy proposals. There are multiple pledges to use Welsh and Scottish legislation to increase regulation, and evidence of proposed policy transfer (as parties cite policy developments elsewhere and call for them to be implemented in their polity). Intra-party divergence on animal welfare is also notable. This refers to 'statewide' parties (that is, parties that operate across the UK or GB) holding different policy positions on animal welfare in the different nations. This is because, in the wake of devolution, to varying degrees, the Labour Party, Liberal Democrats, and the Conservative Parties have moved to (quasi-)federal structures with a degree of autonomy for the Welsh and Scottish wings of these UK parties. Strikingly, on live animal transportation, the right-wing Conservative and Unionist Party in Scotland rejected the policies of its governing administration at Westminster by stating: "In doing so, we will take account of Scotland's geography and established farming models. We do not believe that the scheme as proposed in England is right for Scotland" (Scottish Conservative and Unionist Party 2021, pp. 42–43) [87].

Following the 2021 elections, the parties forming government in Wales and Scotland—the Welsh Labour Party (in Partnership agreement with Plaid Cymru) and the Scottish National Party (SNP) in Scotland—are in the process of implementing their manifesto pledges. For example, the Animal Welfare Strategy for Wales 2021–2026 [89] reasserts the commitments to the introduction of mandatory registration for animals in private keeping; improving the qualifications of animal welfare inspectors; ensuring CCTV in all slaughterhouses, and restricting the use of cages for farmed animals. As these measures are implemented, the associated policy consultations in each nation provide further opportunities for the territorial policy lobbies of civil society organizations to shape the evolving animal welfare regimes. This can be a fractious process, as typified by the Welsh Government's current public consultation on stricter regulation on the release of gamebirds [90]. This has been welcomed by wildlife and animal welfare NGOs, yet it has led opponents to claim it is the backdoor route to a shooting ban [91].

6. Conclusions

This study's findings are significant in a number of key regards. First, the empirical data confirm the importance of devolution to civil society advocacy of animal welfare. They reveal increasing policy divergence and the territorialization of animal welfare regimes with distinctive laws and policies in Wales and Scotland. Moreover, compared to Westminster, civil society activists point to the devolved political spaces as being more structurally allowing and responsive to their claims. This corresponds to what social movements theory calls "political opportunity structures". These are mechanisms that allow civil society organizations to press their policy claims on government (also described in the academic literature on neo-institutionalism, civil society, and multi-level governance).

In addition, our findings show how CSOs' lobbying for policy transfer means that gains (in the form of new laws and policies) in one devolved jurisdiction may subsequently be used in campaigning for similar reforms elsewhere. This is evident in calls in Scotland to introduce a legal ban on electronic dog collars, as introduced in Wales. A further example is hunting with dogs; with bans in force on public land in many areas of Wales, there are calls to devolve powers over hunting with dogs to the Welsh Parliament in order to introduce a full ban, as in Scotland [92]. A further consequence of devolution is that improvements in welfare can take place in one jurisdiction whilst being resisted in others (e.g., the use of snares banned in Wales but lawful in England, and stricter regulation of the release of game birds in Wales). There is also evidence of cross-party working on animal welfare at the devolved level (e.g., 'We will continue to work with other parties across the Scottish Parliament to strengthen protections for animals in Scotland by bringing forward a new Animal Welfare Bill. . .' (Scottish Conservative and Unionist Party 2021, pp. 42–43) [87], as well as intra-party policy divergence in statewide parties (this is the case with all examples of distinctive animal welfare policies advanced by statewide parties in one nation that are not introduced in every polity).

As noted, this study's relevance and rejoinder to cross-cultural studies of animal welfare is that policy and law are not the exclusive domain of central government. Sole emphasis on statewide governments can result in an over-simplification or misrepresentation of the true situation. The political culture and institutional arrangements for governing the modern state are complex and result in contrasting animal welfare policies and laws in constituent nations, states, and territories. Many of the advances seen in Wales and Scotland would not be possible under the pre-devolution "one-size-fits-all" mode of lawmaking from Westminster. This is not to argue that devolution is automatically a panacea for animal welfare. Arguably, one reason why it is easier to get laws on animal welfare in Wales and Scotland is that Westminster plays the role of two legislatures (England and UK), so it is more difficult to get a time slot there than it is in the Senedd or Scottish Parliament, and there are fewer Members of the *Senedd* (MS) (60) in Wales and Members of the Scottish

Parliament (MSPs) (129) than in Westminster (660), so it is easier to lobby those legislatures. Moreover, only certain issues have been devolved (e.g., no defence or foreign policy), so arguably, it is easier to get discussion on issues that have been devolved (e.g., animal welfare) in those legislatures. It is also the case that all the distinctive animal welfare policies and practices outlined in this study may be subject to future reversal. This will be determined by the vagaries of elections, campaigning, and party politics over the following decades. Yet it is argued that to date, at least, the present findings suggest multi-level—or devolved—governance in the UK has generally been beneficial to animal welfare, for it presents opportunities for innovation and, through the process of policy transfer, may lead to a process of 'levelling-up' of welfare standards. The present findings also underline how devolved elections and associated civil society lobbying of parliamentarians are driving policy divergence as parties seek to respond to local demands and public opinion in their efforts to be elected. It is also the case that, in the UK at least, there is increasing evidence that animal welfare policy has moved from the margins to the mainstream of political issues. Finally, it is worth reemphasizing that, as in the present case, future cross-cultural animal welfare studies need to adopt a multi-level governance lens in order to fully understand the way that policy and law develop and diverge within unitary states.

Author Contributions: Each author (P.C., I.R.J., N.N.) is jointly responsible for: conceptualization, methodology, formal analysis, writing—original draft preparation, project administration, and funding acquisition. All authors have read and agreed to the published version of the manuscript.

Funding: This research was funded by the Economic and Social Research Council under Award No. ES/S012435/1.

Institutional Review Board Statement: This study's protocol was approved by the Ethics Committee of the School of Social Sciences, Cardiff University (SREC reference: 267), 10 November 2022.

Informed Consent Statement: Informed consent was obtained from all subjects involved in this study.

Data Availability Statement: Data supporting reported results will be deposited with the ESRC data archive https://www.data-archive.ac.uk/ (accessed 18 December 2023).

Conflicts of Interest: The authors declare no conflicts of interest.

References

1. Paez, A. Grey literature: An important resource in systematic reviews. *J. Evid. Based Med.* 2017, *ahead of print*. [CrossRef]
2. Bonafos, L.; Simonin, D.; Gavinelli, A. Animal welfare: European legislation and future perspectives. *J. Vet. Med. Educ.* **2010**, *37*, 26–29. [CrossRef] [PubMed]
3. Veissier, I.; Butterworth, A.; Bock, B.; Roe, E. European approaches to ensure good animal welfare. *Appl. Anim. Behav. Sci.* **2008**, *113*, 279–297. [CrossRef]
4. Rahman, S.A.; Walker, L.; Ricketts, W. Global perspectives on animal welfare: Asia, the Far East and Oceania. *Rev. Sci. Tech. Off. Int. Des Épizooties* **2005**, *24*, 597. [CrossRef]
5. Rodríguez-Pose, A.; Gill, N. The Global Trend towards Devolution and its Implications. *Environ. Plan. C Gov. Policy* **2003**, *21*, 333–351. [CrossRef]
6. Chaney, P.; Jones, I.R.; Fevre, R. Sentience and salience—Exploring the party politicization of animal welfare in multi-level electoral systems: Analysis of manifesto discourse in UK meso elections 1998–2017. *Reg. Fed. Stud.* **2022**, *32*, 115–140. [CrossRef]
7. Rhodes, R.A.W. *Understanding Governance*; Open University Press: Buckingham, UK; Philadelphia, PA, USA, 1997.
8. Rhodes, R.A.W. Understanding Governance: Ten Years On. *Organ. Stud.* **2007**, *28*, 1243–1264. [CrossRef]
9. Rhodes, R.A.W.; Weller, P. Westminster Transplanted and Westminster Implanted: Explanations for Political Change. In *Westminster Legacies: Democracy and Responsible Government in Asia, Australasia and the Pacific*; Patapan, H., Wanna, J., Weller, P., Eds.; University of New South Wales Press: Sydney, Australia, 2005; pp. 1–12.
10. Jessop, B. *The State: Past, Present, Future*; Polity Press: Cambridge, UK, 2015.
11. Jessop, B. *Putting Civil Society in Its Place: Governance, Metagovernance and Subjectivity*; Policy Press: Bristol, UK, 2020.
12. Rhodes, R.A.W. Understanding Governance: 20 Years On. 2017. Available online: https://www.raw-rhodes.co.uk/wp-content/uploads/2017/07/National-Governance-Review.pdf (accessed on 22 November 2023).
13. Cohen, A.; Arato, A. *Civil Society and Political Theory*; MIT Press: Cambridge, MA, USA, 1992.
14. Eto, M. Reframing Civil Society from Gender Perspectives: A Model of a Multi-layered Seamless World. *J. Civ. Soc.* **2012**, *8*, 101–121. [CrossRef]
15. Salamon, L.; Anheier, H. In Search of the Nonprofit Sector I: The Question of Definitions. *Volunt. Int. J. Third Sect. Res.* **1992**, *3*, 125–151.

16. Edwards, M. *Civil Society*; Polity Press: Cambridge, UK, 2004.
17. De Tocqueville, A. *Democracy in America*; Heffner, R.D., Ed.; Mentor: New York, NY, USA, 1835.
18. Gramsci, A. *Selections from Prison Notebooks*; International Publishers: New York, NY, USA, 1971. First published [1948].
19. Habermas, J. *The Structural Transformation of the Public Sphere: An Inquiry into a Category of Bourgeois Society*; MIT Press: Cambridge, MA, USA, 1962.
20. Putnam, R.; Leonardi, R.; Nanetti, R. *Making Democracy Work: Civic Traditions in Modern Italy*; Princeton University Press: Princeton, NJ, USA, 1993.
21. Hindess, B. Power, Government, Politics' Chapter 4. In *The Blackwell Companion to Political Sociology*; Nash, K., Scott, A., Eds.; Blackwell: Malden, MA, USA, 2004; pp. 40–49.
22. Fondahl, G. The Difference Place Makes: Regional Legislative Approaches to Territories of Traditional Nature Use in the Russian North. *Arct. Rev. Law Politics* **2021**, *12*, 108–133. [CrossRef]
23. Margaras, V.; Vinci, C. Pre-Legislative Briefing Document—Animal Welfare. 2023. Available online: https://www.europarl.europa.eu/RegData/etudes/BRIE/2023/754644/EPRS_BRI(2023)754644_EN.pdf (accessed on 22 November 2023).
24. Government of Canada. *Briefing: Provincial and Territorial Legislation Concerning Welfare of Livestock*; GoC: Ottawa, ON, USA, 2022. Available online: https://inspection.canada.ca/animal-health/terrestrial-animals/humane-transport/provincial-and-territorial-legislation-concerning-/eng/1358482954113/1358483058784 (accessed on 22 November 2023).
25. Rootes, C. Political Opportunity Structures-promise, problems and prospects. *Lett. Maison Française Oxf.* **1999**, *10*, 71–93.
26. Della Porta, D. Political opportunity/political opportunity structure. In *The Wiley-Blackwell Encyclopedia of Social and Political Movements*; Wiley: Hoboken, NJ, USA, 2013; pp. 1–7.
27. Peters, B.G. *Institutional Theory in Political Science*, 4th ed.; Edward Elgar Publishing: Cheltenham, UK, 2019.
28. Almond, G. Comparative Political Systems. *J. Politics* **1956**, *18*, 391–409. [CrossRef]
29. Almond, G.; Verba, S. *The Civic Culture*; Princeton University Press: Princeton, NJ, USA, 1963.
30. Almond, G.; Verba, S. (Eds.) *The Civic Culture Revisited*; Little, Brown: Boston, MA, USA, 1980.
31. Chilcot, S. Defining Political Culture. *West. Political Q.* **1988**, *41*, 419–445.
32. Gardiner, M. *The Cultural Roots of British Devolution*; Edinburgh University Press: Edinburgh, UK, 2004.
33. Birrell, D.; Carmichael, P.; Heenan, D. *Devolution in the UK: Politics, Powers and Policies*; Bloomsbury Publishing: London, UK, 2023.
34. Rothstein, H.R.; Hopewell, S. Grey literature. In *The Handbook of Research Synthesis and Meta-Analysis*; Cooper, H., Hedges, L.V., Valentine, J.C., Eds.; Russell Sage Foundation: New York, NY, USA, 2009; pp. 103–125.
35. Available online: https://bills.parliament.uk/bills/2880 (accessed on 22 November 2023).
36. Available online: https://www.gov.scot/publications/scottish-animal-welfare-commission-terms-of-reference/ (accessed on 22 November 2023).
37. Available online: https://animalwelfarewales.org/ (accessed on 22 November 2023).
38. Available online: https://www.gov.wales/sites/default/files/publications/2020-09/animal-welfare-establishments-code-of-best-practice.pdf (accessed on 22 November 2023).
39. Chaney, P.; Fevre, R.W. Inclusive governance and "minority" groups: The role of the third sector in Wales. *Volunt. Int. J. Volunt. Nonprofit Organ.* **2001**, *12*, 131–156. [CrossRef]
40. Available online: https://www.legislation.gov.uk/ukpga/2004/37/notes/contents (accessed on 22 November 2023).
41. Available online: https://www.legislation.gov.uk/ukpga/2004/37/section/2 (accessed on 22 November 2023).
42. Available online: https://www.legislation.gov.uk/asp/2002/6/contents (accessed on 22 November 2023).
43. Available online: https://www.legislation.gov.uk/asp/2023/1/notes/division/2 (accessed on 22 November 2023).
44. Available online: https://www.uspca.co.uk/hunting-with-dogs#:~:text=Northern%20Ireland%20remains%20the%20only,towards%20this%20so%20called%20sport (accessed on 22 November 2023).
45. Available online: https://www.badgertrust.org.uk/post/badger-trust-welcomes-end-of-selective-badger-cull-in-wales-and-rethink-of-btb-policy (accessed on 22 November 2023).
46. Available online: https://www.legislation.gov.uk/asp/2011/6/contents/enacted (accessed on 22 November 2023).
47. Available online: https://www.legislation.gov.uk/ukpga/1992/51/contents (accessed on 22 November 2023).
48. Available online: https://www.bornfree.org.uk/news/plans-to-cull-badgers-in-northern-ireland-rejected-by-court/ (accessed on 22 November 2023).
49. Available online: https://senedd.wales/media/nnejc1vu/sub-ld14061-em-r-e.pdf (accessed on 22 November 2023).
50. Available online: https://www.legislation.gov.uk/sdsi/2021/9780111048474 (accessed on 22 November 2023).
51. Available online: https://www.legislation.gov.uk/uksi/2018/486/contents/ma (accessed on 22 November 2023).
52. Available online: https://www.gov.uk/government/news/lucys-law-spells-the-beginning-of-the-end-for-puppy-farming (accessed on 22 November 2023).
53. Available online: https://www.legislation.gov.uk/nia/2011/16/pdfs/nia_20110016_en.pdf (accessed on 22 November 2023).
54. Available online: https://www.uspca.co.uk/lucys-law (accessed on 22 November 2023).
55. Available online: https://www.legislation.gov.uk/wsi/2021/416/contents/made (accessed on 22 November 2023).
56. Available online: https://www.legislation.gov.uk/asp/2020/14/contents/enacted (accessed on 22 November 2023).
57. Available online: https://www.legislation.gov.uk/ukpga/2006/45/notes/contents (accessed on 22 November 2023).

58. Available online: https://www.legislation.gov.uk/nia/2011/16/contents (accessed on 22 November 2023).
59. Available online: https://www.legislation.gov.uk/wsi/2010/943/regulation/2/made (accessed on 22 November 2023).
60. Available online: https://www.gov.scot/publications/report-use-handheld-remote-controlled-training-devices-e-collars-dog-training-scottish-animal-welfare-commission/pages/7/ (accessed on 22 November 2023).
61. Available online: https://www.legislation.gov.uk/ukdsi/2023/9780348247305#:~:text=Regulation%202(1)%20makes%20it,cat%20or%20dog%20in%20England (accessed on 22 November 2023).
62. Available online: https://business.senedd.wales/mgIssueHistoryHome.aspx?IId=39968 (accessed on 22 November 2023).
63. Available online: https://www.wwf.org.uk/sites/default/files/2023-05/open-letter-agriculture-bill.pdf (accessed on 22 November 2023).
64. Available online: https://www.wwf.org.uk/wales/land-of-our-future (accessed on 22 November 2023).
65. League Against Cruel Sports. Campaign E-mail to Subscribers Help Us Ban Snares throughout the Rest of the UK, 26 June 2023. Available online: https://www.league.org.uk/ (accessed on 22 November 2023).
66. Cairney, P. The role of ideas in policy transfer: The case of UK smoking bans since devolution. *J. Eur. Public Policy* **2009**, *16*, 471–488. [CrossRef]
67. Available online: https://www.rspca.org.uk/getinvolved/campaign/slaughterhouses (accessed on 22 November 2023).
68. Available online: https://www.theyworkforyou.com/senedd/?id=2023-06-14.3.511921 (accessed on 22 November 2023).
69. Available online: https://www.openseas.org.uk/news/win-for-environmental-justice/ (accessed on 22 November 2023).
70. Available online: https://www.gov.uk/government/news/lobsters-octopus-and-crabs-recognised-as-sentient-beings (accessed on 22 November 2023).
71. Available online: https://www.gov.wales/responses-consultation-scallop-dredging-cardigan-bay (accessed on 22 November 2023).
72. Available online: https://www.gov.wales/written-statement-introduction-vessel-tracking-devices-welsh-scallop-fishery-and-proposed-removal (accessed on 22 November 2023).
73. Available online: https://www.legislation.gov.uk/wsi/2010/269/note (accessed on 22 November 2023).
74. Available online: https://www.gov.wales/king-scallop-fisheries-management-plan (accessed on 22 November 2023).
75. DEFRA. *Summary of 2022 Badger Control Operations*; DEFRA: London, UK, 2023. Available online: https://www.gov.uk/government/publications/bovine-tb-summary-of-badger-control-monitoring-during-2022/summary-of-2022-badger-control-operations (accessed on 22 November 2023).
76. Badger Trust. The True Toll of the Badger Cull. 2022. Available online: https://www.badgertrust.org.uk/post/the-true-toll-of-the-badger-cull (accessed on 22 November 2023).
77. Tuberculosis Eradication (Wales) Order. 2009. Available online: https://www.legislation.gov.uk/wsi/2009/2614/contents/made (accessed on 22 November 2023).
78. Badger Trust v Welsh Ministers EWCA Civ 807. 2010. Available online: https://vlex.co.uk/vid/r-badger-trust-v-793712973 (accessed on 22 November 2023).
79. Welsh Government. *Programme for Government*; Welsh Government: Cardiff, UK, 2021.
80. DEFRA. National Statistics—Figures to March 2022 (Published on 15 June 2022). 2022. Available online: https://www.gov.uk/government/statistics/historical-statistics-notices-on-the-incidence-of-tuberculosis-tb-in-cattle-in-great-britain-2022-quarterly/figures-to-march-2022-published-on-15-june-2022#key-points (accessed on 22 November 2023).
81. Welsh Labour Party. *Moving Wales Forward*; Welsh Labour Party: Cardiff, UK, 2021.
82. Scottish National Party. *Scotland's Future, Scotland's Choice*; SNP: Edinburgh, UK, 2021.
83. Plaid Cymru. *Let Us Face the Future*; Plaid Cymru: Cardiff, UK, 2021.
84. Welsh Conservative Party. *A Plan for Recovery Change: Let's Build a Better Wales*; Welsh Conservatives: Cardiff, UK, 2021.
85. Scottish Labour Party. *Scottish Labour's National Recovery Plan*; Scottish Labour Party: Edinburgh, UK, 2021.
86. Welsh Liberal Democratic Party. *Put Recovery First*; Welsh Liberal Democratic Party: Cardiff, UK, 2021.
87. Scottish Conservative and Unionist Party. *Rebuild Scotland*; Scottish Conservative and Unionist Party: Edinburgh, UK, 2021.
88. Chaney, P. Public Policy for Non-Humans: Exploring UK State-Wide Parties' Formative Policy Record on Animal Welfare, 1979–2010. *Parliam. Aff. J. Represent. Politics* **2014**, *67*, 907–934. [CrossRef]
89. Available online: https://www.gov.wales/sites/default/files/publications/2021-11/animal-welfare-plan-wales-2021-26.pdf (accessed on 22 November 2023).
90. Available online: https://ymgynghori.cyfoethnaturiol.cymru/evidence-policy-and-permitting-tystiolaeth-polisi-a-thrwyddedu/nrw-s-proposed-approach-to-regulating-the-release/ (accessed on 22 November 2023).
91. Dixon, H. Shooting 'Ban by the Backdoor' Set for Wales after Government Advised to Limit Release of Gamebirds. *Daily Telegraph*, 12 November 2023. Available online: https://www.telegraph.co.uk/news/2023/11/12/shooting-ban-by-backdoor-wales-government-gamebirds-limit/ (accessed on 22 November 2023).
92. BBC. Hela: Roger Williams yn Galw am Ddatganoli Pwerau [Welsh–Translation: Hunting: Roger Williams Calls for Devolving Powers]. *BBC*, 15 February 2023. Available online: https://www.bbc.co.uk/cymrufyw/31478199 (accessed on 22 November 2023).

animals

MDPI

Article

A Survey of Public Opinion on Community Cats' General Health and Relationship Quality with Residents in Urban China

Xuan Gu [1,2,*], Zilin Zhang [2], Guo Peng [2,3], Anru Ni [1,2], Bo Wang [2,4], Xiufan Xiong [1], Yujie Liu [1] and Li Wang [1]

[1] Department of Social Work, School of Philosophy and Social Development, Shandong University, Jinan 250100, China; nielsen@sdu.edu.cn (A.N.); emilyxiong1874@outlook.com (X.X.); yujie13627960491@outlook.com (Y.L.); echo19960112@gmail.com (L.W.)

[2] Center for Animal Protection Studies, Shandong University, Jinan 250100, China; zilin1126@outlook.com (Z.Z.); guopeng@sdu.edu.cn (G.P.); wangbojqk@gmail.com (B.W.)

[3] Department of Philosophy, School of Philosophy and Social Development, Shandong University, Jinan 250100, China

[4] Department of Applied Psychology, Guangdong Peizheng College, Guangzhou 510832, China

* Correspondence: guxuan@sdu.edu.cn

Simple Summary: This study addresses the concern of managing and coexisting with community cats in urban areas. The study was conducted in urban China through a survey involving 5382 participants. It explores the residents' perceptions of the general health of community cats and the relationships between humans and these feline inhabitants. The findings show that about 70% of participants considered all or most of the community cats to be in good health, and about 60% reported harmonious or non-conflict coexistence. It was found that residents' relationships with community cats were associated with socio-demographic factors such as education, marital status, and income level. The study highlights the importance of customized approaches to cat management in various cities and communities. This provides useful information for stakeholders to create efficient policies and interventions for the welfare of the community cats and the people who live alongside them.

Abstract: The management and coexistence of community cats in urban areas is a growing concern amid global urbanization. Through a survey-based investigation, we examine the residents' perceptions of the general health of community cats and human-cat relationships in urban China. The data from 5382 participants revealed that approximately 70% of participants perceived community cats as being in good health, and 60% reported harmonious or non-conflict coexistence between residents and these cats. Around 45% of the participants rescued or helped community cats, 38% expressed their intention to adopt, and 18% complained about the issues of community cats to management staff. Linear, logistic, and multilevel-logistic regressions were employed to examine the associations between the types of cities and communities or the participants' socio-demographics and the perceived well-being of community cats or human-cat relationships. The results show that the cats in fourth-tier cities (e.g., county-level cities) had poorer living conditions than in first-tier cities (e.g., Beijing), while the cats in urban village communities (e.g., villages in the city) were less likely to exhibit good health than in ordinary commercial housing communities. The results also show that socio-demographic variables, such as educational attainment, marital status, and income level, predicted participants' relationships with community cats. This study is the first of its kind. It provides valuable insights for stakeholders to develop effective policies and interventions on cat management, emphasizing the need for tailored strategies in diverse urban settings and populations.

Keywords: animal welfare; community cat; China; free-roaming cats; human-animal relationship; stray cat; unowned cat; urban ecology

1. Introduction

Domestic cats (*Felis catus*) are prevalent companion animals worldwide [1–3], playing a pivotal role in human lives as they coexist with humans in diverse settings, including families, animal shelters, and communities [4–6]. Each domestic cat occupies a place on a spectrum of behavior and lifestyle, ranging from completely independent-living individuals with minimal or no direct human contact to those residing in homes relying on human care [7]. Community cats are positioned in the intermediate range of this spectrum for *Felis catus*. In addition to the term "community cats", the domesticated cats that are unowned and free-roaming are also referred to as "stray cats" [8,9], "free-roaming cats" [6,10], "feral cats" [11,12], or "outdoor cats" [13,14]. The term "community cats" is widely employed, irrespective of their sociability, gaining popularity as a replacement for the terms stray cats or abandoned cats. This shift acknowledges that the cats are a commensal species highly valued by many residents [15]. Community cats often denote solitary individuals or in a group, maintaining some degree of direct human contact and tolerance, and are provided with food and care to some extent [7]. In this study, we chose to use the term "community cats" (社区猫) to denote unowned and free-roaming cats.

A contentious matter revolves around the terminology that best refers to community cats. When designating them as feral cats or free-roaming cats, researchers often assume a wildlife perspective and adopt strategies from wildlife management or conservation conflict approach [16]. Conservation conflicts occur because the ecological needs and behaviors of wildlife negatively affect the goals of humans or when the goals of humans negatively affect wildlife [17]. In this approach, community cats are regarded as an urgent and pressing ecological threat [12,18,19]. However, some scholars have challenged the findings about the impacts of community cats on wildlife and biodiversity decline. They discussed boundary conditions, considered the role of community cats in rodent control, and noted that anthropogenic activities pose a greater threat to biodiversity than feline behaviors [20–22]. Moreover, categorizing animals in urban settings into a binary classification of wild or domesticated oversimplifies the complex realities many city-dwelling species face. According to the definitions by Donaldson and Kymlicka [23], wild animals refer to "those living relatively free of direct human management and meeting their own needs for food, shelter, and social structure" (p. 156), and domesticated animals are "created by human labor to meet specific requirements or whims and are adapted to the conditions of continuous care and solicitude people maintain for them" (p. 74). Rats, for instance, are considered wild animals, yet they are non-domesticated creatures that have adapted to life among humans. As we discussed here, whether the community cats in cities are domestic or wild animals may be viewed differently by different residents or researchers. The domestic/wild dichotomy ignores the vast numbers of wild animals living amongst us, even in the city's heart. Donaldson and Kymlicka [23], in their work *Zoopolis*, refer to these urban and peri-urban animals as "liminal" animals, that is wild animals living in close association with humans, which signifies their status as neither wilderness nor domesticated animals. Considering the domestication process of cats and their longstanding interdependence with humans [24], we propose a paradigm shift towards an urban ecology perspective instead of a wildlife perspective when investigating the roles and functions of community cats in cities. Urban ecology, conceptualizing cities as social-ecological systems, represents an applied and transdisciplinary field addressing the social, economic, and environmental dimensions of sustainable development [25]. Therefore, urban ecology considers the ecological niche of community cats and explores the natural and humanistic ecology of interactions among humans, other animals, and the natural environment [21,22].

In China, the relationship between humans and cats is distinctive, which is recorded in ancient accounts and cat domestication history. Archaeological research suggests that approximately 5300 years ago, cats were drawn to the ancient Chinese village of Quanhucun by small animals, such as rodents, thriving on the grain cultivated, consumed, and stored by farmers [26]. The study indicates a commensal relationship between humans and cats. Although these cats were not fully domesticated, the evidence affirms their proximity to farmers and the mutually beneficial nature of the association. In contrast to dogs, horses, and cattle, the reliance of cats on humans was limited as recorded in ancient Chinese literature. For example, the Li Ji (also known as Book of Rites) is a core text of the Confucian canon and articulates this sentiment: "The ancient wise men had appointed all these representatives, and it was felt necessary to make this return to them. They met the (representatives of the) cats because they devoured the rats and mice (which injured the fruits) of the fields and the (representatives of the) tigers because they devoured the (wild) boars (which destroyed them). They met them and made offerings to them". (The Chinese text is based on a translation by Donald Sturgeon in Chinese Text Project: A dynamic digital library of premodern Chinese, Digital Scholarship in the Humanities 2019). (古之君子, 使之必报之。迎猫, 为其食田鼠也。迎虎, 为其食田豕也, 迎而祭之也。) As the poet Liu Zongyuan (773–819) conveyed, "Cats and tigers are worshipped, dogs and horses are buried under drapery". (猫虎获迎祭, 犬马有盖帷。) In summary, due to the needs of agricultural society, a mutually beneficial relationship developed between humans and cats in China, yet cats exhibited both independence and dependency.

From the 1950s to the 1970s in China, keeping companion animals in urban areas was neither encouraged nor permitted [27]. Companion animals at that time were mainly confined to rural settings, serving utilitarian functions such as cats for rodent control and dogs for home guarding. These rural companion animals had owners but existed in a free-roaming state. Since the 1980s, dog management regulations emerged in Chinese cities such as Beijing. However, there are no regulations for cats or other companion animals in China. Due to the increasing population density and a shift towards vertical living in high-rise structures during urbanization and industrialization after 1979, urban-dwelling cats and dogs as companion animals became predominantly indoor-kept, and most cities in China have fewer outdoor activity space for companion animals [28]. Therefore, there is minimal overlap between household and community cats in Chinese cities. The household status of community cats in Chinese and Western countries is noteworthy and requires comparative investigation.

Returning to contemporary China, cats were the most popular companion animals among urban Chinese residents in 2022, with approximately 65.4 million cats in urban households, surpassing the 51.2 million dogs [29]. However, the dark side of China's companion animal boom lies in the vast number of cats abandoned by individuals or breeding companies [30,31], contributing to a staggering increase in the community cat population. On average, China witnesses an annual growth of approximately 40 million stray cats [32]. The term "stray cats" (流浪猫) became popular in Chinese discourse to denote unowned cats residing in urban communities [9]. This nomenclature underscores societal concern for these cats' homeless and abandoned status, whose needs are directly or indirectly addressed by humans. In response to the low welfare of community cats, spontaneous rescuing and helping efforts arise in urban areas, such as community residents' self-governing organizations dedicated to managing cats and dogs [33]. Meanwhile, people participate in the illicit black market of cat theft and meat trade [34]. Against the above background, it is necessary to investigate the community cats' survival situations in urban China and their relationship quality with residents. The answers to these questions will facilitate the formulation of regulations on community cats and improve the welfare of community cats and residents. Regrettably, scientific research exploring the interactions between humans and community cats and the urban ecology shaped by this interaction in contemporary Chinese cities is notably limited.

Three dimensions of urban living hold significant ecological importance: (a) Human needs, social patterns, and economics, (b) domestic animals, and (c) pests and public health [35]. (Forman [35] defined pests as organisms that annoy humans, including fungi, bacteria, invertebrate pests (e.g., cockroaches, mosquitoes, ticks), and vertebrate pests (e.g., raccoons and coyotes in the USA). For the vertebrate pests, Forman particularly noted that all of the species are much appreciated by urban residents and also play important positive ecological roles in urban habitats or urban regions as a whole.) The studies on community cats and their relationships with community residents in urban areas constitute an increasingly important intersection of the first and second dimensions. Taking an urban ecology perspective, we initiated a social survey study to investigate the present situation of community cats in urban communities in China. Specifically, our investigation focuses on community cats' living and health conditions, community residents' relationships with the cats, and their attitudes towards these felines and relevant policies. Addressing and elucidating these questions contributes to understanding the situations and values of community cats in Chinese cities. Furthermore, the insights will guide practices of cat care and management, legislation, and scientific researches that aim to enhance these feline inhabitants' welfare.

The data presented is part of the "Social Research Project on Community Unowned Cats in Urban China", conducted between mid-November 2021 and early February 2022. In the project, we developed a questionnaire to investigate two main issues: (a) The perceived health of community cats and the quality of human-cat relationships in the communities, and (b) the public attitudes towards the community cats and the way of managing the cats (e.g., trapping and killing the cats). The current study focuses on the first issue and explores three research questions. First, what is the general health status of community cats in urban areas? Second, what is the relationship between community cats and residents in urban areas? Third, how do the types of cities and communities and residents' socio-demographic information predict community cats' health and the human-cat relationships?

2. Materials and Methods

2.1. Sampling and Data Collection Procedure

We used a stratified sampling method and defined three strata: Provinces/regions, cities, and urban communities. Firstly, based on economic development and geographical representation, we selected Beijing City (China's capital and national center), Zhejiang Province, and Shandong Province in the first strata. Secondly, the cities within the provinces or regions were classified into different city tiers. The city tier system is a comprehensive indicator used to measure the socio-economic development level and future potential of different cities. It has been widely applied to urban and regional studies in China [36–39]. Cities in mainland China are classified into different tiers based on their performance in business, transportation, urbanism, lifestyle diversity, and development potential, using big data and user behavior data [39]. In this study, the type of city is set as a nominal variable with four tiers of cities. The first-, second-, third-, and fourth-tier cities in our sample varied in GDP and population size, as shown in Table 1. Beijing was chosen as the representative of the first-tier cities. Our sample included five districts in Beijing: Chaoyang, Haidian, Xicheng, Fengtai, and Tongzhou. The second-tier cities included the provincial capitals, Jinan and Hangzhou in the current study, which are the capital cities of Zhejiang Province and Shandong Province, respectively. The third-tier cities were prefecture-level cities (地级市) in the two provinces, and the fourth-tier cities were county-level cities or counties (县级市或县城).

Table 1. Number of sampled cities, communities, and participants (N = 5382).

Type of City	Number of Cities	Number of Communities	Number of Valid Participants	GDP, N × 10^8 CNY #	GDP% of the National GDP	Population, N × 10^4
First-tier city: Beijing	1	10	764	40,270	3.5	1414
Second-tier cities: Jinan	1	3	392	11,432	0.99	817
Second-tier cities: Hangzhou	1	3	308	18,109	1.58	835
Third-tier cities in Shandong Province	6	10	1106	39,123	3.41	4991
Third-tier cities in Zhejiang Province	4	8	753	30,679	2.67	1994
Fourth-tier cities in Shandong Province	8	9	1179	4721	0.42	606
Fourth-tier cities in Zhejiang Province	5	9	880	4836	0.41	438
Total	26	52	5382			

The GDP and population size were based on the provincial statistical yearbook of 2021. Source: [40]. The sum GDP and population size of the third-tier or fourth-tier cities were reported.

The investigators received training before distributing and collecting questionnaires. They also joined supervision provided by the project team to ensure the sampling complied with scientific standards. They were required to confirm participants as community residents, avoid distributing questionnaires to people with specific preferences (such as cat lovers), and prohibit proxy or guided responses. Investigators who completed the investigation received compensation and a certificate of practice.

A total of 150 questionnaires were distributed for large communities with over 5000 residents and 100 questionnaires were distributed for medium-sized communities with 1000 to 5000 residents. The investigators distributed the questionnaires both online and offline. An online questionnaire was set up on the Wenjuanxing platform, a survey service provider in China (also known as Sojump), and distributed through WeChat groups of residents in a community. The investigators also set up booths offline or intercepted people on the roadside. Participants could get access to the online questionnaire by scanning a QR code. For participants who may not be proficient in using smartphones or are reluctant to complete online questionnaires (such as older adults), investigators provided an alternative paper-based questionnaire. Each community had a unique questionnaire link to facilitate tracking the data collection status and each smartphone was limited to one participant's response. We asked each participant to enter a unique password to fill out a questionnaire, but we did not obtain any identifiable information and it was an anonymous survey.

After completing the online questionnaire, participants received Red Packets by WeChat Pay, in which they got a random amount with an average of CNY 10. Those who completed the paper-based questionnaire received gifts worth approximately CNY 10.

2.2. Participants

Our study followed the Declaration of Helsinki and the ethical guidelines for research involving human participants. Prior to answering the questionnaire, all participants provided their informed consent. They were informed that their participation was voluntary and that their privacy would be protected. According to Chinese national policy, the current study did not require ethical approval because it used a survey method and did not collect and report scientific research information on people involved in life sciences and medical issues.

Table 1 presents the number of sampled cities, urban communities, and valid participants. Of the 5728 participants who responded to the questionnaire, we excluded participants who met one of the following criteria: (a) Spent a short time on completing the questionnaire, i.e., less than 120 s; (b) were under the age of 18 or over the age of 86; (c) provided inconsistent information, e.g., retired but with a young age, having children or married under the legal age; (d) selected "other" for gender (N = 12). Ultimately, there were 5382 valid participants available for data analyses.

2.3. Measures

We developed a Chinese questionnaire to assess the perceived health status of community cats and the relationship between community cats and participants. We conducted pilot tests to evaluate and improve the understandability of questionnaire items and the smoothness of the data collection process before developing the formal questionnaire and collecting data. We also collected participants' socio-demographic information. The questionnaire items reported in the current study can be found in the Appendix A.

2.3.1. Community Variables

The community variables included city type and community type. Participants reported their types of community, but the final type was decided by the researchers who interviewed the investigators in each community about the characteristics of the community and housing.

2.3.2. Socio-Demographic Variables

Participants' socio-demographic information, such as gender, education, age, employment, personal income in the last year (i.e., 2020), and marital status, were collected. To use the nominal variables as predictors in regression models, we converted them into dummy variables. A variable with N categories was converted into N − 1 dummy variables, with the most frequently occurring value of the variable representing 0 and the other value of the variable 1.

2.3.3. Perceived General Health of the Community Cats

Participants were asked how frequently they saw community cats in the community. If they selected "often (more than three days per week)" or "occasionally", they then proceeded to report where they saw most of the cats appearing in the community, how the community cats got food, the cats' health condition, and their living condition.

The perceived general health of the community cats was measured by two indicators, i.e., the cats' health condition and living condition as reported by residents. The living condition was measured using a 5-point Likert scale with 1 = extremely well and 5 = extremely bad. To gauge the health condition of cats, we recoded the responses to the cats' health condition into a binary outcome variable: 1, a group named good health (N = 3258), for options "All are in good health" and "Most of them are in good health, and a few are sick or disabled"; and 0, another group named bad health (N = 241), for options "Most of them are sick or disabled, and a few are in good health" and "All are sick or disabled". The response "I don't notice" was recorded as missing data.

2.3.4. Perceived General Relationship between Residents and Community Cats

Participants rated the general relationship between residents and community cats by answering, "What do you think is the current situation of community residents getting along with unowned cats in the community?" with seven potential responses. We then recoded these responses into a binary outcome variable: 1, a group named coexistence (N = 3215), for options "very harmonious" and "basically coexistent"; and 0, a group named conflict (N = 1332), for options "A small number of residents hate unowned cats in the community", "Residents generally demand the removal of unowned cats from the community", and "Property and other managements remove unowned cats from the community". The responses "I don't know" and "other situations" were recoded as missing data.

2.3.5. Personal Relationship with Community Cats

There were three indicators to measure the relationship between participants and community cats. Our first indicator of the personal relationship quality was rescuing behavior, which asks whether participants rescued or helped community cats (including feeding and remote sponsorship). Our second indicator of the relationship quality was adopting intention, asking whether participants considered adopting a community cat if they could. Our third indicator of the personal relationship quality was the complaint, which asks whether participants complained to management staff about the issues of community cats. Participants' responses to the above three questions were coded similarly with "Yes" as 1 and "No" as 0.

2.4. Statistical Analyses

Descriptive statistics were employed to summarize the variables. Linear, logistic, and multilevel-logistic regressions were conducted to examine the associations between community or socio-demographic variables and the perceived health of cats or the human-cat relationship. As the living condition of community cats reported in a 5-point Likert scale was a continuous variable, we employed linear regression analysis to investigate the associations between community variables and community cats' living conditions. As the health condition of community cats and the general relationship between residents and community cats were nominal variables and converted to binary variables, we performed logistic regression analyses to investigate the associations between community variables and community cats' health condition or general relationship with residents.

Given that the indicators of participant's relationship with community cats were binary and that participants nested within communities, we fitted multilevel logistic regression models when investigating the extent to which socio-demographic variables and community variables may predict personal relationship with community cats. Multilevel regression modeling can accommodate nested data and include group characteristics in statistical models, which improves the estimates of the effects within groups [43,44]. The variables we analyzed were at two levels. Socio-demographic characteristics of participants, such as age, gender, marital status, and employment, are at the individual level (i.e., level 1 in the regression models). Community variables, such as community type and city type, are at the community level (i.e., level 2 in the regression models). Level 1 variables were nested in level 2. We chose two levels instead of adding different cities as a separate level because, in multilevel logistic modeling, a minimum of 50 level 1 and 40 level 2 units are necessary to accurately estimate small fixed effects [45].

As some responses (e.g., "I don't know" and "I don't notice") were recoded as missing data, the sample sizes varied when different variables served as the outcomes in the regressions. Specifically, the actual sample size was 4621 when the living condition of community cats was the outcome, 3499 when the health condition of community cats was the outcome, and 4547 when the general relationship between residents and community cats was the outcome. Except for the above situations, the analyses included the data of all 5382 participants.

Statistics were performed in SPSS 27 (IBM, New York, NY, USA). A two-tailed p-value of < 0.05 was considered statistically significant. Only significant effects are discussed.

3. Results

3.1. Descriptive Statistics

Table 2 describes the socio-demographic and community variables of the participants, the community cats' general health, and the relationship between residents and cats as perceived by participants. Approximately 53% of participants were female, and nearly 47% had received a bachelor's degree or were undergraduate students. The average age was approximately 41 years old. Most participants (approximately 62%) lived in ordinary commercial housing. Other types of communities were presented in a decreasing order:

Urban village community, old town community, luxury housing community, indemnificatory housing community, and campus community.

Table 2. Descriptive statistics of variables among participants (N = 5382).

Variables	Categories	N	Percentage (%)	Mean (SD)	China Census 2020, N × 10^6 #
Type of City	First-tier city	764	14		
	Second-tier cities	700	13		
	Third-tier cities	1859	35		
	Fourth-tier cities	2059	38		
Type of Community	Old town community	749	14		
	Indemnificatory housing community	110	2		
	Ordinary commercial housing community	3322	62		
	Luxury housing community	174	3		
	Urban village community	920	17		
	Campus community	107	2		
Gender	Female	2848	53		223 (50%)
	Male	2534	47		224 (50%)
Education	Elementary school or below	488	9		59 (13%)
	Junior high-school	785	15		152 (33%)
	Secondary or high school	1107	21		101 (22%)
	College or university	2547	47		140 (30%)
	Graduate or beyond	455	8		9 (2%)
Age	18–85	5382		41 (15)	45 (15)
Employment	Full employment	3090	57		
	Part-time employment	276	5		
	Household work	383	7		
	Full-time students	649	12		
	Unemployment	187	3		
	Retirement	797	15		
Personal Income in the Last Year	<¥50,000	2889	54		
	¥50,000–¥100,000	1526	28		
	¥100,000–¥200,000	674	13		
	>¥200,000	293	5		
Marital Status	Unmarried	1401	26		10.5 (22%)
	Married	3778	70		33.4 (71%)
	Divorced or widowed	203	4		3.3 (7%)
Frequency of Seeing Community Cats	Often	1617	30		
	Occasionally	3004	56		
	Never	289	5		
	I don't notice.	472	9		
The Living Condition of Community Cats (N = 4621)	5-point Likert Scale	4621		3.3 (0.96)	
The Location Where Community Cats Were Mostly Seen (N = 4621)	Garbage heaps	1461	32		
	Feeding points	619	13		
	Roadside	2026	44		
	Other places	123	3		
	I don't notice.	392	8		
The Way How Community Cats Got Food (N = 4621)	Some people feed them regularly.	639	14		
	Some people feed them occasionally.	1765	38		
	No one feeds them and they find food themselves.	1323	29		
	I don't notice.	894	19		

Table 2. *Cont.*

Variables	Categories	N	Percentage (%)	Mean (SD)	China Census 2020, N × 10^6 #
The Health Condition of Community Cats (N = 4621)	All are in good health.	1492	32		
	Most of them are in good health; A few are sick or disabled.	1766	38		
	Most of them are sick or disabled; A few are in good health.	219	5		
	All are sick or disabled.	22	0.5		
	I don't notice.	1122	24		
Perceived General Relationship between Residents and Community Cats	Very harmonious.	818	15		
	Basically coexistent.	2397	45		
	A small number of residents hate unowned cats in the community.	1083	20		
	Residents generally demand removal of unowned cats from the community.	173	3		
	The management staff remove unowned cats from the community.	76	1		
	I don't know.	816	15		
	Other situations.	19	0.4		
Personal Rescuing or Helping Behavior	Yes	2445	45		
	No	2937	55		
Personal Adopting Intention	Yes	2064	38		
	No	3318	62		
Personal Complaint	Yes	956	18		
	No	4426	82		

Based on the seventh national census in 2020 in China, we calculated the statistics data of the urban population. Due to the availability of the census data, the age range of the population is 18–84 years old when the statistics data of education were calculated and the age range is 20–84 years old when the statistics data of gender and age were calculated. The blank space indicates no data are available. Source: [46].

Participants often saw community cats near garbage piles or feeding sites (accounting for about 45%) and reported that people fed community cats regularly or occasionally (accounting for about 51%). Notably, 29% of the participants reported that no residents fed the cats. Thus, we can infer that community cats rely on humans, to varying extents, to acquire sustenance, which may include discarded human food or kibbles provided by humans. The community cats that could be observed were generally healthy, with about 32% of participants reporting that they were all in good health and about 38% reporting that most were in good health. However, about 6% of participants did report that most or all cats were sick or disabled.

For the relationship between residents and community cats in a community, approximately 60% of participants thought that the residents lived with community cats in a harmonious or non-conflict manner. Nevertheless, approximately 24% thought that there were conflicts between residents and community cats in the community. For personal relationships with community cats, approximately 45% of participants rescued or helped community cats, 38% intended to adopt one if conditions were met, and 18% complained about the cats to management staff at the community.

3.2. The Associations between Community Variables and the Cats' General Health or the Human-Cat Relationship

Table 3 displays the results of linear or logistic regression models. A linear regression was performed to assess how the city type and community type would predict the living conditions of community cats as perceived by participants. The assumptions for the linear regression were met (Durbin-Watson statistic = 1.8, 1.0 < VIFs < 1.6, 0.63 < Tolerance < 0.96). The overall regression model was significant, F (8, 4620) = 4.0, $p < 0.001$. The community cats inhabiting the communities of first-tier cities had better living conditions than those inhabiting the communities of fourth-tier cities (M = 3.4, SD = 0.86; M = 3.2, SD = 0.98, respectively), $p < 0.001$.

Table 3. Results of a linear model and two logistic regression models examining the effects of community variables on the general health of the community cats and the quality of human-cat relationship in urban China.

		Living Condition of Community Cats N = 4621	Health Condition of Community Cats N = 3499	General Relationship between Residents and Community Cats N = 4547
Community Variables		β [95%CI]	OR [95%CI]	OR [95%CI]
Type of City	Fourth-tier cities (ref.)			
	First-tier city	0.06 ** [0.07, 0.27]	1.12 [0.70, 1.80]	1.17 [0.92, 1.48]
	Second-tier cities	−0.02 [−0.14, 0.05]	0.90 [0.57, 1.41]	1.29 * [1.03, 1.61]
	Third-tier cities	−0.02 [−0.12, 0.03]	0.98 [0.70, 1.38]	1.08 [0.91, 1.27]
Type of Community	Ordinary commercial housing community (ref.)			
	Old town community	−0.003 [−0.09, 0.08]	0.98 [0.66, 1.45]	1.20 [0.98, 1.47]
	Indemnificatory housing community	0.03 [−0.01, 0.37]	2.14 [0.66, 6.94]	1.46 [0.92, 2.32]
	Luxury housing community	0.03 [−0.02, 0.30]	1.54 [0.60, 3.91]	0.98 [0.68, 1.42]
	Urban village community	−0.03 [−0.16, 0.01]	0.69 * [0.48, 0.98]	1.13 [0.94, 1.37]
	Campus community	−0.01 [−0.27, 0.16]	4.05 [0.54, 30.58]	2.05 * [1.15, 3.65]

β is standardized regression coefficient; OR, Odds Ratio; CI, Confidence Interval. * $p < 0.05$, ** $p < 0.01$.

A logistic regression was performed to assess how the city type and community type would predict the likelihood of good health condition of community cats. The overall logistic model was not significant, $\chi^2(8) = 13$, $p = 0.10$. The effect of urban villages was significant, $p = 0.04$, with the community cats inhabiting urban villages being 0.69 times less likely to have a good health condition than those inhabiting ordinary commercial housing communities.

A logistic regression was performed to assess how the city type and community type would predict the likelihood of a coexisting relationship between residents and community cats in a community. The overall logistic model was significant, $\chi^2(8) = 22$, $p = 0.004$. Participants in second-tier cities were 1.3 times more likely to have a coexisting relationship than those in fourth-tier cities. Participants in campus communities were 2.1 times more likely to have a coexisting relationship than those in ordinary commercial housing communities.

3.3. The Associations between Socio-Demographic and Community Variables and Personal Relationships with Community Cats

Table 4 displays the results from the estimation of multilevel logistic regression models. The three indicators of personal relationship with community cats (i.e., rescuing behavior, adopting intention, and complaint) were the outcome variables separately in each model. The level 1 variables (i.e., participants' socio-demographic variables) and level 2 variables (i.e., city type and community type) were predictors.

Table 4. Results of multilevel logistic regression models examining the effects of socio-demographic and community variables on personal relationships with community cats in urban China (N = 5382).

		Rescuing Behavior	**Adopting Intention**	**Complaint**
		(Model 1)	(Model 2)	(Model 3)
		OR [95%CI]	OR [95%CI]	OR [95%CI]
	Fixed effects			
	Intercept	13.88 * [1.22, 158.21]	4.31 [0.53, 35.27]	15.65 * [1.14, 214.82]
Socio-demographic Variables				
Gender	Female (ref.)			
	Male	0.95 [0.84, 1.06]	0.99 [0.88, 1.12]	1.12 [0.97, 1.31]
Education	College or university (ref.)			
	Elementary school or below	0.71 * [0.54, 0.93]	0.81 [0.61, 1.09]	1.02 [0.71, 1.46]
	Junior high-school	0.78 * [0.63, 0.95]	1.20 [0.98, 1.48]	1.29 * [1.01, 1.65]
	Secondary or high school	0.88 [0.75, 1.03]	1.14 [0.97, 1.34]	1.18 [0.96, 1.44]
	Graduate or beyond	0.75 * [0.59, 0.95]	0.87 [0.68, 1.11]	1.02 [0.74, 1.39]
Age	18–85	1.00 [1.00, 1.01]	1.02 *** [1.01, 1.02]	1.02 ** [1.01, 1.02]
Employment	Full employment (ref.)			
	Part-time employment	1.09 [1.00, 1.01]	1.01 [0.78, 1.32]	0.89 [0.64, 1.24]
	Household work	0.88 [0.69, 1.12]	0.86 [0.67, 1.11]	0.77 [0.56, 1.06]
	Full-time students	1.00 [0.78, 1.28]	0.96 [0.75, 1.23]	0.93 [0.68, 1.27]
	Unemployment	0.80 [0.58, 1.11]	0.81 [0.58 1.12]	1.02 [0.70, 1.49]
	Retirement	0.77 * [0.61, 0.98]	0.91 [0.70, 1.17]	0.77 [0.56, 1.06]
Personal Income in the Last Year	<¥50,000 (ref.)			
	¥50,000–¥100,000	0.98 [0.84, 1.13]	0.89 [0.76, 1.03]	0.77 ** [0.64, 0.92]
	¥100,000–¥200,000	1.09 [0.89, 1.33]	0.96 [0.78, 1.18]	0.71 * [0.55, 0.93]
	>¥200,000	0.87 [0.65, 1.15]	0.97 [0.73, 1.31]	0.70 [0.48, 1.01]
Marital Status	Married (ref.)			
	Unmarried	1.27 * [1.04, 1.54]	1.67 *** [1.37, 2.04]	0.80 [0.63, 1.03]
	Divorced or widowed	1.31 [0.97, 1.78]	0.97 [0.70, 1.35]	1.61 * [1.12, 2.32]

Table 4. *Cont.*

		Rescuing Behavior	Adopting Intention	Complaint
Community Variables				
Type of City	Fourth-tier cities (ref.)			
	First-tier city	0.65 [0.41, 1.02]	0.64 * [0.44, 0.95]	0.57 * [0.35, 0.92]
	Second-tier cities	0.59 * [0.36, 0.98]	0.54 ** [0.36, 0.92]	0.90 [0.55, 1.45]
	Third-tier cities	0.62 * [0.43, 0.89]	0.79 [0.58, 1.06]	0.76 [0.54, 1.09]
Type of Community	Ordinary commercial housing community (ref.)			
	Old town community	1.26 [0.82, 1.94]	1.68 ** [1.18, 2.39]	1.74 ** [1.15, 2.65]
	Indemnificatory housing community	0.59 [0.21, 1.68]	0.70 [0.30, 1.65]	1.23 [0.46, 3.30]
	Luxury housing community	1.05 [0.39, 2.87]	0.70 [0.31, 1.57]	1.41 [0.56, 3.53]
	Urban village community	0.97 [0.64, 1.48]	0.96 [0.68, 1.36]	0.98 [0.66, 1.48]
	Campus community	0.90 [0.30, 2.68]	1.14 [0.47, 2.81]	0.28 [0.07, 1.10]
	Random effects			
	Intercept	0.22 *** [0.13, 0.37]	0.13 ** [0.07. 0.24]	0.17 ** [0.09, 0.32]

OR, Odds Ratio; CI, Confidence Interval; * $p < 0.05$, ** $p < 0.01$, *** $p < 0.001$.

The results of Model 1 showed that the overall regression model was significant, F (24, 5357) = 4.5, $p < 0.001$. Compared to participants with a bachelor's degree and undergraduate students, those with an elementary school degree, a junior high-school degree, or a master's degree were 0.71, 0.78, or 0.75 times less likely to rescue or help community cats, respectively. Retired participants were 0.77 times less likely to conduct rescuing or helping behaviors than employed participants. Unmarried participants were 1.27 times more likely than married participants to rescue or help the cats. In addition, participants residing in second-tier or third-tier cities were 0.59 or 0.62 times less likely to rescue or help the cats than their counterparts residing in fourth-tier cities, respectively.

The results of Model 2 showed that the overall regression model was significant, F (24, 5357) = 8.9, $p < 0.001$. Older adults were more likely to have an adopting intention than younger adults. Unmarried participants were 1.67 times more likely to have an adopting intention than married participants. Participants residing in first-tier or second-tier cities were 0.64 or 0.54 times less likely to have an adopting intention than their counterparts residing in fourth-tier cities. Participants living in the old town communities were 1.68 times more likely than participants living in ordinary commercial housing communities regarding their intentions to adopt community cats.

The results of Model 3 showed that the overall regression model was significant, F (24, 5357) = 2.9, $p < 0.001$. Participants with a junior high-school degree were 1.29 times more likely to complain about the issues of community cats than those with a bachelor's degree and undergraduate students. Older adults were more likely to complain about the issues of community cats than younger adults. Participants whose personal income ranged from 50,000 to 100,000 or from 100,000 to 200,000 were 0.77 or 0.71 times less likely to complain about the issues of community cats than the poorer participants whose personal income was lower than 50,000. Divorced or widowed participants were 1.61 times more likely to complain about the issues of community cats than married participants. In addition, participants from first-tier cities were 0.57 times less likely than those from fourth-tier cities to complain about the issues of the cats. Participants living in old town communities were 1.74 times more likely to complain about the issues of the cats than participants living in ordinary commercial housing communities.

Figure 1 presents the numbers and corresponding percentages of participants' responses to assessments regarding personal relationships with community cats. Among the participants who complained about the issues of community cats, which comprised 18% of the sample, 13% reported rescuing or helping the cats, and 10% expressed an intention to adopt a community cat if certain conditions were met. The "issues of community cats" may be understood in two ways. It is possible that participants who filed complaints but rescued or intended to adopt community cats were dissatisfied with the poor or inappropriate management practices related to the care and welfare of the cats. On the other hand, participants who filed complaints and did not care about the community cats may be concerned about issues such as security, public health, or mating calls caused by the cats.

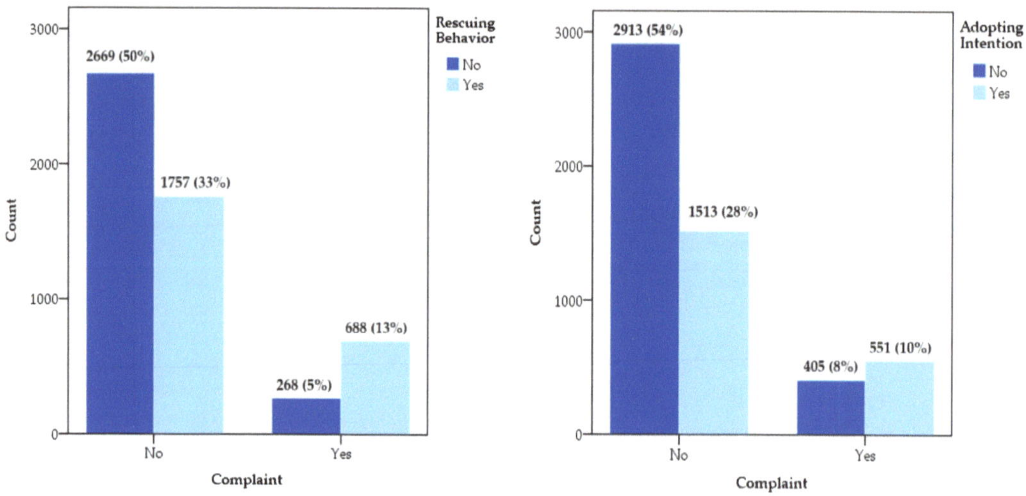

Figure 1. The number and corresponding percentage of options for human-cat relationships.

4. Discussion

The main aim of the current study was to investigate the public opinion on the general health of community cats and the quality of human-cat relationships in urban communities in China. Our survey revealed that 70% of participants reported that all or most community cats were healthy. And 60% perceived a harmonious or non-conflict coexistence between residents and community cats. Moreover, 45% of participants indicated having rescued or helped community cats. The findings support our proposition to use "community cats" as a substitute for the term "feral cats" or "stray cats", particularly in Chinese cities. The findings suggest a high level of awareness and concern among residents for the cats residing in their communities, and most of the cats enjoyed a relatively healthy state and coexistence with residents. These findings offer preliminary support for our conceptualization of an urban ecological research framework for studies on community cats. They suggest that urban animals, exemplified by community cats, fall outside the conventional research categories of wild or domestic animals.

Given the absence of city-level, provincial-level, or national-level laws or regulations governing the management of household cats and community cats in China, our findings offer valuable insights for stakeholders to formulate pertinent documents and intervention strategies concerning the management of cats. We revealed the effects of city type and community type. First, participants reported that the community cats exhibited poorer health and that there were more frequent conflicts with residents in fourth-tier cities than in other types of cities. The residents in fourth-tier cities also had a higher probability of rescuing or helping behavior than those in second- or third-tier cities. The findings indicate that in fourth-tier cities, there are more conflicts between residents and community cats, or more conflicts among residents regarding community cat issues. Second, compared to ordinary commercial housing communities, urban village communities were associated with community cats having an inferior health condition; and residents in old town communities had a higher intention for cat adoption and an increased frequency of complaints. The findings suggest that interpersonal conflicts might occur among residents because of differing opinions on the presence of community cats. In sum, it is imperative to allocate increased efforts and resources, focusing on less-developed cities, urban village communities, and old town communities within cities. Previous studies have demonstrated evidence-based measures in controlling the population of community cats, improving the welfare of the cats, and preventing zoonotic diseases. They include (a) neutering cats, a practice strongly endorsed by veterinary professional bodies in the United States and United Kingdom [47,48]; (b) environmental and behavioral modifications to address feline problem behaviors [49], in order to improve the probability of community cats being adopted or to reduce the probability of household cats being abandoned; (c) providing educational interventions for residents, policymakers, and practitioners to familiarize them with the natural behaviors of cats and the ethical treatment of animals [50,51]; and (d) disseminating knowledge and skills related to trap-neuter-return (TNR) or unity-neuter-vaccinate-return (TNVR) programs [52,53]. Simultaneously, it is crucial to investigate the experiences and practices of exemplary cities and communities where the welfare of community cats is ensured and harmonious coexistence with humans is maintained. The successful modules implemented in these locales hold the potential for replication in other similar regions across China.

When analyzing socio-demographic variables, we found that the participants who were more likely to complain about the issues of community cats tended to have lower levels of educational achievement and income and were more likely to be divorced or widowed. These participants closely align with those belonging to a vulnerable group. Vulnerability means facing a significant probability of incurring identifiable harm while substantially lacking the ability or means to protect oneself [54]. One possible explanation for this phenomenon is that due to limited socio-economic resources, vulnerable groups are more susceptible to the adverse effects that community cats may bring, for example, noises from mating calls, high cat populations, and the chaos caused by community cats rummaging through garbage piles. Moreover, vulnerable groups tend to reside in communities with relatively poor living conditions or low-quality property management [55], which may exacerbate conflicts between the groups and community cats. Furthermore, vulnerable groups may need help accessing scientifically valid information, making it difficult to obtain effective methods for addressing community cat issues or fostering positive interactions with community cats. The findings echo the One Health framework, which asserts that the health of people, other animals, and their shared environment are interdependent [56]. Therefore, it is crucial to involve vulnerable groups as integral stakeholders during discussions about community cat management. Allocating more resources and support will enable vulnerable groups to address and resolve the issues arising from cat overpopulation and conflicts between humans and felines, ultimately promoting both parties' welfare and harmonious coexistence.

The study has two limitations. Firstly, it may be beneficial to inquire whether participants engaged in each behavior separately, such as feeding, TNR, sponsorship, or medical assistance when participants report their rescuing or helping behaviors towards community cats. In a few Chinese communities, residents undertake TNR, medical care, and regular feeding of community cats by the guide of community residents' self-governing organizations [33]. However, residents in more communities may lack the practices of TNR or medical care and merely feed the cats. A study based on social media data revealed instances of irrational feeding practices by Chinese residents, such as offering leftover human food to cats [9]. While this seemingly benevolent feeding behaviors may be construed as caring and rescuing behaviors from the feeder's perspective, they can lead to feline health issues and interpersonal conflicts due to hygiene concerns. Secondly, the questionnaire could include more comprehensive variables at the individual or community level to better understand the variables associated with the health status of community cats and the human-cat relationship. For example, future research could measure whether residents live with companion animals, their attitudes and empathy towards animals [57,58], and whether there are individuals or groups involving in cat rescue activities at the community level. These variables have the potential to affect the well-being of community cats and the attitudes of residents towards them.

5. Conclusions

The current study examines the welfare condition of community cats in urban China and draws three main conclusions. First, the proportion of people complaining about the issues of community cats is relatively small. Instead, more people exhibit positive attitudes, acceptance, and proactive concern towards these animals. This finding contradicts the biased proportion of Chinese media reports portraying the conflicts between humans and community cats [59–61]. Second, in fourth-tier cities, urban village communities, and old town communities, specific interventions and attention are necessary for the health status of community cats and the human-cat relationship. Third, people with lower educational attainment or lower income, or those who are divorced or widowed, are more likely to file complaints. The finding highlights that vulnerable populations may be more susceptible to the adverse effects of community cats' inadequate welfare or management. Therefore, it is crucial to consider the welfare of vulnerable populations and community cats when considering community cat management.

Author Contributions: Conceptualization, G.P., Z.Z., A.N., X.G. and B.W.; methodology, A.N., Z.Z., X.G. and B.W.; investigation, Z.Z., X.X., Y.L. and L.W.; formal analysis, X.G.; writing—original draft preparation, X.G.; writing—review & editing, X.G., Z.Z., B.W. and G.P.; supervision, G.P. and A.N.; project administration, Z.Z.; funding acquisition, G.P. All authors have read and agreed to the published version of the manuscript.

Funding: This research was funded by the Beijing Ta Foundation for Animal Protection.

Institutional Review Board Statement: The study was conducted in accordance with the Declaration of Helsinki. Ethical review and approval were waived for this study as it used a survey method and did not collect or report scientific research information on individuals in life sciences or medical fields in accordance with Chinese national policy.

Data Availability Statement: The data presented in this study are available on reasonable request from the corresponding author.

Acknowledgments: We thank Qing Zhu and Huiyuan Qi for their advice in the sampling stage of the study and Biyu Huang for the suggestions for interpreting data on a draft version of the manuscript.

Conflicts of Interest: The authors declare no conflicts of interest. The funders had no role in the design of the study; in the collection, analyses, or interpretation of data; in the writing of the manuscript, or in the decision to publish the results.

Appendix A The Questionnaire Items Reported in the Study

1. Which community do you currently live in? [Single-choice question]
○ *[The name of the community]* (this community)
○ Other community within the city
○ Other community outside of the city
2. What is your community type? [Single-choice question]
○ Old town community
○ Indemnificatory housing community (e.g., low-rent housing, public rental housing)
○ Ordinary commercial housing community
○ Luxury housing community (e.g., villa area, high-class residential area)

○ Urban villages community (e.g., urban community recently transformed from rural community, "villages in city")

○ Other ________________

3. What is your gender? [Single-choice question]

○ Male

○ Female

○ Other

4. What is your age? (Please fill in a whole number, e.g., 26) [Fill in the blank]

____

5. What is your current highest level of education (including those you are currently enrolled in)? [Single-choice question]

○ Below elementary school

○ Elementary school

○ Junior high school

○ Secondary or high-school

○ College or university

○ Graduate or beyond

6. What is your current employment? [Single-choice question]

○ Full Employment

○ Part-time job

○ Household work (refers to staying at home to take care of other family members, commonly known as "housewife" or "househusband")

○ Full-time students

○ Unemployment

○ Retirement

7. What was your personal gross income in 2020? [Single-choice question]

○ Less than 10,000 CNY

○ 10,000–20,000 CNY

○ 20,000–50,000 CNY

○ 50,000–100,000 CNY

○ 100,000–200,000 CNY

○ 200,000 RMB and above

8. What is your current marital status? [Single-choice question]

○ Unmarried

○ Married

○ Divorced

○ Widowed

9. How frequently do you see community unowned cats (commonly known as "stray cats") in the community where you live? [Single-choice questions]

○ Often (more than three days a week)

○ Occasionally

○ Never (Please skip to Question 14)

○ I don't notice. (Please skip to Question 14)

10. Where do you mostly see unowned cats in the community? [Single-choice questions]

○ Garbage heaps

○ Feeding points

○ Roadside

○ Others ________________

○ I don't notice.

11. What is the pattern of food seeking of unowned cats in your community? [Single-choice question]

○ Some people feed them regularly.

○ Some people feed them occasionally.

○ No one feeds them and cats find food their own.
○ I don't notice.

12. What is the health condition of unowned cats in your community? [Single-choice question]

○All are in good health.
○ Most of them are in good health, and a few are sick or disabled.
○ Most of them are sick or disabled, and a few are in good health.
○ All of them are sick or disabled.
○ I don't notice.

13. With 5 being extremely good living situations and 1 being extremely poor living situations, how would you rate the living situations of unowned cats in your community? [Single-choice question]

○ 1
○ 2
○ 3
○ 4
○ 5

14. Would you consider adopting an unowned cat from the community if you had the conditions to do so? [Single-choice question]

○ Yes
○ No

15. Have you rescued an unowned cat in your community? (Note: Feeding and remote sponsorship are also considered as rescuing behaviors) [Single-choice question]

○ Yes
○ No

16. Have you complained about the issues of unowned cats to management staff in the community? [Single-choice question]

○ Yes
○ No

17. What do you think is the current situation of community residents getting along with unowned cats in the community? [Single-choice question]

○ Very harmonious
○ Basically coexisting
○ A small number of residents hate unowned cats in the community.
○ Residents generally demand the removal of unowned cats from the community.
○ Property and other managements remove unowned cats from the community.
○ I don't know
○ Other situations________________

References

1. Walker, J.K.; Bruce, S.J.; Dale, A.R. A survey of public opinion on cat (Felis catus) predation and the future direction of cat management in New Zealand. *Animals* **2017**, *7*, 49. [CrossRef]
2. Aegerter, J.; Fouracre, D.; Smith, G.C. A first estimate of the structure and density of the populations of pet cats and dogs across Great Britain. *PLoS ONE* **2017**, *12*, e0174709. [CrossRef] [PubMed]
3. Lynn, W.S.; Santiago-Ávila, F.J.; Stewart, K.L. Outdoor cats: An introduction. *Soc. Anim.* **2022**, *30*, 693–702. [CrossRef]
4. Irvine, L.; Cilia, L. More-than-human families: Pets, people, and practices in multispecies households. *Sociol. Compass* **2017**, *11*, e12455. [CrossRef]
5. Dolby, N. Multispecies families: Pets, community, and social action. *Fam. Soc. J. Contemp. Soc. Serv.* **2023**, *104*, 397–403. [CrossRef]
6. Spehar, D.D.; Wolf, P.J. Back to school: An updated evaluation of the effectiveness of a Long-Term Trap-Neuter-Return program on a university's free-roaming cat population. *Animals* **2019**, *9*, 768. [CrossRef]
7. Sparkes, A.H.; Bessant, C.; Cope, K.; Ellis, S.L.; Finka, L.; Halls, V.; Hiestand, K.; Horsford, K.; Laurence, C.; MacFarlaine, I. ISFM guidelines on population management and welfare of unowned domestic cats (Felis catus). *J. Feline Med. Surg.* **2013**, *15*, 811–817. [CrossRef]
8. Crawford, H.M.; Calver, M.C.; Fleming, P.A. A case of letting the cat out of the bag—Why Trap-Neuter-Return is not an ethical solution for stray cat (Felis catus) management. *Animals* **2019**, *9*, 171. [CrossRef]

9. Xu, J.; Jiang, A. Public opinions on stray cats in China, Evidence from social media data. *Animals* **2023**, *13*, 457. [CrossRef]
10. Miller, P.S.; Boone, J.D.; Briggs, J.R.; Lawler, D.F.; Levy, J.K.; Nutter, F.B.; Slater, M.; Zawistowski, S. Simulating free-roaming cat population management options in open demographic environments. *PLoS ONE* **2014**, *9*, e113553. [CrossRef]
11. Aeluro, S.; Buchanan, J.M.; Boone, J.D.; Rabinowitz, P.M. "State of the Mewnion": Practices of feral cat care and advocacy organizations in the united states. *Front. Vet. Sci.* **2021**, *8*, 791134. [CrossRef] [PubMed]
12. Loyd, K.A.T.; Miller, C.A. Influence of demographics, experience and value orientations on preferences for lethal management of feral cats. *Hum. Dimens. Wildl.* **2010**, *15*, 262–273. [CrossRef]
13. Wald, D.M.; Jacobson, S.K.; Levy, J.K. Outdoor cats: Identifying differences between stakeholder beliefs, perceived impacts, risk and management. *Biol. Conserv.* **2013**, *167*, 414–424. [CrossRef]
14. Boone, J.D.; Briggs, J.R.; Hiby, E.; Lawler, D.F.; Levy, J.K.; Miller, P.S.; Nutter, F.B.; Slater, M.R.; Zawistowski, S. Improving and evaluating trap-neuter-return (TNR) management for outdoor cats on the human landscape. In Proceedings of the Vertebrate Pest Conference, Waikoloa Village, HI, USA, 3–6 March 2014.
15. Wolf, P.J.; Hamilton, F. Managing free-roaming cats in US cities: An object lesson in public policy and citizen action. *J. Urban Aff.* **2022**, *44*, 221–242. [CrossRef]
16. Lohr, C.A.; Lepczyk, C.A. Desires and management preferences of stakeholders regarding feral cats in the Hawaiian Islands. *Conserv. Biol.* **2014**, *28*, 392–403. [CrossRef] [PubMed]
17. Madden, F. Creating coexistence between humans and wildlife: Global perspectives on local efforts to address human–wildlife conflict. *Hum. Dimens. Wildl.* **2004**, *9*, 247–257. [CrossRef]
18. Li, Y.; Wan, Y.; Shen, H.; Loss, S.R.; Marra, P.P.; Li, Z. Estimates of wildlife killed by free-ranging cats in China. *Biol. Conserv.* **2021**, *253*, 108929. [CrossRef]
19. Elizondo, E.C.; Loss, S.R. Using trail cameras to estimate free-ranging domestic cat abundance in urban areas. *Wildl. Biol.* **2016**, *22*, 246–252. [CrossRef]
20. Turner, D.C. Outdoor domestic cats and wildlife: How to overrate and misinterpret field data. *Front. Vet. Sci.* **2022**, *9*, 1087907. [CrossRef]
21. Badenes-Pérez, F.R. The impacts of free-roaming cats cannot be generalized and their role in rodent management should not be overlooked. *Conserv. Sci. Pract.* **2023**, *5*, e12861. [CrossRef]
22. Zhang, Z.; Qi, H.; Guo, P. Doubtful estimates of wildlife killed by free-ranging cats in China. A response to Y. Li et al. Biological Conservation 253 (2021) 108929. *Biol. Conserv.* **2021**, *260*, 109200. [CrossRef]
23. Donaldson, S.; Kymlicka, W. *Zoopolis: A political Theory of Animal Rights*; Oxford University Press: New York, NY, USA, 2011.
24. Hulme-Beaman, A.; Dobney, K.; Cucchi, T.; Searle, J.B. An ecological and evolutionary framework for commensalism in anthropogenic environments. *Trends Ecol. Evol.* **2016**, *31*, 633–645. [CrossRef] [PubMed]
25. Verma, P.; Singh, R.; Singh, P.; Raghubanshi, A. Urban ecology–current state of research and concepts. In *Urban Ecology*; Elsevier: Amsterdam, The Netherlands, 2020; pp. 3–16.
26. Hu, Y.; Hu, S.; Wang, W.; Wu, X.; Marshall, F.B.; Chen, X.; Hou, L.; Wang, C. Earliest evidence for commensal processes of cat domestication. *Proc. Natl. Acad. Sci. USA* **2014**, *111*, 116–120. [CrossRef] [PubMed]
27. Tan, C.K.; Liu, T.; Gao, X. Becoming "pet slaves" in urban China: Transspecies urban theory, single professional women and their companion animals. *Urban Stud.* **2021**, *58*, 3371–3387. [CrossRef]
28. Mao, X.; Liu, J.; Zhao, F.; Yuan, S.; Wang, J. The study on pet activity space in modern cities. *Archit. Cult.* **2021**, 159–162. (In Chinese) [CrossRef]
29. Ma, Y. Number of Cats in China 2018–2022. Available online: https://www.statista.com/statistics/992420/china-number-of-cats/ (accessed on 15 March 2023).
30. Zhu, J.; Sun, Q.; Xia, L.; Lu, J. Causes and hazards of stray pets and their prevention and control measures. *Shanghai J. Anim. Husb. Vet. Med.* **2015**, 47–49. Available online: https://qikan.cqvip.com/Qikan/Article/Detail?id=664904929&from=Qikan_Search_Index (accessed on 1 February 2024). (In Chinese).
31. Niu, J. Analysis and reflection on the current situation of conflict of interest between stray dogs and cats and human populations *China Anim. Health Insp.* **2021**, *38*, 55–57. (In Chinese)
32. Wang, J. China's Pet Industry Is Booming. So Are Its Horrific Breeding Mills. Available online: https://www.sixthtone.com/news/1011091 (accessed on 4 September 2022).
33. Gu, Q. The "Haitian Sample" of community self-governance in building animal civilization. *Democr. Leg. Syst.* **2022**, 13–17. (In Chinese)
34. Lau, C. 1000 Cats Rescued in China from Being Slaughtered and Sold as Pork, Mutton. Available online: https://edition.cnn.com/2023/10/24/china/china-police-save-cats-slaughter-intl-hnk/index.html (accessed on 24 October 2023).
35. Forman, R.T. *Urban Ecology: Science of Cities*; Cambridge University Press: Cambridge, UK, 2014.
36. Wu, L.; Bian, Y. Housing, consumption and monetary policy: How different are the first-, second- and third-tier cities in China? *Appl. Econ. Lett.* **2018**, *25*, 1107–1111. [CrossRef]
37. Zhang, H.; Li, L.; Hui, E.C.-M.; Li, V. Comparisons of the relations between housing prices and the macroeconomy in China's first-, second- and third-tier cities. *Habitat Int.* **2016**, *57*, 24–42. [CrossRef]
38. Wong, G.K.M.; Yu, L. Consumers' perception of store image of joint venture shopping centres: First-tier versus second-tier cities in China. *J. Retail. Consum. Serv.* **2003**, *10*, 61–70. [CrossRef]

39. Wang, C.; Shen, J.; Liu, Y. Administrative level, city tier, and air quality: Contextual determinants of hukou conversion for migrants in China. *Comput. Urban Sci.* **2021**, *1*, 6. [CrossRef]
40. Population Statistics Yearbook. China Economic and Social Big Data Research Platform. Available online: https://data.cnki.net/ (accessed on 21 January 2024).
41. Bray, D. Building 'community': New strategies of governance in urban China. In *China's Governmentalities*; Routledge: London, UK, 2009; pp. 100–118.
42. He, Y. *The Road to People's Cities*; People's Publishing House: Beijing, China, 2022.
43. O'Dwyer, L.M.; Parker, C.E. A Primer for Analyzing Nested Data: Multilevel Modeling in SPSS Using an Example from a REL Study. Available online: http://ies.ed.gov/ncee/edlabs (accessed on 14 December 2022).
44. Sommet, N.; Morselli, D. Keep calm and learn multilevel logistic modeling: A simplified three-step procedure using Stata, R, mplus, and SPSS. *Int. Rev. Soc. Psychol.* **2017**, *30*, 203–218. [CrossRef]
45. Schoeneberger, J.A. The impact of sample size and other factors when estimating multilevel logistic models. *J. Exp. Educ.* **2016**, *84*, 373–397. [CrossRef]
46. Office of the Leading Group of the State Council forthe Seventh National Population Census (OLG). *China Census Yearbook 2020*; China Statistics Press: Beijing, China, 2022.
47. Wongsaengchan, C.; McKeegan, D.E. The views of the UK public towards routine neutering of dogs and cats. *Animals* **2019**, *9*, 138. [CrossRef]
48. Chu, K.; Anderson, W.M.; Rieser, M.Y. Population characteristics and neuter status of cats living in households in the United States. *J. Am. Vet. Med. Assoc.* **2009**, *234*, 1023–1030. [CrossRef]
49. Halls, V. Tools for managing feline problem behaviours: Environmental and behavioural modification. *J. Feline Med. Surg.* **2018**, *20*, 1005–1014. [CrossRef]
50. McLeod, L.J.; Hine, D.W.; Driver, A.B. Change the humans first: Principles for improving the management of free-roaming cats. *Animals* **2019**, *9*, 555. [CrossRef]
51. Finka, L.R.; Ripari, L.; Quinlan, L.; Haywood, C.; Puzzo, J.; Jordan, A.; Tsui, J.; Foreman-Worsley, R.; Dixon, L.; Brennan, M.L. Investigation of humans individual differences as predictors of their animal interaction styles, focused on the domestic cat. *Sci. Rep.* **2022**, *12*, 12128. [CrossRef]
52. Kreisler, R.E.; Cornell, H.N.; Levy, J.K. Decrease in population and increase in welfare of community cats in a twenty-three year trap-neuter-return program in Key Largo, FL: The ORCAT program. *Front. Vet. Sci.* **2019**, *6*, 7. [CrossRef]
53. Boone, J.D.; Miller, P.S.; Briggs, J.R.; Benka, V.A.; Lawler, D.F.; Slater, M.; Levy, J.K.; Zawistowski, S. A long-term lens: Cumulative impacts of free-roaming cat management strategy and intensity on preventable cat mortalities. *Front. Vet. Sci.* **2019**, *6*, 238. [CrossRef]
54. Schroeder, D.; Gefenas, E. Vulnerability: Too vague and too broad? *Camb. Q. Healthc. Ethics* **2009**, *18*, 113–121. [CrossRef]
55. Kuran, C.H.A.; Morsut, C.; Kruke, B.I.; Krüger, M.; Segnestam, L.; Orru, K.; Nævestad, T.O.; Airola, M.; Keränen, J.; Gabel, F. Vulnerability and vulnerable groups from an intersectionality perspective. *Int. J. Disaster Risk Reduct.* **2020**, *50*, 101826. [CrossRef]
56. World Health Organization. One Health. Available online: https://www.who.int/news-room/questions-and-answers/item/one-health (accessed on 1 December 2023).
57. Gu, X.; Bexell, S.M.; Wang, B. Attitudes toward nonhuman animals during the coronavirus disease (COVID-19) outbreak in China. *Anthrozoös* **2022**, *35*, 219–235. [CrossRef]
58. Gu, X.; Xie, L.; Bexell, S.M. The link between attitudes toward animals and empathy with humans in China: Mediation of empathy with animals. *Anthrozoös* **2023**, *37*, 1–14. [CrossRef]
59. Public Legal Service of China. Legal aid Provided by Shanghai Songjiang District Legal Aid Center to Gao Moumou in the Case of Animal Infringement Dispute. Available online: https://alk.12348.gov.cn/LawSelect/Detail?dbID=46&dbName=FYGL&sysID=27726 (accessed on 21 January 2024).
60. Shanghai Legal News. Stray Dogs and Cats Hurt Homeowners, Is the Feeder Responsible? This Court Collected Cases to Tell You. Available online: https://m.thepaper.cn/newsDetail_forward_25429982 (accessed on 21 January 2024).
61. Sun, Y. Community Coordinates to Find New Homes for More than 20 Stray Cats after Love Feeding Caused Nuisance. Available online: https://news.bjd.com.cn/2022/09/13/10156481.shtml (accessed on 21 January 2024).

MDPI AG
Grosspeteranlage 5
4052 Basel
Switzerland
Tel.: +41 61 683 77 34

Animals Editorial Office
E-mail: animals@mdpi.com
www.mdpi.com/journal/animals